T0263519

Guide to Satellite TV

Guide to Satellite TV

Installation, reception and repair
Fourth Edition

D. J. Stephenson
BA, IEng, FIEIE

Newnes
An imprint of Butterworth-Heinemann
Linacre House, Jordan Hill, Oxford OX2 8DP
225 Wildwood Avenue, Woburn, MA 01801–2041
A division of Reed Educational and Professional Publishing Ltd

Ⴁ A member of the Reed Elsevier plc group

OXFORD   AUCKLAND   BOSTON
JOHANNESBURG   MELBOURNE   NEW DELHI

First published 1990
Second edition 1991
Third edition 1994
Fourth edition 1997
Reprinted 1999

British Library Cataloguing in Publication Data
A catalogue record for this book is available from the British Library

Library of Congress Cataloging in Publication Data
A catalogue record for this book is available from the Library of Congress

ISBN 0 7506 3475 8

Printed and bound by Antony Rowe Ltd, Eastbourne
Typeset by Keyword Typesetting Services Ltd, Wallington, Surrey

Contents

Preface

The first edition was received well by the technical press way back in 1990. A lot has changed since then in this fast developing field. This new fourth edition is revised and updated to include topics relevant to digital video broadcasting, which is expected to replace analogue systems entirely sometime during the early 21st century owing to its flexibility and impressive reduction in bandwidth.

As is traditional with *Newnes Guide to Satellite TV*, the pitch is set between that of a simple installation guide and that of an involved theoretical textbook. New sections are added to cover: multi-feed antennas capable of receiving two or more satellites from a single fixed dish; universal wideband LNBs suitable for analogue and digital reception; simplified downlink budget calculations for specifying digital receiving equipment; and finally an overview of the widely adopted DVB/MPEG-2 system.

Newnes Guide to Satellite TV is intended to provide the necessary information to specify, install and maintain both fixed and polar mount antenna systems along with small IF distribution systems for small blocks of flats and hotels. Regrettably, it is not possible in a book of this size to include every facet of the fast developing satellite TV industry, but it is hoped the choice of included material is sufficient to grasp at least the essential working details an installer needs to know.

Finally, it is fit I acknowledge the work of the many engineers who have contributed to the vast pool of knowledge from which much material in this book is ultimately 'borrowed'.

D. J. Stephenson
Merseyside

1 Overview of satellite TV

Introduction

Direct-to-home satellite TV, although not intended as such, has been with us now for a number of years. A few enthusiasts gathered together the rudiments of a satellite receiving system and eavesdropped on the programme material destined for cable operators via low power general telecommunications satellites. Equipment, mainly from outside Europe, was imported to supply this relatively small demand. This was usually expensive, cumbersome and indescribably ugly. Reception was often poor in heavy rain or cloudy conditions, so it was not surprising that there was little public interest. Today, Ku-band antennas of monstrous proportions are all but designated to the scrapheap. Antennas above about 1.5 m are rarely seen in Europe, since newer and more powerful replacement satellites have mainly replaced the early low power craft. Most interest continues to be in FSS band satellites where popular services have been available for some time. Major satellite operators have adopted a variety of methods to extend transponder capacity. The cluster approach is common, where a group of perhaps, 2, 3, 4 or 5 satellites are co-located in the same orbital slot (e.g. Astra). Another alternative is to operate satellites in a variety of different orbital slots but beam services to same area (e.g. Eutelsat II series). There are also cases where operators combine a mixture of the two approaches such as the co-location of Eutelsat II F6 (Hotbird 1) with Hotbird 2 at 13°E. Obviously co-location has a significant advantage in that only a single, low-cost fixed dish is needed to receive signals from a number of satellites.

Basic terms and concepts

For those new to telecommunications, who are unfamiliar with some of the basic terms and concepts used, here follows a brief preparatory section of basic principles necessary for the understanding of satellite reception. Trained technicians may like to skip this section.

Sinusoidal electromagnetic waves (e/m waves)

All radio and television signals consist of electrical and magnetic fields which, in free space, travel at the speed of light (approximately 186 000 miles per second or 3×10^8 metres per second). These waves consist of an *electric field (E)*, measured in volts per metre and a *magnetic field (H)*, measured in amps per metre. The *E* and *H* field components are always at right angles to each other and the direction of travel is always at right angles to both fields. The amplitudes vary sinusoidally as they travel through space. In fact, it is impossible to produce a non-sinusoidal e/m wave! (The importance of this statement will be grasped more easily when modulation is discussed).

The sine wave (see Figure 1.1)

Cycle: One complete electrical sequence.
Peak value (V_p): Maximum positive or negative value – also called the amplitude.
Period (t): Time to complete one cycle.
Frequency (f): Number of cycles per second measured in Hertz (Hz). (One hertz = one cycle per second). It follows that period and frequency are reciprocals of each other:

$$t = 1/f$$

Commonly used multiples of the hertz are:

Kilohertz (kHz)　　= 10^3 Hz = 1000 Hz
Megahertz (MHz) = 10^6 Hz = 1 000 000 Hz
Gigahertz (GHz)　= 10^9 Hz = 1 000 000 000 Hz

RMS value: This is 0.707 of the peak value and, unless otherwise stated, any reference to voltage or current in technical literature is normally taken to mean this value. For example, the supply mains in the UK is a sinusoidal variation, stated to be '240 volts' so the peak value is 240/0.707 = 339 volts.

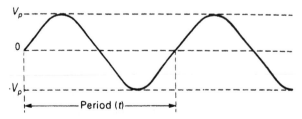

Figure 1.1   *The sine wave*

Angular velocity (ω)

This is an indirect way of expressing the frequency:

$\omega = 2\pi f$ radians per second.

Instead of considering the number of complete cycles, angular velocity is a measure of how fast the vector angle is changing. The voltage equation of a sine wave, which gives the instantaneous value (v) of a sine wave at any point in the cycle is given by:

$v = V_p \sin \theta$

where V_p is the peak voltage and θ is an angle measured in radians (not degrees). There are 2π radians in a circle and, since the sine wave can be visualized as a vector rotating in a circle, the above equation can be written in terms of frequency and angle:

$v = V_p \sin 2\pi ft$

For convenience and brevity, the $2\pi f$ part is often lumped together and given the title of angular velocity (ω). Using this notation, the equation of the sine wave can be written as:

$v = V_p \sin \omega t$

Wavelength

Since e/m waves travel at a known velocity and vary sinusoidally, it is possible to consider how far a wave of given frequency (f) would travel during the execution of one cycle. Denoting the speed of light as (c), the wavelength (W) is given by:

$W = c/f$

From this, it is clear that the higher the frequency, the shorter the wavelength. Satellite broadcasting employs waves in the order of 10 GHz frequency so the order of wavelength can be calculated as follows:

$W = (3 \times 10^8)/(10 \times 10^9)$
$W = 3 \times 10^{-2} = 3$ cm

In practice, the frequencies used are not necessarily a nice round figure like 10 GHz. Nevertheless, the wavelengths in present use invariably work out in terms of centimetres – they are, in fact, known as 'centimetric waves'. It is pertinent at this stage to question why such enormously high frequencies are used in satellite broadcasting? Before this can be answered, it is necessary to understand some fundamental laws relating to broadcasting of information, whether it be sound or picture information.

Carrier frequency

For simplicity, assume that it is required to transmit through space a 1000 Hz audio signal. In theory, an electrical oscillator and amplifier could be rigged up and tuned to 1000 cycles per second and the output fed to a piece of wire acting as a primitive aerial. It is an unfortunate fact of nature that, for reasonably efficient radiation, a wire aerial should have a length somewhere in the order of the 'wavelength' (W) of 1000 Hz. Using the equation given above:

$$W = c/f = (3 \times 10^8)/(10^3) = 3 \times 10^5 \text{ metres}$$
$$= 300\ 000 \text{ metres which is about 188 miles!}$$

Apart from the sheer impracticality of such an aerial, waves at these low frequencies suffer severe attenuation due to ground absorption. Another important reason for using high frequencies is due to the considerations of bandwidth which is treated later.

The solution is to use a high frequency wave to 'carry' the signal but allow the 'intelligence' (the 1000 Hz in our example) to modify one or more of its characteristics. The high frequency wave is referred to as the carrier (f_c) simply because it 'carries' the information in some way. The method of impressing this low frequency information on the carrier is called modulation. There are two main types, amplitude modulation (AM) and frequency modulation (FM).

Amplitude modulation (see Figure 1.2)

The low frequency modulating signal is made to alter the amplitude of the carrier at the transmitter before the composite waveform is sent to the aerial system. If the amplitude of the modulating signal causes the carrier amplitude to vary between double its unmodulated height and zero, the modulation is said to be 100 per cent. Terrible distortion results if the modulation amplitude is ever allowed to exceed 100 per cent.

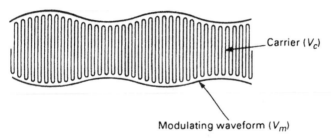

Carrier (V_c)

Modulating waveform (V_m)

Figure 1.2 *Amplitude modulation*

Modulation factor (*m*)

This is the ratio of modulation amplitude (V_m) to carrier amplitude (V_c):

$m = V_m/V_c$

When $m = 1$, the modulation is 100 per cent. Although 100 per cent modulation is an advantage, it is too dangerous in practice because of the possibility of overmodulation, so 80 per cent ($m = 0.8$) is normally considered the safe limit.

Sidebands

Although the modulating signal in Figure 1.2 is shown as a simple sinusoidal waveform, in practice it will be more complex. Thus the envelope of the waveform will be non-sinusoidal. Bearing in mind that only sinusoidal waveforms can be sent through space, there is clearly something odd to explain. This is where a little school maths comes in handy.

The unmodulated carrier sine wave has the instantaneous form:

$v = V_p \sin \omega_c t$

But the amplitude of this wave (V_p) is made to vary by the modulating frequency which causes V_p to have the form:

$V_p = V_m \sin \omega_m t$

Substituting this expression in the first equation gives:

$v = V_m \sin \omega_m t \sin \omega_c t$

In school, we were told that one of the trigonometrical identities is:

$\sin A \sin B = 1/2 \cos (A - B) - 1/2 \cos (A + B)$

So it follows that the modulated carrier waveform splits up in space into three pure sinusoidal components:

1 The carrier frequency.
2 A frequency equal to the sum of the carrier and modulating frequencies. This is called the 'upper sideband'.
3 A frequency equal to the difference between the carrier and modulating frequency. This is called the 'lower sideband'.

Taking a simple numerical example, if the carrier frequency is 1 000 000 Hz and the modulating frequency is 1000 Hz, then the upper sideband is a 1 001 000 Hz sine wave and the lower sideband is a 999 000 Hz sine wave. In practice, the modulating frequency will seldom be anything as simple as a 1000 Hz sine wave but, more probably, may consist of speech or picture information which contains a complex mixture of frequencies. This does not invalidate the former reasoning. It just means that instead of single frequency upper and lower sidebands, there will be, literally, a band of sinusoidally varying frequencies either side of the carrier. For example, the music frequency spectrum extends from about 20 Hz to about 18 kHz so, to transmit high quality sound, the upper sidebands would have to contain a spread of frequencies extending from 20 Hz to 18 kHz above the carrier, and the lower sidebands, frequencies 20 Hz to 18 kHz below the carrier. Television transmission is more difficult because pictures have a far greater information content than sound. The sidebands must extend several MHz either side of the carrier. The wider the sidebands of a transmission, the greater space it will occupy in the frequency spectrum, so broadcast stations geographically close together must operate on frequencies well away from each other in order to prevent interference from their respective sidebands. Since television stations occupy several MHz in the spectrum, carrier frequencies are forced into ever higher and higher frequencies as the number of stations fight for space. There are several novel solutions to the over-crowding problem. For example, it is not essential to transmit both sidebands since all the required information is contained in one of them, providing of course the carrier is sent with it. Such transmissions are called SSB (single sideband). An even more drastic curtailment is to reduce the carrier amplitude at the transmitter to almost zero and use it to synchronize a locally generated carrier at the receiving end, a technique known as 'single sideband vestigial carrier' transmission.

Frequency modulation (FM)

Whereas amplitude modulation alters the envelope in the 'vertical plane', frequency modulation takes place in the 'horizontal plane' (see Figure 1.3). The amplitude of the carrier is kept constant but the frequency is caused to deviate proportional to the modulating amplitude.

Frequency deviation

The maximum amount by which the carrier frequency is increased or decreased by the modulating amplitude is called the frequency deviation. It is solely dependent on the amplitude (peak value) of the modulating voltage. In the case of satellite broadcasting, the signal

Constant amplitude carrier

Figure 1.3 *Frequency modulation*

beamed down to earth has a typical frequency deviation of about 16 MHz/V and the bandwidth occupied by the picture information is commonly about 27 MHz.

Modulation index (*m*)

This is the ratio of the frequency deviation (f_d) to the highest modulating frequency (f_m):

$$m = f_d/f_m$$

In contrast with amplitude modulation, the modulation index is not necessarily restricted to a maximum of unity.

Johnson noise

Any unwanted random electrical disturbance comes under the definition of noise. Such noise is all-pervading and is the worst enemy of the electronic designer. It begins in conventional circuitry, particularly with the apparently harmless resistor because, at all temperatures above zero kelvin (0 K), a minute, but not always negligible, e.m.f. (called Johnson noise) appears (and can be measured) across the ends. This is due to random vibration of the molecules within the body of the resistor and nothing whatever can be done to stop it. Although the following equation for Johnson noise is not particularly important in this text it is worth examining if only to grasp the strange connection between noise e.m.f.s and temperature.

RMS value of Johnson noise $= (4k\ tBR)^{\frac{1}{2}}$

where $t = 0$ absolute temperature kelvin (ι ..m temperature may be taken as around 290 K)

k = Boltzman's constant $= 1.38 \times 10^{-23}$
R = the resistance in ohms
B = the bandwidth of the instrument used to measure the e.m.f.

Those with sufficient zeal to work out the noise from a one megohm resistor at room temperature would come up with a value of about 0.4 millivolts! This may seem small but it is relative, rather than absolute values that are important. If the wanted signal is of the same order as this (in practical cases it could be much smaller) then the noise will swamp it out. Note from the equation, which incidentally is not restricted to man-made materials, the noise depends on the temperature, and the bandwidth of the 'instrument used to measure it'. Such an 'instrument' includes a broadcast receiving station! A high quality transmission has wide sidebands so the receiving installation must also have a wide bandwidth in order to handle the information in the sidebands.

The occurrence of this form of noise entering the chain can seriously limit the quality of reception. Although Johnson noise has been used as an example, there are many other forms of noise (including ground and the man-made variety) which are treated in other parts of this book.

Signal to noise ratio (S/N ratio)

This is the ratio of the desired signal e.m.f. to any noise e.m.f. present and should be as high as possible. If this ratio falls to unity or below, the signal is rendered virtually useless. (It is possible, but expensive, to use computer generated 'signal enhancement' techniques in some cases, but for domestic satellite broadcasting this is out of the question.)

Comparison of FM and AM

There are two features of AM which, in the past, have been responsible for its popularity:

1 The demodulation circuitry in the receiver, called 'rectification', is simple, requiring only a diode to chop off one half of the composite waveform and a low pass filter to remove the carrier remnants.
2 The sidebands are relatively narrow so the transmission does not occupy too much space in the available frequency spectrum.

The most serious criticism of AM is that noise, at least most of it, consists of an amplitude variation. That is to say, any noise e.m.f.s present ride on the top of the envelope as seen in Figure 1.4. So, apart from meticulous design techniques based on increasing the S/N ratio, nothing much can be done about reducing noise without degrading signal quality by crude methods such as bandwidth reduction. FM, on the other hand, is often stated to be 'noise free'. This is not true! A FM transmission is as vulnerable to noise pollution as AM but, due to the manner in which the information is impressed on the carrier, much of the noise can be removed by the receiver circuitry. Since noise rides on

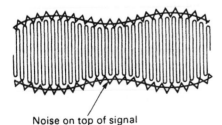

Noise on top of signal

Figure 1.4 *Noise on AM signals*

the outside of an FM waveform, it is possible to slice off the top and bottom of the received waveform without destroying the information (remember that the information is inside the waveform rather than riding on the top and bottom). The slicing-off process is known as 'amplitude limiting'. A disadvantage of FM is the wide bandwidth required. FM is only possible if the carrier frequencies are relatively high. Fortunately, satellite broadcasting is well above 1 GHz so this is a trivial disadvantage. It cannot be denied that the circuitry required to extract the information from an FM carrier is, to say the least, awkward! The circuitry which performs this function is called an 'FM demodulator' which often takes bizarre forms. Among the various circuits that have been developed for FM demodulation are discriminators, ratio detectors and phased locked loops. This latter type is the most often used method and will be explained in Chapter 4.

Decibels (dB)

Decibels provide an alternative, and often more convenient, way of expressing a ratio between two powers. Instead of the actual ratio, the logarithm to base 10 of the ratio is used as shown below:

$$dB = 10 \log P_1/P_2$$

The sign of the result is positive if P_1 is greater than P_2 and negative if P_1 is less than P_2. To avoid the trouble of evaluating negative logarithms, it is a good plan always to put the larger of the two powers on top and adjust the sign afterwards in accordance with the above rule.

Examples: If $P_1 = 1000$ and $P_2 = 10$ then, dB = 10 log 1000/10 = 10 log 100 = +20 dB. (If P_1 was 10 and P_2 was 1000, the absolute value in decibels would be the same but it would be written as −20 dB). There are several advantages of using dBs instead of actual ratios:

1 Because the human ear behaves logarithmically to changes in sound intensity, decibels are more natural than simple ratios. For example, if the power output of an audio amplifier is increased from 10 watts to 100 watts, the effect on the ear is not 10 times as great.
2 Decibels are very useful for cutting large numbers down to size. For example, a gain of 10 000 000 is only 70 dB.
3 The passage of a signal from the aerial through the various stages of a receiving installation is subject to various gains and losses. By expressing each gain in terms of positive dBs and each loss in negative dBs, the total gain can be easily calculated by taking the algebraic sum.

Example: $(+5) + (−2) + (+3) + (−0.5) = 5.5$ dB.

A few of the more commonly used dB values are as follows:

Decibels (dB)	Relative power increase
0	1.00
0.5	1.12
1.0	1.26
2.0	1.58
3.0	1.99
6.0	3.98
12.0	15.85
15.0	31.62
18.0	63.09
21.0	125.89
50.0	100 000
100.0	10 000 000 000

Voltage dB

Although dBs are normally used in conjunction with power ratios, it is sometimes convenient to express voltage ratio in dB terms. The equation in these cases is:

$$dB = 20 \log V_1/V_2$$

The use of 20 instead of 10 is because power is proportional to the square of the voltage so the constant is 20 instead of 10.

Ku-band satellite TV

The European nations have almost exclusively adopted Ku-band (10.95 to 14.5 GHz) for the transmission of satellite TV signals. Likewise, the contents of this book are entirely biased toward Ku-band reception.

The Clarke belt

Back in 1945, Arthur C. Clarke, the famous scientist and science fiction novelist, predicted that an artificial satellite placed at a height of 35 803 km directly above the equator would orbit the globe at the same speed with which the earth was rotating. As a result, the satellite would remain stationary with respect to any point on the earth's surface. This equatorial belt, rather like one of Saturn's rings, is affectionately known as the *Clarke belt*. Any satellite within this belt is termed *geostationary*, and is placed in a subdivision known as an *orbital slot*. Signals are sent up to a satellite via an *uplink*, electronically processed and then re-transmitted via a *downlink* to earth receiving stations. Figure 1.5 illustrates the points made.

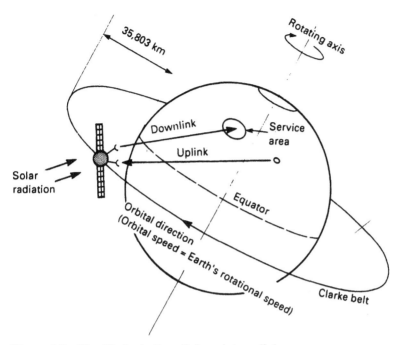

Figure 1.5 *The Clarke belt, uplink and downlink*

The uplink

The uplink station is a fairly complex affair because not only do the signals have to be sent but they have to be sent at a differing frequency, usually in the higher 14 GHz band, to avoid interference with downlink signals. Another function performed by the uplink station is to control tightly the internal functions of the satellite itself (such as station keeping accuracy), although these technicalities need not concern us here. Uplinks are controlled so that the transmitted microwave power beam is extremely narrow, in order not to interfere with adjacent satellites in the geo-arc. The powers involved are several hundred watts.

The downlink

Each satellite has a number of *transponders* with access to a pair of receive/transmit antennas and associated electronics for each channel. For example, in Europe, the uplink sends signals at a frequency of about 14 GHz, these are received, down-converted in frequency to about 11/12 GHz and boosted by high power amplifiers for re-transmission to earth. Separate transponders are used for each channel and are powered by solar panels with back up batteries for eclipse protection. The higher the power of each transponder then the fewer channels will be possible with a given number of solar panels, which in turn, is restricted by the maximum payload of launch vehicles as well as cost. Typical power consumption for a satellite such as ASTRA 1A is 2.31 kW with an expected lifetime of 12.4 years. Satellites are conveniently categorized into the following three power ranges:

1   *Low power* – These have transponder powers around the 20 W mark and are primarily general telecommunication satellites. Due to the low transmission power of each transponder they can support many channels with the available collected solar energy. Many of these transponders relay programme material for cable TV operators across Europe. Small numbers of enthusiasts eavesdrop on these broadcasts but, unfortunately, receiving dishes of monstrous proportions are necessary for noise free reception, often in excess of 1 metre. This state of affairs is clearly not too popular with the general public who consider them, quite understandably, as dinosaurs of a past age. Even so, domestic TV reception is not the primary reason for the existence of such high channel capacity satellites. Transponder bandwidths can vary.

2   *Medium power* – These satellites have typical transponder powers of around 45 W, such as those on board Astra 1A. Such satellites are now commonly termed semi-DBS (*direct broadcast service*) and represent the first serious attempt to gain public approval by offering the prospect of dustbin-lid-sized dishes of 60 cm diameter. About sixteen transponders are average for this class at the present time. Medium power European satellites usually operate in the frequency band 10.95 GHz to 11.70 GHz and form the *fixed satellite service* (FSS). The transponder bandwidths are commonly 27 MHz or 36 MHz. Some medium power satellites, such as the Eutelsat II series, also have a number of transponders that can be active in the 12.5 GHz to 12.75 GHz band, originally termed the business band service (BBS) by the International Telecommunication Union (ITU).

3   *High power* – These pure DBS satellites have transponder powers exceeding 100 W and have a correspondingly reduced channel capacity of around four perhaps five channels. The specified dish size is minimal, about 30 to 45 cm in the central service area. These are perhaps the ideal size as far as the public are concerned and interest in satellite TV is expected to blossom as these come on stream. European transponder frequencies are in the band 11.70 to 12.50 GHz which is known as the *DBS band*. It has been agreed that the transponder bandwidths are 27 MHz.

Microwaves and the receiving site

The medium used to transmit signals from satellite to earth is *microwave electromagnetic radiation* which is much higher in frequency than normal broadcast TV signals in the VHF/UHF bands. Microwaves still exhibit a wave-like nature but inherit a tendency to severe attenuation by water vapour or any obstruction in the line of sight of the antenna. The transmitted microwave power is extremely weak by the time it reaches earth and unless well designed equipment is used, and certain installation precautions are taken, the background noise can ruin the signal. A *television receive only* (TVRO) site consists of an antenna designed to collect and concentrate the signal to its focus where a *feedhorn* is precisely located. This channels microwaves to an electronic component called a *low noise block* (LNB) which amplifies and down-converts the signal to a more manageable frequency for onward transmission, by cable, to the receiver located inside the dwelling.

Between the feedhorn and the LNB, a *polarizer* may be located, the function of which will be explained shortly. The complex of feedhorn, polarizer and LNB is often referred to collectively as the *head unit*. Figure 1.6 shows a typical downlink from a medium power satellite to domestic premises.

The antenna

The antenna or 'dish' is concerned with the collection of extremely weak microwave signals and bringing them to a focus. The surface must be highly reflective to microwaves and is based on a three-dimensional geometric shape called a *paraboloid* which has the unique property of bringing all incident radiation, parallel to its axis, to a focus as shown in Figure 1.6. There are two main types of antenna, one is called *prime focus* and the other *offset focus*. Briefly, a prime focus antenna has the head unit mounted in the central axis of the paraboloid whereas the offset focus configuration, as shown in Figure 1.6, has the head unit mounted at the focal point of a much larger paraboloid of which the observable dish is a portion. Antennas are normally made from steel, aluminium or fibreglass with embedded reflective foil. Antennas will be treated in more detail in Chapter 2.

Antenna mounts

The purpose of a mount is to point the antenna rigidly at any chosen satellite. There are two main types of mount, the *azimuth/elevation* (AZ/EL) mount which has simple horizontal and vertical adjustments and the *polar mount* which allows the dish to be tracked across the

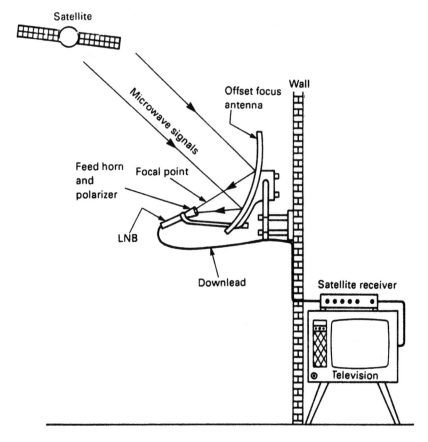

Figure 1.6   *Typical downlink configuration*

entire visible geo-arc, stopping at any chosen satellite. The first type, AZ/EL, is used mainly for fixed single satellite reception although a degree of multi-satellite reception can be provided in certain cases, such as the Astra plan to put five or more satellites in the same orbital slot. A single fixed dish will be able to receive signals from all three proposed spacecraft. Polar mounts are usually motorized and remotely controlled by an indoor positioner and are capable of receiving a fair number of satellites.

The feedhorn

The feedhorn, positioned at the focal point of an antenna, is a device which collects reflected signals from the antenna surface whilst rejecting any unwanted signals or noise coming from directions other than that parallel to the antenna axis. These are carefully designed and precision engineered to capture and guide the incoming microwaves to a *resonant probe* located at the front of the LNB. A feedhorn is really a *waveguide* whose basic theory has been known since the early days of radar, during the Second World War. They normally consist of rectangular or circular cross-section tubes and exhibit two important properties, dictated by waveguide theory. First, signals having wavelengths longer than half the internal dimensions are severely attenuated as the signal progresses down its length. Secondly, wavelengths shorter than the waveguides designed dominant mode become rapidly attenuated; thus the feedhorn behaves, in effect, like a band pass filter. The reason for the fluted horn is to match the free space impedance of the air with that of the waveguide. Feedhorns will be covered in a little more detail in Chapter 3.

Polarization

Current polarization techniques are classified as either linear or circular and are utilized for the following main reasons:

1 *Linear polarization* – A method to extend the number of channels that can occupy a given bandwidth, by using either horizontal polarization (*E* field horizontal to the ground) or vertical polarization (*E* field vertical to ground). This effectively doubles the number of channels that can be provided by a satellite since two channels can share the same frequency, providing they have opposite polarizations. In reality, these channels are staggered to minimize crosstalk (interference) between the two. Two jargon phrases which may cause confusion with regard to polarization are *co-polarized channels*, meaning channels of the same polarization and *cross-polarized channels* meaning they are of opposite polarization. Figure 1.7 shows the two types of linear polarization.

2 *Circular polarization* – This method involves spinning the *E* field of the microwave signal into a spiral or corkscrew as shown in Figure 1.8. The two opposite polarizations this time are:
   (a) Clockwise or right hand circular polarization (RHCP);
   (b) Anticlockwise or left hand circular polarization (LHCP).

Although circular polarization can be used in much the same way as linear polarization, to extend the number of channels, it is more frequently used in high power DBS satellites for a different reason. DBS satellites usually have all their channels fixed at a single polarization either LHCP or RHCP. There is no need to extend the channel capability because this is limited more by power considerations than the numbers of channels. Adjacent DBS satellites in the geo-arc, due to their high power output, usually have opposite polarizations to reduce interference between signals on their earthward journey. Cross-polarization leads to an equivalent suppression in interference in excess of 20 dB and is not noticeable to the viewer.

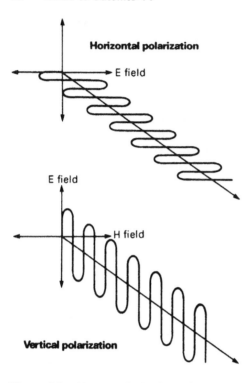

Horizontal polarization

E field

E field

H field

Vertical polarization

Figure 1.7   *Linear polarization of signals*

Polarizers

Polarizers are fitted either between the feedhorn and the LNB or inside the feedhorn itself and fall into three main categories.

1 *The V/H switch type* – These are simply a pair of probes positioned 90 degrees apart. A solid state switch can select the output from one or the other depending on the selected polarization sense. This type is restricted to single satellite systems.
2 *Mechanical polarizer* – This type mechanically rotates a lightweight metal polarizer probe to lie in the plane of the required incoming electric field, that is to say the polarizer probe is vertical for receiving vertical polarized signals and horizontal for receiving horizontally polarized signals. The servo motor automatically positions the polarized probe according to the channel polarity selection stored in the receiver's memory. These polarizers, because mechanical movement is involved, have become less popular recently due to their inherent wear and subsequent unreliability. They are also liable to seizure in very cold weather and are often relatively slow in operation.

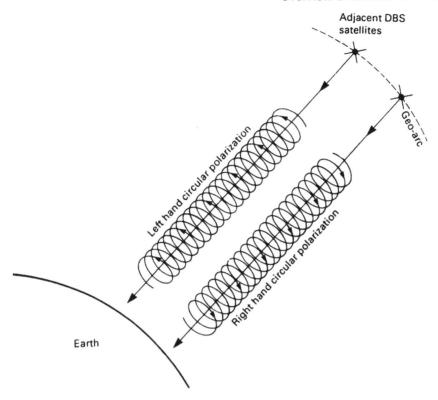

Figure 1.8 *Circular polarization from adjacent DBS satellites*

3 *Magnetic polarizers* – This is the favoured replacement for the mechanical type of polarizer; it consists of a ferrite former wound with copper wire, into which a remotely controlled current is passed. The flow of this current generates a magnetic field which twists the incoming waves, depending on the polarization sense selected, to the orientation required for reception. This type of polarizer causes a slight attenuation of the incoming signal in the region of 0.3 dB. Because magnetic polarizers have no moving parts they are, in the main, reliable. The polarization reference plane is sometimes marked on the casing.

The low noise block (LNB)

The function of a LNB is to detect the weak incoming microwave signals via an internal tuned resonant probe, provide low noise amplification, and finally down-convert the whole block of frequencies to one suitable for cable transmission. It is common nowadays for the combination of feedhorn, polarizer and LNB to be manufactured as a single sealed unit. The entire assembly is often referred to as an LNB, for convenience, but it should be remembered that this is not strictly the case. The Marconi

'LNB' matched with the majority of low cost Astra receivers in the UK have this particular feature. All the components of a LNB are hermetically sealed against moisture.

Satellite receivers

The purpose of a satellite receiver is the selection of a channel for listening, viewing, or both, and transforming the signals into a form suitable for input to domestic TV and stereo equipment. Down-converted signals of about 1 GHz are fed by coaxial cable from the LNB to the input of the receiver. The various subsections of a receiver, which will be treated in Chapter 4, are listed below:

1 Power supply.
2 Second down-conversion and tuner unit.
3 Final IF stage.
4 FM video demodulator.
5 Video processing stages.
6 Audio processing stages.
7 Modulator.

It will be increasingly common to find TV sets with built-in satellite receivers designed to cover both the FSS and DBS band. These will probably be the norm in the years to come. It is hoped that the familiar set top receiver, multiple decoders and spaghetti wiring may gradually disappear as the industry develops into the 1990s. If not, we may need a stepladder in the lounge to operate all the add on boxes of equipment.

Effective isotropic radiated power (EIRP) and footprint maps

An *isotropic radiator* is defined as one which radiates uniformly in all directions. For purposes of illustration it is perhaps better to use a lightbulb analogy. Imagine a 40 W lightbulb suspended from a ceiling so as to be in line with a keyhole. An observer looking through the keyhole would see just a 40 W isotropic radiator. If a parabolic reflector from an old car headlamp is placed directly behind it, then the energy from the bulb will be reflected and magnified in one general direction, toward the keyhole, similar to a car's headlamp on main beam. To an observer, with a restricted field of view, the light source will appear as an isotropic radiator of much higher power. In other words, the effective power appears much higher than the actual power. This effect is somewhat similar to that which occurs with a parabolic transmitting antenna of a satellite. To a distant observer, which in this case is the receiving site antenna, the radiated power appears much higher than that of an isotropic radiator because the transponder antenna has a parabolic reflector and the receiving site antenna ('eye at the keyhole') has a restricted view of the transmitted beam. We know that the EIRP of the Astra 1A satellite is 52 dBW in the central service area, and that the transponder power is 45 W, therefore we can calculate the effective isotropic radiated power in watts as seen by the antenna.

$$\text{EIRP} = 10 \log (\text{effective power})$$
$$\text{effective power} = 10^{(\text{EIRP}/10)}$$
$$= 10^{(52/10)}$$
$$= 158\,489 \text{ W or } 158.5 \text{ kW}$$

From this we can calculate the magnification factor of the transponder's transmitting antenna:

$$\text{magnification} = 158\,489/45$$
$$= 3522 \text{ times}$$

Repeating the calculation for a typical DBS satellite with a transponder power output of 110 W and an EIRP value of 61 dBW in the central service area we get:

$$\text{effective power} = 10^{(\text{EIRP}/10)}$$
$$= 1\,258\,925 \text{ W or } 1.25 \text{ MW}$$
$$\text{magnification} = 1\,258\,925/110$$
$$= 11\,445 \text{ times}$$

As with the lamp analogy the intensity of the beam will fall off as the distance from the main axis increases, since the beam will naturally diverge, in a conic fashion, with distance. A satellite *EIRP footprint* map is constructed by linking contours or lines through points of equal EIRP in the service area. The values will decrease away from the centre, as can be seen in Figure 1.9 which shows footprint maps for the four beams generated by the Astra 1A satellite. The above calculations show that large and unwieldly numbers start to emerge when we talk in effective power terms; this is why EIRP is measured in logarithmic decibel units relative to 1 watt. Remember that a 3 dB increase corresponds to a doubling of power. Therefore the apparent small increases in the values seen on footprint maps correspond to large changes in power levels. In this way relatively small numbers can be used to describe large power changes. Most footprint maps have this characteristic circular shape with EIRP levels falling off linearly away from the main service area.

Free space path loss

As the radiated signal of a transponder travels towards earth it loses power by spreading over an increasingly wide area thus diluting the signal strength. This effect is known as the *free space path loss* and

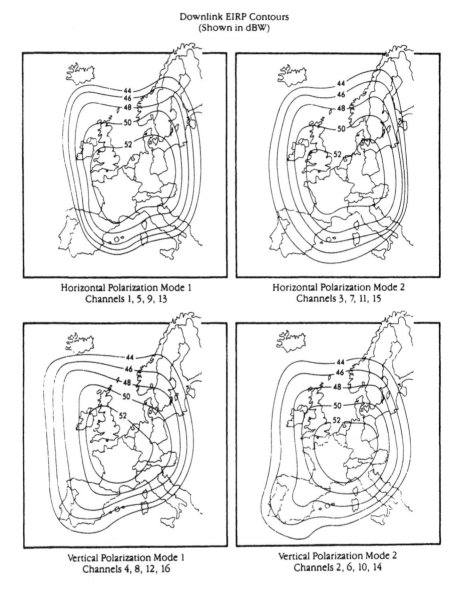

Figure 1.9 *Astra 1A footprint map*
(Source: Société Européenne des Satellites)

the greater the distance the receiving site from the satellite the more it increases. Contributory factors include absorption of microwaves by gases and moisture in the atmosphere. The power density of signals, measured in watts per square metre, finally arriving at earth are extremely weak.

Rain attenuation

One of the major problems with satellite reception is rain, and to a lesser extent snow and hail. The weak incoming microwave signals are absorbed by rain and moisture, and severe rainstorms occurring in thundery conditions can reduce signals by as much as 10 dB (reduction by a factor of 10). Not many installations can cope with this order of signal reduction and the picture may be momentarily lost. Even quite moderate rainfall can reduce signals by 2 to 3 dB which is enough to give noisy reception on some receivers. Another problem associated with rain is an increase in noise due to its inherent noise temperature which is similar to that of the earth. In heavy rain depolarization of the signal can also occur resulting in interference from signals of the opposite polarization but same frequency. This effect is more noticeable with circular polarization.

Noise and its effects

Any body, above the temperature of 0 K or −273°C has an inherent *noise temperature*. Only at absolute zero temperature does all molecular movement or agitation cease. At higher temperatures molecular activity causes the release of wave packets at a wide range of frequencies some of which will be within the required bandwidth for satellite reception. The warmer the body the higher the equivalent noise temperature it will have, resulting in an increase in noise density over the entire spectrum of frequencies. The warm earth has quite a high noise temperature of about 290 K and consequently rain, originated from earth, has a similar value. The characteristic appearance of noise on FM video pictures can be either black or bright white tear drop or comet shaped blobs ('sparklies') that appear at random on the screen. It is subjectively far more annoying than the corresponding snowy appearance of noise on terrestrial AM TV pictures. Video cassette recorder pictures, also frequency modulated, display annoying sparklies as a result of worn/dirty heads or faulty head amplifiers. Only relatively small amounts of FM noise can be tolerated.

Downlink frequency allocations

The ITU has split the world up into three regions. The approximate frequency allocations above 10 GHz are as follows.
Region 1: Europe, CIS, Africa and Middle East
 Fixed satellite service (FSS) band: 10.70–11.70 GHz
 12.50–12.75 GHz
 17.70–21.20 GHz

Direct broadcast service (DBS):   11.70–12.50 GHz
Broadcast satellite service (BSS):   21.40–22.00 GHz
                                     (from 2007)

Region 2: The Americas and Greenland
    Fixed satellite service (FSS) band: 11.70–12.20 GHz
                                         17.70–21.20 GHz
    Direct broadcast service (DBS):   12.20–12.70 GHz
    Broadcast satellite service (BSS):   17.30–17.80 GHz
                                         (from 2007)

Region 3: India, Asia, Australasia and the Pacific
    Fixed satellite service (FSS) band: 11.70–12.75 GHz
                                         17.70–21.20 GHz
    Direct broadcast service (DBS):   11.70–12.75 GHz
    Broadcast satellite service (BSS):   21.40–22.00 GHz
                                         (from 2007)

Travelling wave tube amplifier

On board the satellite, weak signals received from the uplink Earth station are translated to a lower frequency carrier to reduce interference, then retransmitted on the downlink. Ku-band satellites commonly use devices known as travelling wave tube amplifiers (TWTAs) as the final power amplification stage within a transponder because of their wide bandwidth at Ku-band frequencies. Efficiencies as high as 70% or more are easily obtainable. Figure 1.10 shows a simplified schematic diagram of a TWTA. The intermediate anodes and the focusing arrangement have been left out because they contribute little to the understanding.

There are three major components:

1   A long helix of wire through which the RF signal is passed.
2   An electron gun which fires a focused beam of electrons though the helix.
3   An anode held at a critical voltage for determining the beam velocity.

The RF input signal is fed to one end of the helix and the amplified signal emerges at the other end. The signal progresses through the wires at the velocity of light so the main purpose of the helix is to slow down the

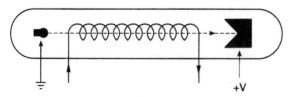

Figure 1.10 *Simplified TWTA schematic*

axial velocity in order for the average *beam velocity* to be synchronous with the RF signal.

Velocity modulation

The action of a TWTA depends on the concept of *velocity modulation.* In the absence of an RF signal on the helix, the electron density of the beam is constant during its journey from the cathode to the anode. However, depending on the phasing, an RF signal will interact with the beam, causing the velocity to increase at certain points within the helix and decrease at other points. Some beam electrons will be slowed down and others will be speeded up. This gives rise to 'bunching', because electrons which have been slowed down will be overtaken by those which have been speeded up, causing the electron density to increase at certain points along the helix. In between these bunches of high density will be troughs of low density. If the RF field increases the beam velocity then it gives up some of its energy, but if the RF field decreases the beam velocity then energy is transferred from the beam to the RF signal. By ensuring that the average beam velocity is comparable with the axial velocity of the RF wave, and the bunches are *slowed down* at the far end of the helix, a travelling wave tube becomes an effective amplifier. The axial velocity of the helix depends on the ratio of circumference to pitch and is, typically, arranged to be around one-tenth the speed of light.

Digital video broadcasting (DVB)

This section introduces concepts of digital video broadcasting, a topic which is covered in more detail in Chapter 11.

As the current millennium comes to end, there will be a rapid transition to digital video broadcasting (DVB) using the MPEG-2 (ISO/IEO 13818) international standard of data compression and digital (phase) modulation of the compressed signals. These standards have been adopted in Europe and many other countries for digital TV transmission via satellite and cable. Owing to the economy on bandwidth and the higher tolerance to noise, there are plans to extend it to terrestrial broadcasting.

The first generation of DVB consumer receivers will be the set-top integrated receiver/decoder (IRD). The receivers will have standard RF and SCART interfaces to connect to antenna, cable and TV/VCR. The up-market models will have a personal computer interface that will be useful for multimedia and Internet use. It is expected that World Wide Web page delivery would be ideally suited to satellite since the band-width is higher.

The bit rate used for a video broadcast can be selected depending on the quality required by the broadcaster. Good VHS video quality can be obtained with a bit rate of 2 Mbit/s. Standard PAL/SECAM/NTSC quality can be obtained with bit rates in the range 4–6 Mbit/s. Studio quality D2MAC and PAL+ can be obtained with bit rates of 8 Mbit/s. A maximum data rate of 15 Mbit/s would be needed for high definition television (HDTV).

For audio encoding the MPEG 11 layer 2 algorithm is used (ISO/IEO 13818-2), which is based on a MUSICAM sub-band coding system. Bit rates in the region of 192 kbit/s are needed to provide near CD quality audio.

The MPEG-2 standard allows for many video, audio and data streams to be combined into one single transport stream for uplinking. This multiplexing technique thus transmits many different programs via a single 38.01 SM bit/s stream on a single satellite transponder. The transport stream used in Europe may be composed as follows:

Data type	Bit rate (kbit/s)	Number of programs
Video	2000	18
Video	9000	4
Audio	192	190
Data	64	590
Data	2.4	15800

The primary modulation scheme adopted for satellite transmission is QPSK (quaternary phase shift keying) and for cable 64-QAM (quarternary amplitude modulation).

2 Antennas

Introduction

The job of the antenna is to concentrate weak incoming microwave signals from a distant satellite to a point where the signals can be collected. It should do this with the maximum efficiency and reject unwanted signals and noise. Satellite TV operates principally in the S, C, Ku and Ka bands. Table 2.1 shows the microwave bands and their approximate frequency ranges.

Table 2.1 *Microwave bands*

Band	Frequency range (GHz)
P	0.2–1.0
L	1.0–2.0
S	2.0–3.0
C	3.0–8.0
X	8.0–10.0
Ku	10.0–15.0
Ka	17.0–22.0
K	26.0–40.0

The vast majority of satellite TV antennas are *parabolic reflectors* and are based on a particular shape known as a parabola, shown in two-dimensional form in Figure 2.1. There are exceptions of course; a yagi array antenna may be used for S-band reception and flat plate antennas for Ku- and Ka-band reception. Other possible methods include lenses.

Parabolic reflectors

A parabolic reflector (dish) has the property of focusing incoming wavefronts, parallel to the main axis, AB, in phase to a single focal point. At the focal point is fitted a head unit which is said to *illuminate* the dish.

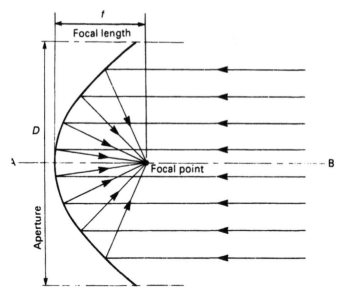

Figure 2.1   *Parabolic reflector*

Any waves entering at angles other than parallel with the main axis, or *boresight*, are reflected so as to miss the focal point altogether as shown in Figure 2.2.

It is often convenient to use optical analogies when dealing with parabolic reflectors since they appear to behave in a similar way to their optical counterparts. Terms such as 'feeds', 'illumination' and 'radiation patterns' often appear to be more relevant to transmission

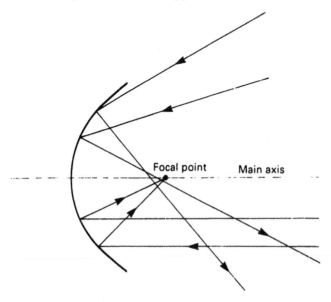

Figure 2.2   *Reflection of incoming ways*

than reception, but since reception is basically the reverse process of transmission, albeit with lower power levels involved, we can loosely say that what is true for transmission is equally true, in reverse, for reception. With this in mind, descriptions of antenna theory often tend to switch, at a whim, between transmission and reception terminology.

The basic equation for a parabola curve is $y^2 = 4fx$, so if we let y be the radius of the antenna (i.e. diameter/2) and x be the depth of the dish and rearrange, the focal length, f, of any parabolic antenna may be expressed by:

$$f = \frac{D^2}{16d} \tag{2.1}$$

where D is the dish diameter, and d is the dish depth.

Since the depth and diameter of a dish can be measured quite accurately, it is always possible to calculate the position of its focal point.

Gain of a parabolic reflector

The antenna gain increases with the effective antenna size which takes into account the efficiency (p), its gain can be expressed by:

$$\text{Antenna gain}\,(G_a) = 10 \log \left(\frac{(\pi d)^2 p}{100\lambda^2} \right) \text{ dB} \tag{2.2}$$

where d = the antenna diameter (m)
$\quad\quad p$ = the percentage antenna efficiency (60–80% typically)
$\quad\quad \lambda$ = wavelength (m)

Note: the antenna efficiency may be specified as a normalized value less than 1 (e.g. 0.67 or 0.80) rather than a percentage. In such cases delete the 100 term in the denominator and substitute the normalized factor for p.

Table 2.2 represents the gain of a parabolic reflector at 11 GHz and shows that for a given frequency the gain increases as the diameter is increased. The gain also increases as the efficiency is increased.

Factors affecting the performance of an antenna

There are many undesirable effects that affect the overall performance of an antenna as illustrated in Figure 2.3 and the following list.

1 Waves, from whichever angle they enter the aperture, are diffracted at the rim and scattered. Some unwanted signals, perhaps from other satellites, may converge on the focal point. Conversely, some wanted signals may be diffracted and miss the focal point.

Table 2.2   *Gain at 11 GHz for a range of dish diameters and efficiencies*

Dish diameter (metres)	Antenna efficiency					
	55%	60%	65%	70%	75%	80%
0.30	28.17	28.55	28.90	29.22	29.52	29.80
0.40	30.67	31.05	31.40	31.72	32.02	32.30
0.50	32.61	32.99	33.34	33.66	33.96	34.24
0.60	34.20	34.57	34.92	35.24	35.54	35.82
0.62	34.48	34.86	35.21	35.53	35.83	36.11
0.65	34.89	35.27	35.62	35.94	36.24	36.52
0.85	37.22	37.60	37.95	38.27	38.57	38.85
0.95	38.19	38.56	38.91	39.23	39.53	39.81
1.00	38.63	39.01	39.36	39.68	39.98	40.26
1.20	40.22	40.59	40.94	41.26	41.56	41.84
1.50	42.15	42.53	42.88	43.20	43.50	43.78
1.80	43.74	44.12	44.46	44.78	45.08	45.36
2.00	44.65	45.03	45.38	45.70	46.00	46.28
2.20	45.48	45.86	46.21	46.53	46.83	47.11
2.50	46.59	46.97	47.32	47.64	47.94	48.22
3.00	48.17	48.55	48.90	49.22	49.52	49.80
3.50	49.51	49.89	50.24	50.56	50.86	51.14
5.00	52.61	52.99	53.34	53.66	53.96	54.24
10.00	58.63	59.01	59.36	59.68	59.98	60.26

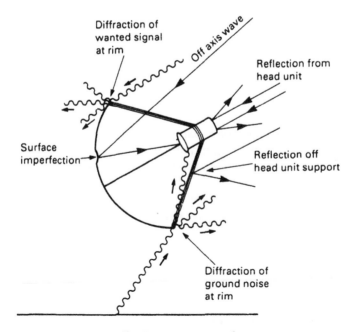

Figure 2.3   *Factors affecting antenna performance*

2 Surface irregularities will cause reflection errors.
3 The head unit and support structure may block and reflect incoming signals.
4 The head unit, occupying a larger space than the focal point will pick up off-axis signals.
5 Ground noise from the relatively warm earth will enter the antenna by diffraction and enter the system via antenna side lobes.
6 Galactic noise will enter the main lobe, although this is a minor contributor below 1 GHz (2.7 K typical at Ku-band which is the background radiation from the 'big bang').
7 Signals may be absorbed by the reflective surface of the antenna.
8 The antenna illumination method adopted by the feed.

Lobe patterns and beamwidth

Ideally, an antenna should exhibit a sharp pencil beam in the direction of the satellite. Unfortunately, since the wavelengths involved are small compared with the aperture (diameter), a fixed focal point is not truly practical. This gives rise to a slightly divergent main beam and some undesirable collection of 'off-axis' signals. The resulting polar diagram consists of a thin pencil beam called the *main lobe* and a series of lower amplitude side lobes as shown in Figure 2.4. Since a polar diagram is often difficult to interpret the Cartesian co-ordinate form is preferred. The normalized theoretical signal response for a uniformly illuminated 0.65 cm antenna at 11 GHz is shown in Figure 2.5. In reality, the other factors, listed above, will contribute to a much more ragged version but it should still be possible to distinguish the overall shape shown.

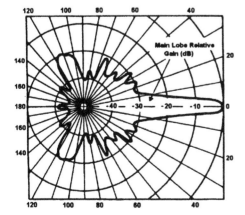

Figure 2.4 *Typical radiation pattern of a parabolic reflector (polar co-ordinates)*

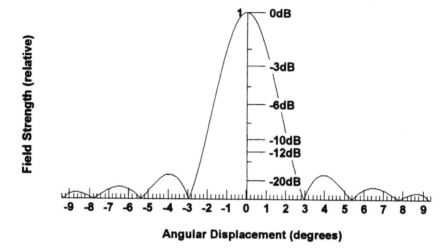

Angular Displacement (degrees)

Theoretical values assume uniform illumination
Antenna Aperture = 0.65 metres
Frequency = 11.000000 GHz

Figure 2.5 *The lobe pattern of a 65 cm parabolic reflector (uniform illumination)*

Ground noise enters the antenna system principally through side lobes so these are arranged to be as low as possible in relation to the main lobe amplitude. A uniformly illuminated antenna theoretically produces the first and largest of these side lobes at about −17.6 dB down on the main lobe maximum.

The illumination is rarely uniform in practice. The exact distribution being dependent on the type of feed adopted. This leads to the concept of the effective area or the efficiency of an antenna system. In other words, most signal power is collected from the central area of the dish and is tapered off to the outside edge. The wasted aperture can be thought of as a shield against ground noise.

Under-illuminating the dish reduces the first side lobe level to less than −20 dB thus decreasing ground noise pick-up. This seems ideal, but there are a number of draw backs, one being a reduction in gain and another is a corresponding increase in beamwidth. The half-power beamwidth is taken as the width of the main lobe at a point −3 dB down. The equations used to calculate the beamwidth at any specified level of the main lobe are fairly complex and tedious to perform, but the −3 dB beamwidth, first side lobe amplitudes and position of the first null, depending on the illumination adopted can be calculated easily from the expressions given in Table 2.3. As can be seen from the table, the cosine distribution is close to the average if the illumination method adopted is not known and may be used as a first approximation of the −3 dB beamwidth.

Table 2.3

Ilumination method	 ◄─ D ─► Aperture distribution	−3 dB Beamwidth (degrees)	1st sidelobe amplitude (dB)	1st null (degrees)
Uniform		$\dfrac{58.4\lambda}{D}$	− 17.6	$\dfrac{69.9\lambda}{D}$
Cosine		$\dfrac{72.8\lambda}{D}$	− 24.6	$\dfrac{93.4\lambda}{D}$
Cosine²		$\dfrac{84.2\lambda}{D}$	− 30.6	$\dfrac{116.3\lambda}{D}$
Pedestal		$\dfrac{66.5\lambda}{D}$	− 26.5	$\dfrac{86.5\lambda}{D}$

Antenna efficiency

The antenna efficiency is the percentage of the incoming signal finally arriving to be collected at the focal point. Efficiencies of antennas generally lie in the range 60–80%. The main factors which determine efficiency are given below.

1 Some of the parallel-to-axis wanted signals will be diffracted at the rim of the dish and will be lost from the focal point. This is governed by physical laws.
2 Surface imperfections will cause reflection errors. Manufacturing quality can reduce this effect by reducing the so-called *RMS deviation* of the surface.
3 With prime focus dishes, with a centrally mounted head unit, both the head unit and the support will block incoming signals. This can be reduced by using an offset focus design.
4 The surface of the antenna will absorb a certain percentage of the incoming signal.
5 The feed illumination method adopted can significantly affect the efficiency. The practice of tapering the amount of signal collected toward the edge of the dish reduces ground noise pick-up but unfortunately also reduces the efficiency.

Antenna noise

Any signal received is combined with an element of noise which degrades the overall performance:

Signal = wanted signal + noise

Obviously, the noise component must be kept as small as is possible, taking into account cost and available technology. Noise can come from many sources and is produced by the thermal agitation of atoms and molecules above absolute zero (−273°C or 0 K; note that the degree sign is not used on the kelvin scale). This is why noise is said to have an equivalent noise temperature. The noise temperature of the earth is normally standardized at 290 K (17°C). There are three main sources of noise in the environment:

1 *Extraterrestrial noise sources* – This is wide bandwidth radiation caused by the energy conversion in stars and the residual background radiation of the 'big bang'. This tends to taper off at 1 GHz and settles to that of the residual background noise alone which is taken as 2.7 K. Above 2 GHz, there are only a few isolated points of very strong non-thermal noise, principally from Cygnus A, Cygnus X, Cassiopeia A and the Crab nebula. There is also a narrow band of increased noise from the Milky Way. The Sun is an enormous source of noise at around 10 000 K at 12 GHz and the Moon at about 200 K. This noise enters the antenna mainly via the main lobe.
2 *Man-made noise* – This noise emanates from microwave pollution due to man's electrical activities and principally enters the antenna via the side lobes.
3 *Ground noise* – In the long term, this is the major component of noise incident on the antenna aperture, and depends mainly on the antenna diameter, antenna depth, and elevation setting. The smaller the diameter of the dish the wider and more spread out will be the side lobes, so more noise will enter from the warm earth. The noise temperature also increases as the elevation angle decreases, since lower elevation settings will pick up more ground noise due to side lobes intercepting the ground (diffraction effects at the antenna rim). This may be reduced by various methods of feed illumination. The design of the antenna itself also plays a part. A deep dish picks up less ground noise at lower elevations than do shallow ones, also prime focus mounted head units will add to noise since it is 'seen' at the same temperature as the Earth. Inclining the head unit away from the earth and towards the cool sky as happens in the case of an offset focus design can also improve things. This practice tends to counteract the negative effects of increased beamwidth for small antennas set at low elevation angles. Table 2.4 gives approximate values for ground noise (see Chapter 5 for an equation for calculating the approximate antenna noise).

Illumination, noise and *f/D* ratio

A parabolic reflector can be shallow or deep depending on the slice of parabola envisaged during manufacture. Referring to Figure 2.6 it is

Table 2.4 *Typical ground noise v antenna diameter at various elevation settings (K)*

Antenna size (m)	Elevation angle (°)			
	15	20	25	>30
3.7	35	31	29	29
3.0	36	32	31	30
2.3	40	36	35	34
1.8	45	41	39	38
1.3	51	45	42	41
0.9	59	56	54	53
0.6	76	73	71	70

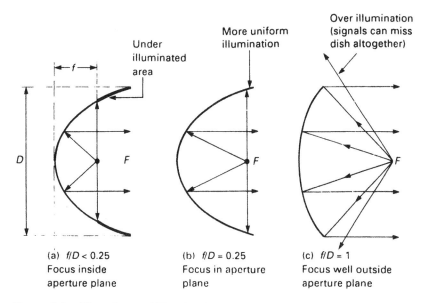

(a) f/D < 0.25
Focus inside
aperture plane

(b) f/D = 0.25
Focus in aperture
plane

(c) f/D = 1
Focus well outside
aperture plane

Figure 2.6 f/D *ratios and illumination*

difficult to get uniform illumination over such a wide angle with configuration (a); the effect is known as 'under-illumination'. On the other
hand, it is difficult to collect a high proportion of signal with configuration (c); the effect is known as 'over-illumination'. The condition giving
maximum gain is configuration (b) where the dish is uniformly illuminated. Figure 2.6(b) is where the focal point is in the aperture plane and
parabola geometry dictates that this is when the focal distance/diameter
ratio, f/D, is 0.25. Deep dishes with low f/D ratios tend to have lower
amplitude side lobes at the expense of gain so are more shielded from

ground noise entering beyond the rim. In practice, the f/D ratio is normally greater than 0.25 but less than 1 for aesthetic and material usage reasons, so the feed is designed to counter the effects of over-illumination. In other words, the feed is designed to compensate with a narrow beamwidth so as not to over-illuminate the dish. Typical f/D ratios for prime focus dishes are 0.25 to 0.6 and for offset fed dishes 0.6 to 0.7.

Prime focus antennas

The typical 'Jodrell Bank' type parabolic dish is called 'prime focus'. The head unit is positioned at the focal point at the centre of the dish as in Figure 2.3. Large C band antennas often have this method of feed, due to its inherent mechanical stability. For medium to high power Ku-band reception its use is rare.

Offset focus antennas

The offset focus configuration is most often used for small dishes designed for use with medium to high power semi-DBS and DBS (direct broadcast service) satellite services operating in the Ku and Ka bands. Figures 2.7 and 2.8 show that the offset focus dish is really a section cut out of a much larger parabola (this does not imply that they are manufactured this way!). The main advantages are that the head unit is out of the way of incoming signals, thus eliminating blockage, and the head

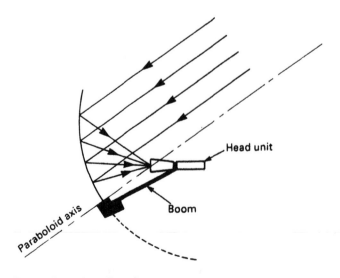

Figure 2.7   *An offset focus antenna*

unit is inclined away from the earth toward a cool sky. A further advantage is that snow and other debris can slide off the dish more easily. The actual elevation setting of the dish itself is reduced by the angle of offset. As can be seen from Figure 2.8, the lobe pattern for the minor diameter would have a wider main lobe than that of the major diameter, so a flattened elliptical beam as shown in Figure 2.9 would be expected. From the cut-out shape shown, where the dish is higher than it is wide, it would be expected that the elevation is more critical to adjust than is the azimuth. Sometimes the dish can be made wider than it is high to minimize interference pick-up from adjacent satellites. This can be envisaged by rotating the ellipse in Figure 2.9 through 90°.

The Cassegrain antenna

Figure 2.10 shows another possible configuration for domestic satellite reception, called the Cassegrain antenna. However, it is little used because of the added expense of the hyperbolic subreflector although it does have certain advantages. The profile may be slimmer, since the subreflector intercepts reflected waves before their normal prime focal point and re-reflects them back to a rear mounted head unit. The main disadvantage is that the subreflector blocks some of the incoming signal, however this may be overcome by using an offset design. The hyperbolic subreflector has to be in the order of 10 wavelengths diameter in order to be efficient. As a general rule, if the diameter of the main reflector is greater than 100 wavelengths, the Cassegrain antenna becomes a real contender. For smaller sizes the traditional front-fed design is normally preferred.

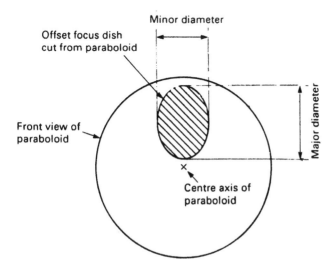

Figure 2.8   *Offset dish cut from a larger parabola*

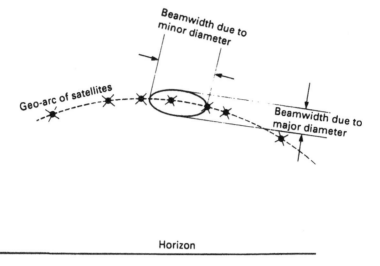

Figure 2.9   *The elliptical beam of an offset focus antenna*

The Gregorian antenna

The Gregorian antenna differs from the Cassegrain antenna in that the hyberbolic sub-reflector is replaced by an elliptical surface. Again, the antenna can be prime focus or offset fed. Figure 2.11 shows the offset focus example. This configuration can inherit high efficiency since the advantages of offset design are incorporated with those attributed to rear mounting of the low noise block (LNB).

1   The LNB is shielded from the weather and extra noise introduced from a hot sun is diminished.

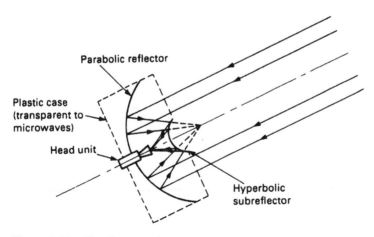

Figure 2.10   *The Cassegrain antenna*

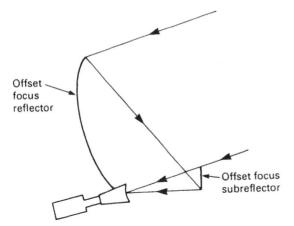

Figure 2.11 *The Gregorian antenna*

2   The feed can be made to 'see' the sub-reflector more efficiently as sky noise, rather than ground noise, is the result of over-illumination (see Chapter 3).
3   The sub-reflector can be made to see the edge of the dish more efficiently.

The backfire antenna

Following the Gregorian antenna principle and sharing its advantages is the backfire antenna (Figure 2.12).

Sub-reflectors are usually in the order of 5 wavelengths in diameter or approximately 10% of dish size. The main disadvantage of this system is the blocking of the incoming signal by the positioning of the deflector plate.

The flat antenna

Instead of concentrating the signal power to a single focal point as with parabolic reflectors or lenses, it is possible to collect it in an array of smaller discrete aerial elements, in the form of slots or holes, distributed over a flat surface. The aerial elements in the outer surface can be connected by a series of equal length printed circuit transmission lines to a common output where a LNB is positioned. The substrate which keeps the printed circuit transmission lines apart is specially designed to minimize losses, although some designs may be air spaced. Secondary arrays or parasitic array elements may be incorporated to suppress unwanted polarizations. The practice of tapering down signal sensitivity toward the outside edges of the aperature, reduces side lobes. Size for

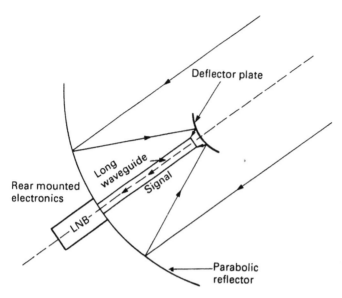

Figure 2.12 *The backfire antenna*

size, flat antennas tend not to be so efficient as parabolic reflectors, and they also inherit larger side lobes. On the plus side, feedhorns are redundant, and overall they tend to be more aesthetically pleasing.

Dual view dishes

Dual view dishes are a novel idea for capturing signals from satellites in two different regions of the geo-arc with a single fixed mounting. These two faced or kipper dishes usually provide for the fitting of two separate LNBs on a single fixed boom. A portion of the available antenna area is dedicated to collect signals from one satellite while the other portion collects signals from the other. Although these creations are perhaps cheaper than two separate dishes, it is important to perform a site survey since the number of possible mounting sites may be restricted due to blocking of signals by trees, buildings, etc.

Lenses

The parabolic reflector can be replaced by a lens which has similar properties in focusing incoming radiation. An analogy can be made with the optical telescope. At optical frequencies there are two major classes of telescope: the reflector and the refractor. A parabolic dish can be loosely thought of as a parabolic mirror (reflector) concentrating incoming radiation to a focus where an eyepiece (feed) is positioned. Likewise an optical refracting telescope using a lens rather than a mirror has its centimetric wave equivalent. Lenses operate by having a refractive index greater than unity, and when used at centimetric wavelengths are often made from dielectric material which has the property of slowing down travelling waves so bringing them to a focus. Figure 2.13 shows the general dielectric lens action and the corresponding ray diagram.

Gains and radiation patterns for lenses are similar to those for parabolic reflectors and an illumination taper is needed for the same reasons. It is not easy to achieve uniform illumination of a lens and it emerges that those with long focus are more evenly illuminated than those with short focus. Also, long focus lenses are lighter and thinner (Figure 2.14). However, lenses can be stepped to reduce weight and bulk without affecting performance too much.

Feeds for lenses fulfil a similar function as those for reflectors. In practice, lens focal length is chosen to be of the order of the lens aperture dimension. This is greater than that usually associated with a parabolic reflector and consequently the feed needs to be more directional. Horns are commonly used and, to avoid stray radiation entering the horn, sides of the horn are conveniently extended right up to the lens

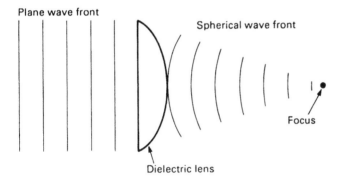

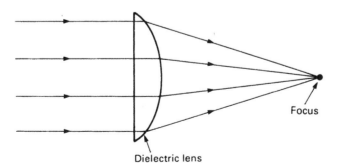

Figure 2.13 *Dielectric lens action*

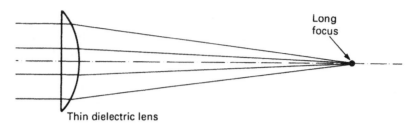

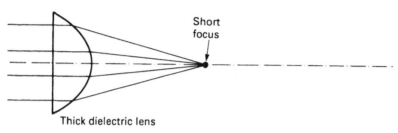

Figure 2.14 *Long and short focus lenses*

as illustrated in Figure 2.15. Although the illustration appears to super-ficially resemble a conventional feedhorn in appearance, actual size is at least a magnitude greater to work as an antenna in its own right. The lens aperture can approach the diameter of the corresponding reflector size for a given band. The refractive index of dielectric lenses can also be altered by introduction of a volume distribution of metal spheres, rods or discs, and thus artificial dielectric materials can be manufactured.

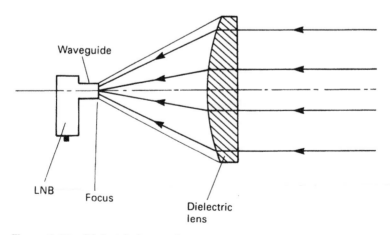

Figure 2.15 *Dielectric lens antenna*

S-band Yagi arrays

There is no clear dividing line between aerial type for one frequency band and another, so inevitably there will be some overlap. For S-band reception in the range 2.5 GHz to 2.7 GHz it is common to use a Yagi array similar to that used for UHF TV reception. The gains of such arrays vary from about 24 dBi to 28 dBi or more which is equivalent to a 3.0 m dish. The main advantages of using a Yagi over a parabolic reflector is its size, wind loading and ease of installation. When receiving central footprint S-band signals from craft, such as Arabsat, it is normally sufficient to use two phased antennas fitted in a special phase-matching harness. Four antennas would be needed for fringe areas such as northern Europe and in some locations in the southern Mediterranean. Figure

Figure 2.16 *A typical S-band Yagi array (Tomira International)*

2.16 shows a typical Yagi used for the reception of Arabsat, the LNB is directly connected to the Yagi and the intermediate frequency (IF) lead from the LNB is then connected to the receiver. To install simply set to the computed elevation and azimuth then peak for maximum signal strength.

Electronically aimed antennas?

The ideal situation for multi-satellite reception, in the future, will possibly move towards a flush-mounted wall antenna which is electronically selective of any required satellite signal viewable in the geo-arc. In other words, virtual directivity of the antenna in both elevation and azimuth might be performed by electronic methods or otherwise. Cumbersome, mechanical means of moving antennas are, by their nature, doomed to eventual extinction. Research is going on in this area at the moment and it may be just a matter of time before such a system is launched on the market with both a usable efficiency and, more importantly, at reasonable cost. Perhaps 'mushroom farms', frowned on today, are only a temporary phase.

Antenna mounts

A mount should rigidly and accurately target an antenna onto the chosen satellite. There are three basic types:

1 AZ/EL fixed mounts.
2 Twin axis AZ/EL robotic mounts (sometimes grouped as horizon-to-horizon mounts).
3 Polar mounts (including single axis horizon-to-horizon mounts).

Fixed AZ/EL mounts

This is the simplest mount of all and is simply clamped to a specific azimuth and elevation setting in order to receive signals from a fixed region of the geo-arc. They are easy to install and are cheap to manufacture. Most satellite installations are of this type; this is encouraged by the tendency of fixed satellite service (FSS) operators to cluster more than a single satellite in an orbital slot. An example is the intended five strong Astra cluster at 19.2°E where five satellites with 16 to 18 active transponders each are all co-located in the same region of sky. Figure 2.17 shows how the azimuth and elevation settings relate to the geo-arc and Figure 2.18 shows a typical fixed AZ/EL mount.

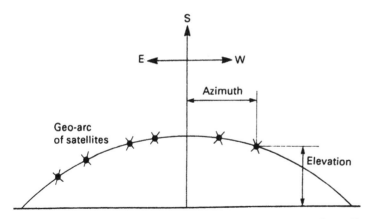

Figure 2.17 *How AZ/EL settings relate to the geo-arc of satellites*

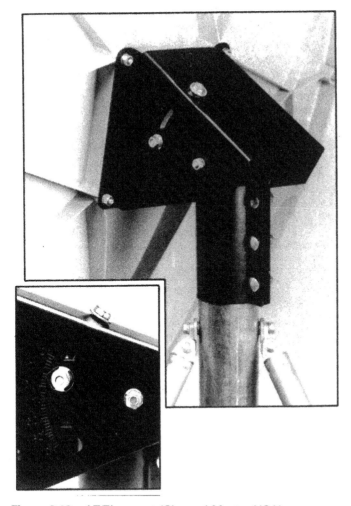

Figure 2.18 *AZ/EL mount (Channel Master USA)*

The modified polar mount

The typical low cost method of tracking the geo-arc of satellites is the modified polar mount. Movement is restricted to a single axis in such a way as to make the net elevation vary as the dish is swung from horizon to horizon. The lowest cost method is to drive the dish using a motorized screw jack called a *linear actuator* as shown in Figure 2.19. Polar mounts have been used by astronomers for years to compensate for the Earth's rotation while observing celestial bodies.

Traditionally, the polar axis angle is set to the latitude of the site so that rotation around the polar axis can track a distant fixed object as the Earth rotates. The polar axis is parallel to a line intersecting the north and south poles so the antenna will point straight out into space as shown in Figure 2.20, tracing the arc shown in Figure 2.21. An angle, depending on latitude, known as the 'declination offset angle' is then introduced so that the antenna is lowered onto the geo-arc of satellites, as shown in Figure 2.22. The standard declination offset angle is the difference between the apex dish declination (i.e. 90° minus the apex elevation or slope of dish) and the polar axis angle (latitude). The 'apex' meaning the highest observable point of the geo-arc which is due south in the northern hemisphere or due north in the southern hemisphere. A confusing habit which abounds is to abbreviate the term 'declination offset' to simply 'declination'. This could be misleading, since traditionally 'declination' has been reserved to describe the angle corresponding to the slope of the dish face.

Figure 2.19   *Polar mount assembly (Channel Master USA)*

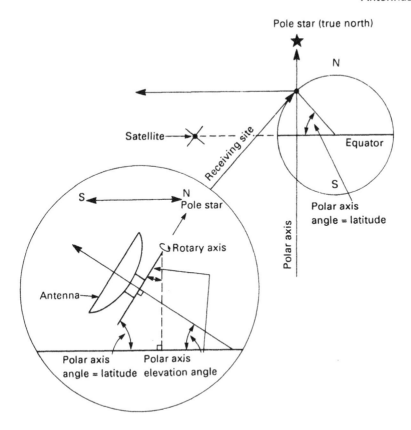

Figure 2.20   *The polar mount with 0° declination offset*

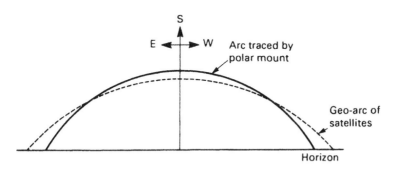

Figure 2.21   *Arc traced out by the polar mount with 0° declination offset*

Due to the relative closeness of the geo-arc to the Earth's surface, the circular arc traced out by the above set-up does not exactly match the more elliptical geo-arc as illustrated in Figure 2.23. A modification attributed to the English engineer, S. J. Birkhill, back in the late 1970s, consisted of introducing a small, yet significant, correction factor to counteract this tracking inaccuracy. This angle is always less than

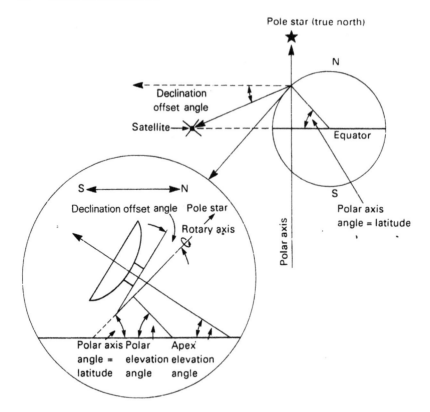

Figure 2.22 *Polar mount with declination offset*

0.75°. In effect, the polar axis angle is increased by tilting it slightly forward thus reducing the previously described declination offset angle by a similar amount. Since then, the modified polar mount setting has been universally agreed and adopted. Equation 2.3 may be used to calculate directly the modified declination offset (α), and Equation 2.4 may be used to calculate the small correction factor (β) which is added to the polar axis.

$$\alpha = 90° - tan^{-1}\left(\frac{\sqrt{m^2 - \cos^2 \phi}}{\sin \phi}\right) \text{(degrees)} \tag{2.3}$$

$$\beta = tan^{-1}\left(\frac{\sqrt{m^2 - \cos^2 \phi}}{\sin \phi}\right) - tan^{-1}\left(\frac{m - \cos \phi}{\sin \phi}\right) \text{(degrees)} \tag{2.4}$$

where: ϕ = latitude of site (degrees)
   α = modified declination offset angle (degrees)
   β = the small correction factor to the polar axis (degrees)
   m = ratio of the geostationary orbit to the radius of the Earth (i.e. 6.61).

The graph of Equations 2.3 and 2.4 are plotted for various latitudes in Figures 2.24 and 2.25, respectively. The angles calculated are accurate to within ±0.05°. If you tabulate or plot out the graphs of Equations 2.7 and 2.8 for various latitudes, you will see that they correspond, with negligible error, to those produced using Equations 2.5 and 2.6. You will find the latter equations considerably easier to calculate. There will be more discussion regarding the installation of modified polar mounts in a later chapter together with reliable angle setting tables for various latitudes. It is sufficient here to indicate how the angles α and β relate to the overall angle terminology of polar mounts using the symbols defined above.

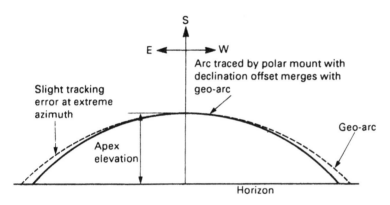

Figure 2.23 *Arc traced by polar mount with declination offset*

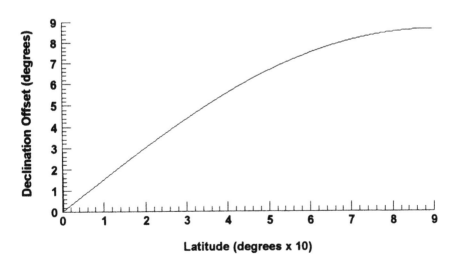

Figure 2.24 *How the modified declination offset angle, α, varies with latitude. This is the reduced declination offset angle resulting from the slight forward tilt of the polar axis in a modified polar mount*

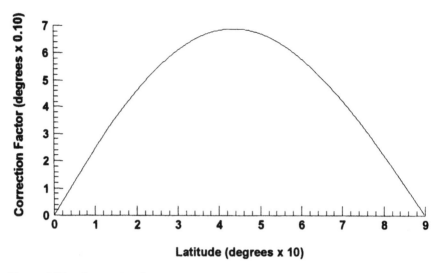

Figure 2.25 *Correction factor, β, applied to polar mount geometry. This correction factor is added to the polar axis angle (latitude) to enable accurate tracking of the geo-arc*

Apex declination (due south slope of dish face) $= \alpha + \beta + \phi$
Apex elevation $= 90° - (\alpha + \beta + \phi)$ or ($90° -$ apex declination)
Polar axis $= \phi + \beta$
Polar elevation $= 90° - (\phi + \beta)$ or ($90° -$ polar axis)

$$\alpha \approx 8.60 \sin \phi \quad \text{(degrees)} \tag{2.5}$$
$$\beta \approx 0.69 \sin 2\phi \quad \text{(degrees)} \tag{2.6}$$

A typical low cost polar mount assembly uses a screw jack similar to that shown in Figure 2.19. Actuators come in a variety of sizes to suit particular dish sizes and the motor is driven mostly from 36 volt or occasionally 24 volt or 12 volt supplies from the positioner control unit. The limit of travel of a screw jack restricts the tracking of the geo-arc to about ±50° (i.e. 50° to the east and 50° to the west). To prevent the jack from being overdriven in either direction, adjustable limit switches are often provided.

Horizon-to-horizon mounts

The term 'horizon-to-horizon', or H–H, mounts, is used to distinguish them from simple jack operated polar mounts with restricted travel and are mainly used with small antennas up to 1 metre or so, but a few models are available for dishes as large as 2.5 metres. Single axis H–H mounts are polar mounts, but there are also two-axis versions which are really motorized AZ/EL mounts (see next section). Single-axis H–H mounts are directly driven using a motor and gearbox and can span an azimuth of 180°, hence the origin of their name. Figure 2.26 shows a

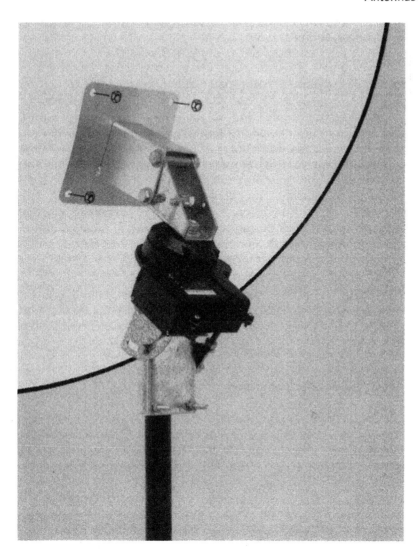

Figure 2.26 *A typical single axis H–H antenna mount (Oakwood International Corporation)*

typical single-axis version; the polar elevation is set from the elevation angle plate and the apex delination angle from the upper declination bracket. Although there are many mechanical arrangements according to manufacturer, the above two adjustments still have to be made to the mount. (See Chapter 6 for details of setting all polar mounts). The number of pulses per degree from the position sensor varies from about 3 pulses per degree to 4.5 pulses per degree, depending on the manufacturer. Versions can be obtained with reed switch, optical or potentiometer position sensors. Adjustable limit switches are incorporated to prevent damage to the dish. The setting range for the polar elevation of H–H mounts is typically 0 to 60° or 20° to 70°, and that for the apex declination is 0 to 40°.

Robotic AZ/EL mounts (twin axis H–H mounts)

These multi-satellite mounts are two-axis motorized in both azimuth and elevation, and are capable of targeting any chosen satellite in the geo-arc with unsurpassed accuracy. Mounts of this type are often grouped as a variant of H–H mounts, but it is important to remember that they do not model a polar mount in operation as do single-axis H–H mounts. Sophisticated dual digital servo mechanisms acting under microcomputer control drive a pair of geared motors which can target the antenna to any pre-programmed AZ/EL co-ordinates stored in memory. In conjunction with a suitably programmed electronic control system, they can be made to track satellites in inclined geosynchronous orbits. This would be impossible with a single-axis polar mount. The cost is high compared with actuator type polar mounts and single-axis H–H mounts, so this may account for their comparative rarity. A more elaborate type of positioner unit would be needed to drive a twin-axis system.

Motorized positioning systems

Actuators and H–H mounts are controlled remotely by a unit known as a *positioner*. An antenna positioning system, in general, works like this. On selecting a particular satellite, the current position count is compared with that programmed into the memory and the motor drives the antenna in the appropriate direction. The instantaneous position count is continuously compared with the value in the memory until such time as the two counts match, at which point the supply to the motor is cut. The position sensor or transducer can be a reed switch, Hall effect transistor or an optical counter. The positioner also has inbuilt safeguards to stop the actuator or H–H motor from being driven beyond certain software selectable limits to prevent damage to the dish. Motor supplies are typically 36 volts. A simplified schematic of a position count feedback system is given in Figure 2.27.

Multi-feed antennas

Another method of enabling multi-satellite reception using a single antenna is to use a dual or multi-feed arrangement with a fixed dish. If the feed of a parabolic antenna is displaced laterally from the focal point, the beam is squinted or scanned away from boresight in the

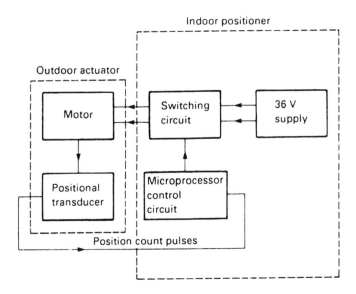

Figure 2.27 *A simplified schematic of a positioner unit*

opposite direction. There is a corresponding gain loss and beam broadening associated with this displacement, together with the suppression of sidelobes in the direction of the scan and the appearance of a trailing coma sidelobe as shown in Figure 2.28. A displacement of θ degrees of the feed results in a beam displacement of between 0.8 θ and 1.0 θ. This figure is sometimes referred to as the beamfactor of the antenna.

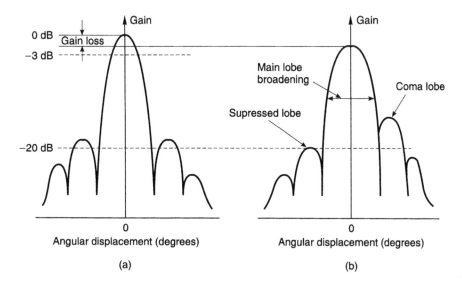

Figure 2.28 *(a) Normal beam pattern; (b) squinted beam pattern*

The gain loss depends on the feed displacement and the antenna's focal distance to diameter ratio, *f/D*. Figure 2.29 shows a universal squint diagram for antennas of any size and shows the gain loss versus beam squint angle in half-power beamwidths (−3 dB). The values are plotted for three different *f/D* ratios. The loss for other *f/D* values may be estimated by interpolation.

For practical purposes the squinting process can be increased, by up to 10 beamwidths, until the gain loss and comma lobe interference produce unacceptable results. Large squint angles are associated with aberrations such as astigmatism, excessive beam broadening and filled-in pattern nulls.

As can be seen from Figure 2.29, antennas with a high *f/D* ratio (*f/D* = 1.0) are more tolerant of feed displacement than those with low *f/D* ratios (*f/D* = 0.25). Specially designed antennas to exploit this effect are available. A typical example is the OA-1600 antenna manufactured by Swedish Microwave which is pictured in Figure 2.30. This is a 1.65 m high efficiency, 13° offset multi-focus dish capable of receiving up to 9 Ku band satellites within a 26° angle. The surface is specially designed from a combination of spherical and parabolic geometry. Antennas of this type would be used by commercial operators and cable companies. Standard domestic antenna designs are also capable of dual feed operation for the domestic market but are less efficient. A typical arrangement in Europe would be a set up designed to receive the Astra/Hotbird combination at 19.2°E and 13°E respectively. The dish would be aimed at

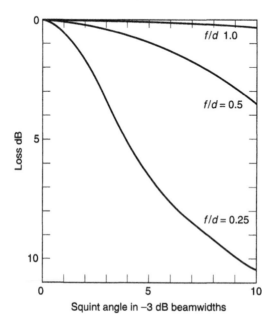

Figure 2.29   *Gain loss versus squint in −3 dB beamwidths for various f/d ratios*

Figure 2.30 *Wide scan antenna showing multi-feed arrangement (Source: Swedish Microwave)*

Astra (boresight) and the second LNB would be displaced laterally to receive signals from Hotbird.

There are also antennas available to receive C-band signals in a similar way but due to the relatively close spacing of C-band satellites the feeds would be mounted so close together that mechanical interference problems may result.

Antenna construction

Most modern antennas are constructed using one of the following three main techniques.

Spinning

Flat sheet metal is placed onto a parabolic mould and the whole assembly spun. A forming tool or roller then gradually changes the shape of the sheet until the familiar dish shape emerges. A large range of dish sizes can be made from the same parabolic mould and are consequently cheap to produce.

Pressing

Flat sheet metal is pressed into the required shape using a hydraulic press with a male and female die. Water is often utilized to help produce a good smooth and reflective surface. The process is similar to that of pressing car body panels. Different pairs of die need to be used for each antenna size.

Foil embedded in fibreglass

This method uses lightweight metal reflective foils embedded in a solid fibreglass paraboloid dish. The foils are placed in a mould and then injection moulding techniques complete the process. Fibreglass antennas are slightly less efficient than their solid metal counterparts since microwaves need to penetrate the fibreglass layer reflect and then re-emerge. Some of these will be absorbed in the process. This method of construction is common with offset feed designs.

Painting

A common enquiry by customers is 'can we paint the dish a different colour?'. Although not recommended, the answer is yes, as long as the same principles are applied as that in manufacture. The paint should be applied as smoothly as possible, any bumps or drips may cause reflection errors. The paint used should not be optically reflective such as metallic paints or gloss finishes otherwise the Sun's radiation may become focused on the head unit and cause problems. Always use 'vinyl matt' finish paints which exhibit lower solar reflection properties. The amount of microwave absorption and reflection errors using this type of paint are minimal in practice and no loss of performance should be noticed.

Coating

A number of spray and brush-on antenna coatings have been developed to improve reception (hence, reduce sparklies) in bad weather. These coatings are offshoots of similar compounds used by the military and are claimed to keep antenna surfaces free of residual rain drips, ice and snow. The compounds are expensive but may be economical if you live, say, in the Pennines or Scottish highlands.

Visual impact of antennas

Satellite antennas vary widely in appearance and general quality. Some degenerate into ghastly rust-stained monstrosities in less than a couple of years, while others of similar age can look as if they were installed yesterday. Mounting brackets in particular often look like they are reclaimed from a sunken battleship. By choosing materials which inherit a certain resistance to corrosion, you can do your bit to improve the environment. The use of glass or aluminium mesh dishes may also be a step in the right direction in this respect.

Glass dishes

A type of glass known as 'metallized armour plate glass' manufactured by Pilkington Glass is known to be 99% reflective at centimetric wavelengths. This material can be formed into tubes, plates and reflectors and can be assembled into various antenna systems. Armour plate glass is reputed to be five times stronger than window glass, and coupled with a clear polycarbonate mounting bracket this would seem the ultimate environmentally friendly antenna. There are certain other advantages. The glass holds its shape better than metal in windy conditions and heating elements similar to those fitted in car rear windows could be incorporated to keep the dish clear of snow and ice where necessary.

Mesh dishes

Mesh or perforated dishes are environmentally an improvement overall, they are certainly less noticeable. A considerable reduction in wind loading is achieved for a given size over its solid metal counterpart. Perforated surfaces are insensitive to polarization and can be regarded as short waveguides, the frequency of which is well beyond the cut-off of the operational frequency. Mesh dishes need careful handling since they are easily deformed, but provided the deviation from a pure parabola is less than one-sixteenth of a wavelength of incoming radiation this is not a major problem.

Solar outages

Around the spring and autumn equinoxes, when the day and night are about the same length, the Sun crosses the equator and traces an arc that is directly behind the geo-arc of satellites. This momentarily disrupts satellite reception and causes a phenomenon known as a *solar* or

Sun outage. The exact date, time and duration of such events depends on the receive site location, the satellite in question, the Earth station antenna beamwidth or focal resolution, the station keeping accuracy of the satellite and, of course, the accuracy of antenna pointing.

The Sun is a powerful broadband microwave transmitter and has a noise temperature well in excess of 25 000 K. As the Sun passes directly behind the satellite, when viewed from Earth, reception may be degraded or sometimes even swamped by the overwhelming noise from the Sun. The observable result on the screen is impulse noise (sparklies).

The outage may last several minutes either side of the peak each day during the season and will last longer the smaller the antenna. Similarly the event will occur for several days both before and after the peak day, and therefore outages will occur at roughly the same time each day and may repeat on a daily basis for a week or more.

Damage to receiving equipment is also possible due to the intense heat from the Sun being focused onto the head unit. Rare stories of protective caps melting and dramatic LNB explosions due to ingressed water vaporization have been reported, so if you have expensive equipment, and a solid white painted antenna, it may be a good idea to shield the head end during these events.

A precise knowledge of when these events occur is useful. If you booked transponder time to relay, say, a news feed and it happened just to coincide with a solar outage on the downlink you might be slightly annoyed. You could use the advance information to book a transponder at an alternative time or on another satellite.

The calculations needed to predict solar outages are complex and require a suitable computer program. One such program is Satmaster Pro outlined in Appendix 2.

3 Head units, cables, line amplifiers and connectors

Head unit

A head unit is a convenient group name for the assembly positioned at the focus of a dish antenna and comprises the following component parts:

1 Feedhorn.
2 Polarizer.
3 Low noise block (LNB).

The basic operation of these components has been outlined in Chapter 1. However, the treatment in this chapter is more concerned with why feedhorns and LNBs are needed at all!

Why do we need a feedhorn and LNB?

Signals passing through a length of coaxial cable suffer attenuation per unit length due to:

1 Dielectric losses in the material necessary to support the inner conductor.
2 Skin resistance due to the finite diameter of the inner conductor.
3 Radiation loss due to the coaxial cable operating as an aerial.

All these losses increase with signal frequency. However, by correct choice of cross-sectional dimensions and materials, the attenuation due to these losses is acceptable providing the signal frequencies are not too high! It is unfortunate that broadcast signals from the current satellites are in the 11 to 12 GHz range and coaxial cables, except in very short lengths, are unable to cope. There are two ways of overcoming the problem:

1 Employ frequency conversion techniques at the aerial head to lower the frequency sufficiently for subsequent handling by coaxial cable.
2 Abandon coaxial cable altogether and pass the signal down the inside of a hollow metal tube – in other words, use waveguides!

Since waveguides can pass gigahertz signals with negligible attenuation, the ideal solution would be to use them exclusively to provide the path between the aerial (the 'dish') on the outside of the wall or roof and the receiver inside the house. Unfortunately, waveguides demand precision engineering and the component cost would be prohibitive. Apart from the cost, it is doubtful if house owners in an up-market estate would like to see a long length of piping snaking down the wall only to disappear through the window frame and continue along the skirting boards of the lounge. The solution, as always when there are conflicting requirements, is a compromise: a short stub of waveguide to pass the satellite signals to the first stage frequency converter (positioned close to the dish), and high grade coaxial cable for the rest of the run, down the wall to the receiver. The frequency converter section is called the LNB and the short waveguide stub and its connection to the focal point of the dish is known as the feedhorn.

Basic waveguide operation

It is not necessary to delve deeply into the horrifying complexities of waveguide theory because they form only a small part of the total transmission path in commercial satellite installations. However, those cursed with an insatiable appetite for knowledge should consult one of the many textbooks on the subject. The following treatment only provides a rough outline of the essential features of waveguides. A word of advice; some texts specify the A dimension of a rectangular guide as the major dimension, but in the UK the larger dimension is often taken as the B dimension. Another point is that modes H_{01}, H_{02} ... etc. are sometimes referred to as TE_{10}, TE_{20} ... etc. This non-standardized terminology can sometimes lead to considerable confusion when studying waveguide theory from a selection of text books.

Rectangular waveguides

A signal passing along a coaxial cable conforms to conventional circuit practice, that is to say, it is described in terms of voltage (V) appearing across the inner and outer conductors and the current (I) flowing along the conductors. In a waveguide, as the name implies, the signal remains in the form of an electromagnetic wave and is therefore more properly considered in terms of an electric field (E) and a magnetic field (H). It is essentially a wave disturbance, trapped within the confining walls of the guide. However, the guide dimensions must be chosen to satisfy the following boundary rules imposed by electric and magnetic fields:

1   An *E* field can never be parallel to a perfectly conducting surface if close to that surface.
2   An *H* field can never be at right angles to a perfectly conducting surface when close to that surface.

Because of these restrictions and because the wall of a waveguide is, for all practical purposes, a perfect conductor, it appears that a wave can never propagate straight down a metal tube but must travel by a series of reflections from wall to wall – it bounces along. (See Figure 3.1(a)).

The bouncing angle depends on the cross-sectional dimensions of the guide. For reasons given later, the longer dimension (B) is the more critical and is normally designed to be greater than a half wavelength and less than a full wavelength. Within these two limits, the B dimension determines the bouncing angle. The longer the B dimension, subject to the upper limit, the less times the wave will bounce (wide bouncing angle). The wave suffers slight attenuation during each bounce, so the fewer the bounces the more efficient the propagation down the guide. (See Figure 3.1(b)). If the B dimension is shortened down towards the half wavelength limit, the number of bounces increase dramatically. In fact when B is exactly a half wavelength, the wave bounces backwards and forwards in the same place and is said to be in an *evanescent* mode; in plain English, it won't come out the other end! (See Figure 3.1(c)). It would seem from the foregoing that it would be a good thing to have the B dimension as near as possible to a full wavelength in order to minimize the number of bounces the wave

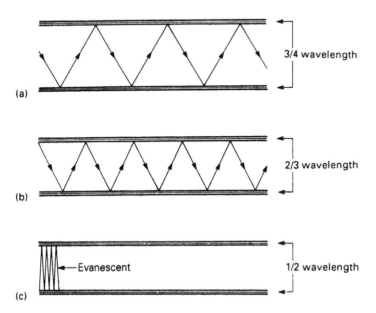

(a)   3/4 wavelength

(b)   2/3 wavelength

(c)   Evanescent   1/2 wavelength

Figure 3.1   *Effect of B dimension on bouncing angle*

has to make during its passage down the guide. Unfortunately, there is another unwanted effect which occurs when the B dimension is too near the full wavelength limit – the danger of allowing so-called 'higher order modes' to be propagated. To ensure predictability, waveguide techniques are directed towards the propagation of a simple wave pattern known as the 'dominant mode'. Such a mode demands the smallest possible guide cross-section and is known as the H_{01} mode. By ensuring the B dimension is not too near the half wavelength limit, only the dominant mode can propagate since higher order modes (H_{02} and beyond) become evanescent. Figure 3.2 illustrates the difference between the magnetic field patterns in the dominant mode and the second high order mode. The conflicting requirements of wide bouncing angle and avoidance of higher order modes leads to a B dimension compromise in the region of three-quarters of a wavelength. The 11 GHz band corresponds to a wavelength of approximately 2.7 cm so the B dimension of a rectangular guide is chosen to be around 2 cm. A waveguide thus has the advantage of filtering out unwanted signals other than the designed wavelength region.

The A dimension is not critical but, by definition, it must be shorter than the B dimension otherwise the wave could slip round and the *E* and *H* field directions would be unpredictable. In the dominant mode, the *E* field lines lie across the guide axis, parallel to the A dimension so if A is too short and the signal strength is exceptionally high, the electric stress could cause arcing across the guide walls. Fortunately, the *E* field strength at a satellite receiving dish is far too weak to worry about arcing. This danger only crops up during the design of high powered centimetric transmitters.

Rectangular waveguides, by their nature, are highly selective of a single linear polarization sense, either horizontal or vertical. A polarizer section is often employed to twist the incoming plane of the selected linear polarization sense to that of the LNB probe. To receive circular

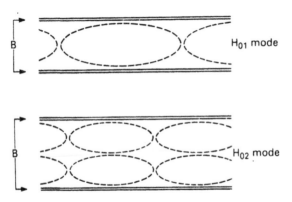

Figure 3.2   *Magnetic field patterns*

polarized signals with a single probe LNB it is necessary to embed a dielectric slab at 45° in the throat of the preceding waveguide. If only one circular polarization sense is broadcast from the satellite, a head unit without a vane or dielectric slab will still resolve the signals but with an attenuation of about −3dB. Ku band waveguides are technically known as WR75 or WG17 (19.05 mm × 9.53 mm) and C band WR 229 or WG 11A (58.17 mm × 29.08 mm). Waveguides can also absorb energy if there are surface irregularities; this is quantified as a waveguide loss in decibels.

Waveguide impedance

As all technicians will know, a coaxial cable has a definite characteristic impedance which is independent of its length and normally fixed at 75 ohms. Technicians will also be aware of the unpleasant effects of a coaxial cable feeding a mismatched load impedance. It is therefore not surprising to learn that waveguides also have a characteristic impedance and require matched loads before they will function correctly. In the dominant mode, the characteristic impedance of a rectangular guide is dependent on the B dimension, and because of the restrictions mentioned above is normally in the region of 600 ohms. At the receiving dish, whenever a section of waveguide is used, however short, there is a problem arising from the inherent mismatch between the waveguide impedance and the impedance of free space. The waves collected at the focal point of the dish if collected by the waveguide must somehow be matched to the waveguide. It is difficult to visualize the concept of the impedance of free space because how can 'nothing' have an impedance? The answer to this seeming paradox is tied up with the permittivity (k_0) and the permeability (μ_0) of free space and the nature of electromagnetic propagation. All such waves, whether radio, television, x-rays or gamma rays have the following properties:

1   They all consist of an *E* field and an *H* field, the amplitudes of which vary sinusoidally as they move through space.
2   The *E* and *H* fields are always at right angles to one another and travel in a direction at right angles to both fields.
3   The ratio of the *E* field to the *H* field is always exactly 120 π ohms or 377 ohms approximately. (See Figure 3.3).

It is this last property which is mysterious and why we can make the following statement:

The characteristic impedance of free space is 377 ohms.

We can understand the reasoning behind this if we first note that the *E* field is a *voltage* concept and is considered in terms of volts per metre while the *H* field is a *current concept* and is considered in terms

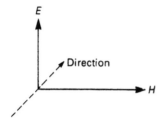

Figure 3.3   *E, H and direction vectors*

of amps per metre. Since the ratio of E to H fields in free space is 377 ohms, then the ratio of V to I must also be 377 ohms. In symbols,

$$E/H = \frac{V/d}{I/d} = V/I = 377 \text{ ohms}$$

Since the ratio of voltage to current has the dimensions of impedance, it is not illogical to conclude that free space has an impedance of 377 ohms. Strictly speaking, the waves in the vicinity of the dish are not in perfectly free space but this is little more than an academic distinction.

The horn aerial

It is clear that a mismatch will occur if the open end of a rectangular waveguide is just left at the focal point of the dish. It will certainly collect power from the dish but there will be a mismatch between free space impedance (377 ohms) and the waveguide impedance (about 600 ohms). This will lead to the partial reflection of the incoming microwave signal back to the antenna. A quantity, known as the voltage standing wave ratio (VSWR), is a measure of how much signal is reflected back and lost due to impedance mismatch. Under matched conditions, the rms voltage should be the same at all points along the length of the transmission path. Mismatching causes voltage nodes (voltage minima) and antinodes (voltage maxima) to appear at fixed points on the path. The ratio of an antinode to a node is called the VSWR. VSWR should ideally approach 1:1 but normally 1.5:1 and below is acceptable in practice. It was stated above that the impedance of a guide is reduced as the B dimension is increased so one way out of the muddle is to widen the tip of the B dimension as shown in Figure 3.4. The extremity of the guide now acts as a horn aerial at the focal point of the dish. There are several exotic ways of terminating the guide at the other end, including a projecting pin aerial or probe placed at a point in the guide corresponding to an E field maxima. This is normally located in the throat of the LNB.

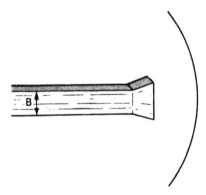

Figure 3.4   *Horn feed*

Circular guides

It is also possible to use guides with circular cross-section although there are some disadvantages, the main one being the difficulty of predicting the plane of the *E* and *H* fields. Because of the symmetry of a circular structure, there is nothing to prevent the wave-pattern slipping round. The wave might start with, say, the *E* field lines at a certain angle, but a few inches along the guide the pattern could twist round a fraction. Good design can only proceed if parameters are predictable so an unpredictable slip in the *E* field is not conducive to good design! However, if the length is short, as indeed it would be in this case, there may not be room for such a slip to occur so circular guides can be used. The dominant mode in a circular guide is the E_{01}. One advantage of circular guides is that they tend to be cheaper to produce. Often a transition from circular to rectangular waveguide is encountered along the path to the aerial probe.

Circular guides, by their nature, can simultaneously propagate both senses of linear polarization (vertical or horizontal) or circular polarized signals. A transition from circular to rectangular may occur after polarization selection. Circular waveguides for the Ku band are known as C120.

Waveguide components

The uniform waveguide, whether circular or rectangular, forms the basis of wave transmission at centimetric wavelengths. The usefulness of such guides can be extended by the further addition of various circuit components which have their equivalents in electronic circuit theory. Examples of these are attenuators, junctions, terminations, matching and transforming elements.

An important principle in waveguide practice is that a wave can be transformed with negligible reflection through changes in waveguide form if the change is made gradually. This can often be seen in head units where a circular waveguide feedhorn is attached to a rectangular waveguide LNB throat: transition from circular to rectangular waveguide is made gradually over a minimum of quarter of a wavelength. Where waveguide components are connected together flanges are commonly used; these are accurately machined so as to keep internal reflections to a minimum and provide a guard against water ingress with compressed O rings.

A common component encountered is the so-called orthomodal transducer (OMT) shown in Figure 3.5. This essentially splits one waveguide into two and has many uses where two LNBs are required for whatever purpose. For example, in SMATV systems for Astra, one LNB is dedicated to horizontally polarized channels while another, offset by 90 degrees, is dedicated to vertically polarized channels. Often, dual-band operation requires two separate LNBs and an OMT is sometimes used to split the signal to each LNB from a common broadband feed/polarizer combination. Each OMT branch is fashioned for the frequency band for which it is designed. Some OMTs have circular waveguides (so-called C120 standard) and some use rectangular waveguides (known as WR75) or a combination of the two. Waveguide adapters can be obtained to make reflection free connections between circular and rectangular guides. If making up your own system from component parts such as LNBs, feedhorns, polarizers and OMTs be careful to pay attention to the waveguide type used in each component part and, where necessary, use appropriate adaptors. In view of the enormous amount of components available from manufacturers it is advisable to check with your supplier about compatibility if at all in doubt.

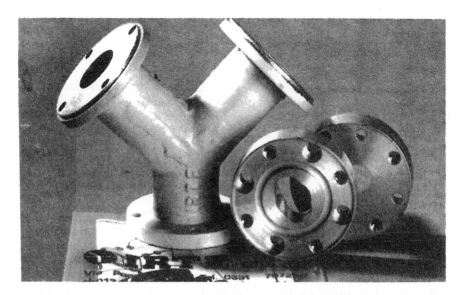

Figure 3.5 *An OMT complete with waveguide adaptors (Micro X)*

For applications where saving of space is important, filling the interior of a waveguide with dielectric material allows its cross-sectional area to be reduced in the ratio of $1/\sqrt{k}$ where k is the dielectric constant of the filling material.

There are many combination feeds for various situations. For example, Figure 3.6 shows an impressive triple-band feed for the S, C and Ku bands combining feedhorn and polarizer sections for multi-band reception. It can resolve both V and H polarization for the S and Ku bands and left- and right-hand circular polarization (LHCP and RHCP) for the C band. For the S band the small offset dipole antenna is physically rotated as necessary to receive either V or H polarizations. Three output ports are provided, the C band LNB is attached to the larger WR229 waveguide flange, the Ku band LNB is attached to the smaller WR75 and an S band LNB may be connected using a standard 'N' connector. Scalar rings are provided to optimize the illumination for both shallow and deep dishes. There are many other special combinations available designed specifically for reception from multi-band hybrid satellites.

Feeds

A centimetric aerial system has two clearly defined parts: the reflector, and the source which feeds it (the head unit or feed). We sometimes say feed because it is convenient, in some cases, to visualize the antenna as a primary radiator (feed) working in conjunction with a secondary radiator (reflector). These terms are used, by convention, from early days of

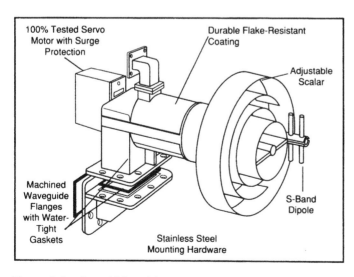

Figure 3.6 *A multi-band feedhorn for S, C, and Ku bands showing scalar rings (Chaparral Communications)*

radar so may appear more relevant to transmission rather than reception.

The illumination or part of the reflector/lens that a feed or feedhorn 'sees' is usually tapered to decrease amplitude of side lobes relative to the main beam. This practice can reduce side lobes to at least 20 dB down, relative to the main beam. Feeds are usually designed to provide an illumination pattern which tapers typically 15 dB towards the outside edge of the dish relative to the centre; thus most of the signal received by any satellite system is from the central portion of the antenna. This optimal illumination taper reduces pick-up of extraneous ground noise from beyond the edge of the dish. Remember from Chapter 2 that ground noise from the warm earth (or wall) is the major noise contributor to total system noise. This distribution pattern is dependent on the feed design which can take a variety of forms designed to work in conjunction with set fixed ranges of f/D ratios. This point can be important when putting together component parts of an aerial system from different manufacturers' parts. The range of f/D values with which a feed provides optimum performance is obtained from manufacturers' literature.

The good design of an antenna system for centimetric waves requires careful experimental work or the use of powerful computers to measure or model amplitude and phase distributions at feed and antenna regions. A uniform phase distribution across the antenna aperture should be obtained to minimise main beam width and reduce side lobe amplitude. The beam width diagram of the feed should also have the required amplitude to illuminate the reflector correctly, in general the following rules apply:

1 A uniform amplitude distribution gives maximum gain.
2 An amplitude distribution tapering from maximum in the centre towards dish edge reduces side lobes, but this is at the expense of gain.
3 An amplitude distribution tapering from maximum at the edge towards the centre gives a sharper main lobe but increases side lobe amplitude and reduces gain. This is known as an inverse taper distribution.

If an open-ended waveguide is positioned at the focus of a reflector then power is certainly collected from the open end but, as with any aerial system with a small aperture, beam width is very wide causing considerable over-illumination. The aperture may be increased by flaring out the waveguide into a fluted horn shape. This flaring tends to illuminate the dish more evenly and aid matching. Flaring open-ended waveguides to about one wavelength normally illuminates the paraboloid adequately. Flanges or scalar rings may be used at the horn mouth in prime focus antennas to adjust the pattern. Although a long feedhorn produces a minimum of side lobes, a compromise length is always adopted. This is because waves have further to travel from the outside of the flute than at the centre thus creating unpredictable phase variations down its length. Very short horns may include lenses to correct phase.

Scalar rings

With prime focus antennas, adjustable scalar rings are commonly used to optimize the illumination for either deep or shallow dishes. A scalar feedhorn, as shown in Figure 3.6, is essentially a waveguide pipe with a series of quarter wavelength deep rings surrounding it. The number of rings can vary, but three to five are common. The rings are commonly adjustable for peaking with a wide range of antenna f/D ratios (0.35 to 0.5). The shallower the dish, the nearer the rings are set to the mouth of the waveguide pipe. Specially manufactured feeds are normally needed for especially deep dishes of $f/D = 0.25$ to 0.32 for the C band.

Figure 3.7   *Corrugated feedhorn with OMT (Channel Master USA)*

Offset feeds

Flared feedhorns for offset focus dishes are carefully dimensioned to ensure E and H fields are detected with equal efficiency. This is more difficult than in design of prime-focus feeds as only a portion of the parabola is illuminated. A more restricted range of f/D ratio dishes can be used with a particular feedhorn design. Dielectric lenses can also be used to replace conventional feedhorns and these can give a more even illumination of the dish (thus a higher gain) and produce a sharper cut off at the edge of the dish (lower side lobes). Another means of reducing reflections to a low level is to progressively step down or corrugate the diameter of the waveguide or feedhorn. Figure 3.7 shows a corrugated feedhorn attached to another design of OMT.

Dielectric antenna feeds

This type of feed, for example Marconi's Polyrod feed, incorporates a dielectric lens which is designed for use with parabolic reflectors. Two examples are given, both with integral current-operated ferrite polarizer. One is optimised for use with prime focus antennas with f/D ratios of 0.35 to 0.5 while the other is for use with offset fed antennas with f/D ratios between 0.60 and 0.70. Both are available with either WR75 or C120 waveguide flanges. Figure 3.8 shows the two types. Several advantages are claimed for the polyrod feed:

1 E and H plane polar illumination is virtually the same which ensures maximum antenna efficiency in any polarization.
2 The beam has a more rectangular shape than that associated with conventional feedhorns which results in a more efficient illumination of the reflector.

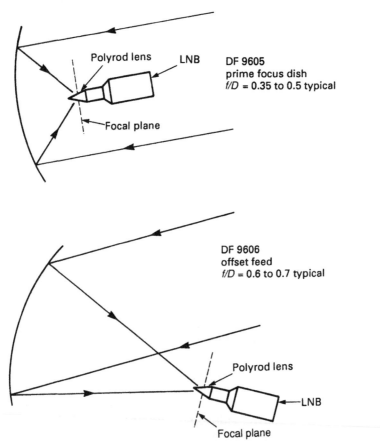

Figure 3.8 *Polyrod fed LNBs, models DF9605 and DF9606 (Marconi Electronic Devices)*

3 The antenna gain/beamwidth product is virtually independent of frequency over the entire 10.95 GHz to 12.75 GHz frequency range.

The Marconi Polyrod feed, with alternative polarizer designs, is also supplied as a compact integrated unit with a fixed single satellite system.

Swinging the main lobe

Sideways movement of a feed in the focal plane of a reflector swings the main lobe to a certain extent. However, if this is pushed too far then defocusing of the beam occurs. A tilt of θ degrees of the feed relative to the paraboloid axis produces a beam displacement of about 0.80 θ as shown in Figure 3.9.

This effect can be put to experimental DIY use if one fixed reflector is needed to capture signals from two close satellites or dual-band operation is required from the same satellite. Provided the satellite spacing is not excessive, two feeds mounted side-by-side and a single parabolic reflector could be used. Various mechanical add-ons, involving an extra LNB, have appeared to exploit this effect as a low cost way of receiving signals from two relatively closed spaced satellites (i.e. $< 4°$ apart). There have been doubts expressed about this practice since many antenna booms are not designed to accept the extra weight and may be liable to distortion or fracture. However, specifically designed antenna systems are available using up to four LNBs in this way.

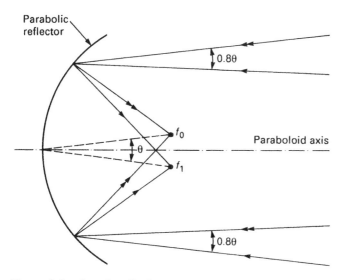

Figure 3.9 *Angular displacement of beam produced by feed position*

Polarization

In Chapter 1 it was shown that electromagnetic radiation, which includes satellite TV signals, are propagated through space with the E field (electric field) at right angles to the H field (magnetic field). The direction of propagation is at right angles to both.

The reference plane for polarization treatment is the E field. Current polarization techniques are classified as either linear or circular. Rain also affects polarization strategies, and heavy bursts can lead to significant depolarization of the fields causing cross-polar interference. Circular polarized signals are affected most by this.

Linear polarizers

Dual linear polarization is used to extend the number of channels that can occupy a given bandwidth, by using either E field horizontal polarization or E field vertical polarization. Its use can effectively double the number of channels that can be provided by a satellite since two channels can share the same frequency, provided they have opposite polarizations.

In reality, channels are staggered to minimize interference or crosstalk between the two. Two jargon terms are commonly encountered in the subject of polarization: 'co-polarized channels', meaning of the same polarization; and 'cross-polarized channels', meaning of the opposite polarization. As a general rule, although there are many exceptions, low to medium power fixed satellite service (FSS) band telecommunications satellites tend to employ linear polarization and direct broadcast service (DBS) band high power craft employ circular polarization.

There are three basic designs of linear polarizer:

V/H switched polarizers

V/H switched polarizers, sometimes called 'solid state probe selection polarizers' consist of a pair of probes set at right angles to each other (orthogonal). These project directly into the waveguide. One is set to detect vertical polarization and the other horizontal. A solid state switching technique is employed (e.g. PIN diode) to select the required output. For example, a LNB supply of 13 volts might select vertical polarization and 17 volts horizontal. These lower cost units tend to be relatively inefficient due to associated losses and noise and can only be used for single satellite or co-located satellite cluster reception. This is because the probes are fixed and no provision for fine tuning of skew is provided. A pair of these units, and an OMT, or a dual output LNB, are ideal for use in head ends of multi-user intermediate frequency (IF) distribution systems.

Mechanical polarizers

An internal motor physically rotates a probe, under remote pulse control, according to the polarization sense selected. They provide good isolation between the two polarities, but the major disadvantage is that the moving parts tend to be unreliable and may seize in cold weather. They are little used for Ku/Ka-band reception in Europe but are commonplace in C-band installations in North America.

Electromagnetic or ferrite polarizers

The basic operation of electromagnetic or ferrite polarizers involves the 'Faraday rotation principle' which twists the E field plane to that of the LNB probe orientation. This is performed by varying the magnitude and polarity of current flowing through the polarizer windings, depending on how much polarization skew is selected. The insertion loss of such a device is about 0.3 dB. Satellite receivers designed to operate with this type of polarizer usually provide some sort of fine control over the amount of rotation of the incoming signal called 'skew' since:

1 The vertical E field may not always be truly vertical, depending on the satellite/receive site longitudinal difference and the satellite offset.
2 The amount of depolarization (twisting) of the signal on its downward journey to earth varies with frequency.

These types of polarizer are very reliable since they have no moving parts and are ideal for multi-satellite reception systems.

Circular polarizers

Circular polarization involves spinning the E field so the radiation is propagated through space in a corkscrew or spiral fashion. Again two separate types emerge, clockwise spinners known as right-hand circular polarization (RHCP) and anti-clockwise spinners known as left-hand circular polarization (LHCP). A left-hand rule can be used to determine spin direction, the outward curl of the fingers of the left hand show the direction of spin, and the outstretched thumb the direction of propagation for LHCP. Conversely, a similar rule applies to the right hand for RHCP.

Although circular polarization can be used in a similar way to linear polarization to increase the number of channels on a given satellite, it is also used for a different reason with DBS high power satellites. Individual DBS satellites usually have all their channels fixed at a single circular polarization sense either RHCP or LHCP. Adjacent DBS satellites, if their footprints are not sufficiently far apart to prevent interference, will utilize the opposite polarization sense to each other. The number of transponders is usually limited by power considerations rather than frequency re-use, although video compression techniques may be used to increase the channel capacity. The use of cross-polarization in adjacent satellites leads to an interference suppression of over 20 dB.

To receive circular polarized signals with a linear polarizer it is necessary to embed a dielectric slab at 45° in the throat of the LNB or waveguide. If only one circular polarization sense is broadcast from the satellite a head unit without a vane or dielectric slab will still resolve the signals but with an attenuation of about 3 dB. More elaborate methods of resolving RHCP and LHCP involve the use of one of a pair of orthogonal probes having its output delayed by a quarter wavelength relative to the other; the resultant output detects one circular polarization sense and reversal of the delay detects the other. Installations for circularly polarized signals are easier and less prone to polarization losses than are those for linear polarization since it is not necessary to align the LNB with the polarization plane of the signal.

Polarization skew

Linearly polarized signals from geostationary satellites usually have the polarization orientation specified with respect to the equator. For a global beam, the satellite's horizontally polarized electric vector is normally parallel to the equator with the vertically polarized vector set orthogonally at the sub-satellite point as shown in Figure 3.10. The angular difference between the orientation of a given polarization sense transmitted and that sense received at the Earth station is called the *polarization tilt angle*, τ, and may be calculated using

$$\tau = \tan^{-1}\left(\frac{\sin B}{\tan A}\right) \text{ (degrees)} \tag{3.1}$$

where: A = the latitude of the Earth station (positive for northern hemisphere, negative for southern hemisphere)
  B = the longitude east of the Earth station minus the longitude East of the satellite

This is not the whole story. The polarization is sometimes intentionally tilted in spot beams so that the vertically polarized signal is oriented to the local vertical at the spot beam centre. This practice reduces adverse depolarization effects caused by rain in the service area. Note the result is equally applicable if the orthogonal horizontal polarization sense is referenced to the local horizontal.

Tilt angles are best interpreted from the satellite's viewpoint. As with all vectors, positive values are interpreted as anti-clockwise with respect to the reference horizontal (or vertical as appropriate), and negative angles as clockwise. Assuming there is no spot beam polarization

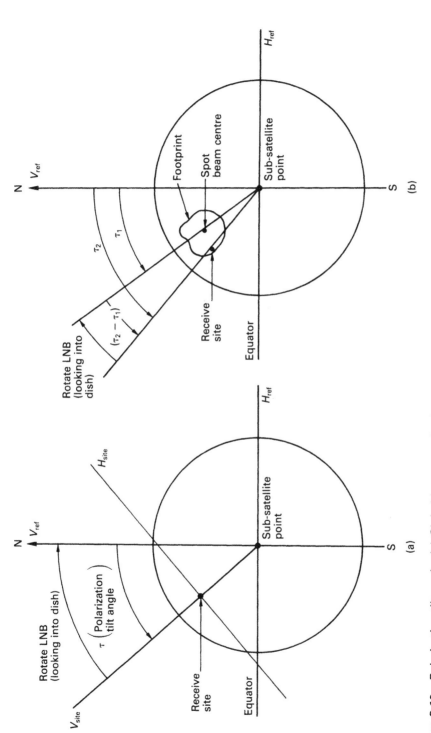

Figure 3.10 *Polarization tilt angle. (a) Global beams and reference planes. (b) The case where the vertical polarization sense is tilted to the local vertical at the spot beam centre*

tilting, the tilt angle is simply calculated using Equation 3.1. To align the vertical planes, the corrective setting of the LNB orientation (looking into the dish) is of the opposite sign (direction) to the calculated tilt angle as shown in Figure 3.10(a).

If a satellite already adopts a tilted linear polarization scheme, as shown in Figure 3.10(b), the difference in the two tilt angles, paying attention to sign, will then give the resultant value at the receive site and consequently the angle of rotation of the LNB in order to correct for it. This is why published values for polarization offset from a given satellite often bear no apparent resemblance to its position in the geo-arc as would initially be expected. The amount of rotation of the LNB to correct for tilt angle is called 'polarization skew'.

The low noise block (LNB)

The low noise block is a fairly complex piece of equipment but its basic operation is as follows: the short stub of waveguide is continued to a resonant probe or aerial located in the LNB throat. At this probe the incoming microwave signals are converted to minute electrical signals which are subsequently amplified and block down-converted to a frequency more suited for onward transmission by coaxial cable. The overall gain of a LNB is typically within the range 50 dB to 60 dB. Efforts to exceed this figure have been found to be influenced by the law of diminishing returns. A word of warning, the internal probe aerial should never be touched or tampered with because it is delicate and easily damaged.

The whole assembly is hermetically sealed against the ingress of moisture. If moisture should get into the unit, corrosion, then subsequent failure, may result. Some head units combine feedhorn polarizer and LNB into a single unit, others have component parts which need to be bolted together. In the latter case silicone rubber 'O' rings are fitted between the connection flanges to prevent the seepage of moisture into the feedhorn and LNB sections. The first IF output from a LNB is generally via an 'F' connector. This should be waterproofed by the installer with either self-amalgamating tape or a rubber weatherproof bolt. The subsections of a simplified LNB are shown in Figure 3.11.

Low noise amplifiers (LNAs)

The low noise input amplifier of an LNB invariably uses some form of gallium arsenide (GaAs) technology which is well suited to microwave frequency use. The reasons for this are four-fold:

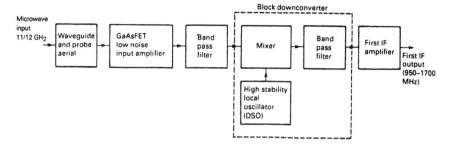

Figure 3.11 *Simplified schematic of a LNB*

1 GaAs can withstand higher working temperatures than can silicon based devices due to its inherently larger band gap. This can be advantageous when the full heat of the Sun is focused on the LNB case at certain times of the year. This property also lends itself to exploitation in high power situations.

2 Electron mobility is several times higher than in silicon devices; this shortens transit times and increases the upper working frequency limit.

3 GaAs field-effect transistors (FETs) when used as amplifiers at typical outdoor temperatures of around 15°C have very low leakage currents and thermal generation. This makes them ideal low noise input devices.

4 The internal method of construction allows for lower parasitic capacitances than is possible using silicon. Even very small values of stray capacitance may be troublesome at microwave frequencies.

High electron mobility transistors (HEMTs) are now standard even in low cost LNBs. These devices have even greater electron mobility, higher upper working frequencies and improved noise performance over traditional GaAs FETs. At the time of writing, 0.8 dB to 1.0 dB noise figures are commonplace for the Ku band and 30 K (0.43 dB) for the C band. Low noise performance is always more difficult to achieve the higher the frequency. Up to a certain degree, a lower noise LNB can compensate for a slightly undersized antenna, although the trade-off cannot be pushed too far.

Noise figure and noise temperature

In manufacturers' specifications for equipment it is normal to quote the noise figure of a LNB in decibels. This can be converted to its corresponding noise factor by the following relationship:

$F_{LNB} = 10^{(NF/10)}$ where NF is the noise figure of the LNB.

This noise factor value can be substituted in the following equation to obtain the equivalent noise temperature, which is an alternative way of specifying the noise performance of a LNB.

$T_{LNB} = 290(F_{LNB} - 1)$ where F_{LNB} is the noise factor of the LNB.

For convenience, a computer generated table, Table 3.1, has been produced showing the relationship between noise figure and noise temperature at an ambient temperature of 290°K. The results are plotted in Figure 3.12.

To obtain the equivalent noise figure from the noise temperature use the following two reversal equations:

$F_{LNB} = 1 + (T_{LNB}/290)$ and NF $= 10 \log (F_{LNB})$

Table 3.1   *Noise figure v noise temperature (ambient temperature 290 K)*

Noise figure (dB)	Noise temperature (degrees K)
0.00	0.00
0.10	6.75
0.20	13.67
0.30	20.74
0.40	27.98
0.50	35.39
0.60	42.96
0.70	50.72
0.80	58.66
0.90	66.78
1.00	75.09
1.10	83.59
1.20	92.29
1.30	101.20
1.40	110.31
1.50	119.64
1.60	129.18
1.70	138.94
1.80	148.93
1.90	159.16
2.00	169.62
2.10	180.32
2.20	191.28
2.30	202.49
2.40	213.96
2.50	225.70
2.60	237.71
2.70	250.01
2.80	262.58
2.90	275.45
3.00	288.63

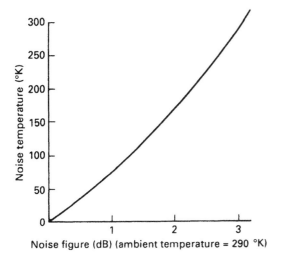

Figure 3.12 *Noise figure v noise temperature (ambient temperature 290 K)*

Block down-converters

Another important function of the LNB is to change the frequency of the incoming signals to a more manageable frequency for onward transmission to the receiver by high quality coaxial cable. A frequency changer or down-converter consists basically of a local oscillator and mixer which is conceptually similar to that used in the common superheterodyne radio receiver. However, the technology required to keep the fixed frequency local oscillator sufficiently stable to shift a whole 750 MHz group of microwave channels to a lower range proved difficult. With the advent of dielectric stabilized oscillators (DSO) this process has become the standard. For the fixed satellite service band in Europe, the group or block of channel frequencies is by convention down-converted to the range 950 to 1700 MHz using a low side local oscillator of 10 GHz. Potential interference from UHF TV transmitters and other civil transmissions dictates the choice of this down-converted band and, unfortunately, this necessitates the use of higher grade cables than standard UHF coaxial. The main advantages of block down-conversion are:

1 More than one receiver, fed from a common antenna and head unit, can independently receive any of the selected co-polarized channels in the group.
2 The channels are less likely to drift off tune since the channel selection circuitry is indoors where ambient conditions are relatively constant.

The output of the frequency changer is known as the 'first IF frequency', or simply 'satellite IF'.

Universal LNBs

Now that DVB broadcasting is becoming commonplace in Europe and elsewhere, modified designs of LNB are available suited to receive both Ku-band analogue and digital transmissions over the entire frequency range 10.7 GHz to 12.75 GHz. Such devices have become to be known as universal LNBs and are well suited to receive all signals from the Astra and Eutelsat satellites.

Universal LNBs span the entire range 10.7 GHz to 12.75 GHz, and down-convert using a switched local oscillator to two separate IF bands which can be processed by the indoor receiver or RD. A 9.75 GHz local oscillator is switched in to select the low band satellite IF (950 to 1950 MHz) and a 10.60 GHz local oscillator is used to select the high band satellite IF (1100 to 2150 MHz) at the head end.

Bearing in mind that 13 V/17 V voltage threshold switching is already used to control the polarizer sense as previously described, an additional switching arrangement is used by the indoor receiver to control the local oscillator frequency. The solution adopted, for low cost units intended for the domestic market, is to arrange the default local oscillator to be 9.75 GHz and use a 22 kHz tone signal to switch in the 10.60 GHz local oscillator as required. The indoor receiver handles all the control signals required for both the polarizer and the LNB band selection and sends them all up the coax cable to the head end, so no additional wiring is required. Twin output versions are also available for feeding two separate receivers and are essentially two complete independently switchable LNBs in the same box. Typical thermal noise figures for these low cost LNBs are around 1.2 dB at present.

To receive all the collocated SES Astra and Eutelsat Hotbird Satellites at 19.3°E and 13°E respectively, a universal LNB is needed which covers the entire FSS (10.7–11.7 GHz), DBS (11.7–12.5 GHz) and Telecom (12.5–12.75 GHz) bands. Astra satellites have 27 GHz transponders and Hotbirds have 33 GHz transponders.

Table 3.2   *SES Astra and Eutelsat Hotbird deployments*

Launch Date	FSS 10.70 GHz	FSS 10.95 GHz	FSS 11.20 GHz	FSS 11.45 GHz	DBS 11.70 GHz	DBS 12.05 GHz	Telecom 12.50–12.75 GHz
1989				Astra 1A			
1991					Astra 1B		
1991		Eut. II F1			Eut. II F1		
1993		Astra 1C					
1994	Astra 1D		Hotbird 1				
1995					Astra 1E		
1996					Hotbird 2	Astra 1F	
1997							Astra 1G
1997						Hotbird 3	Hotbird 4
1998	Hotbird5						

For the commercial market and cable operators, higher quality wide-band dual output devices are used, whereby the LNB provides two separate outputs for the HI and LO bands. These devices have low phase noise so are well suited to the new digital TV standard. An OMT is required for dual polarization reception. A typical example of this type of device is the WDLNB 1000E model manufactured by Swedish Microwave and shown in Figure 3.13. The specifications are as follows:

Input frequency	10.7–12.5 Hz
Noise figure	0.8 dB typical, 1.0 dB max
Output frequency	950–1950 MHz (low band), 1100–2150 MHz (high band)
Gain	52 dB ($\pm$4 dB)
Local oscillator frequencies	9.75 GHz (low band), 10.6 GHz (high band)
Local oscillator phase noise	-75 dBc @ 10 kHz min
Output SWR	2 : 1 max
Image rejection	40 dB min
Mixing products in band	-30 dBm max at output (1700 MHz)
DC power	12–20V/300 mA max (DC feed through either connector)

Figure 3.13 *A professional quality universal LNB (Source: Swedish Microwave)*

Local oscillator phase noise

When high frequency bands are used for digital broadcasts this type of noise becomes an increasing problem. The periodic and thermal noise generated by the local oscillator results in unwanted phase and amplitude modulation effects or phase noise within the local oscillator circuit itself.

Phase noise can corrupt the steep transients of a digital pulse if they happen to coincide, leading to unacceptable bit error performance. The phenomenon is known as 'phase hits' or 'knock outs'. The result of all this is yet another important noise parameter for judging overall LNB performance. LNBs intended for digital reception are designed to operate within industry wide phase noise limits.

The phase noise level is referenced to the carrier level measured in a kilohertz bandwidth at a given frequency offset from that of the local oscillator. The units are dBc/Hz at a given offset frequency (typically 1 kHz, 10 kHz and 100 kHz). The accepted limits of local oscillator phase noise is typically −50 dBc/Hz @ 1 kHz, −75 dBc/Hz @ 10 kHz and −95 dBc/Hz @ 100 kHz or better.

Many older LNBs designed for analogue transmissions may be unsuitable for digital transmissions because of their susceptibility to excessive phase noise.

Practical down-conversion

The Astra system (Europe)

The Astra system is currently envisaged to comprise several collocated satellites at 19.2°E. These are Astra 1A through to 1G. It seems likely that the vast majority of installations in Europe will continue to be 60 cm to 80 cm dishes targeted at the Astra cluster.

By way of an example of practical down-conversion, we will look at just three of the Astra satellites shown in Figure 3.14. A fifty transponder group on board 1A, 1B and 1C extends between 10.70 GHz and 11.7 GHz as shown in Figure 3.13. On 1C, there are also two channels between 10.90 and 10.95 MHz which is the start of the extension into the range 10.70 to 10.95 GHz by the later satellites of the cluster. The lower extension to the 10.95 to 1700 MHz FSS band was not envisaged in the original frequency allocations by the WARC, so many LNBs particularly on older equipment were unable to downconvert and thus receive these additional channels. Traditionally, the FSS group of frequencies is mixed with a local oscillator running at 10 GHz to provide a down-converted group extending from 950 MHz to 1700 MHz which forms part of the satellite IF block of frequencies. Since some terrestrial UHF broadcasts operate in the range 700

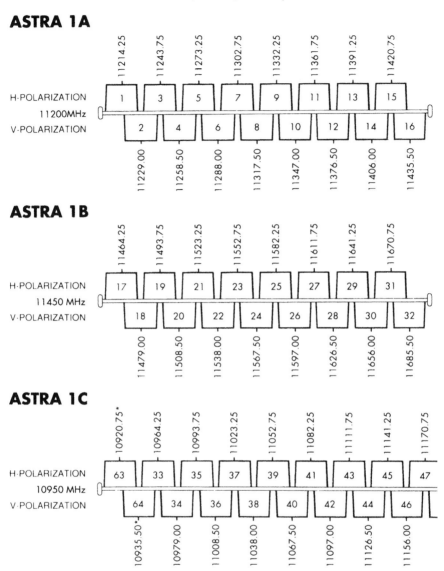

Figure 3.14 *Astra downlink plan for 1A, 1B and 1C only*

MHZ to 860 MHz there would be considerable overlap if this local oscillator frequency were maintained to include sub-10.95 GHz channels. The solution was to reduce the frequency of the local oscillator to 9.75 MHz.

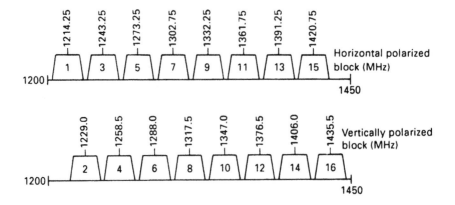

Figure 3.15 *Down-converted frequencies for Astra 1A*

Cables

The down-converted group of satellite channels, commonly known as the 1st IF, occupy a band of frequencies from 950 MHz to 1700 MHz for the fixed satellite service in Europe. Unfortunately, these frequencies are far too high for ordinary UHF TV cable to be used except in very short runs of a few metres. In practice, cable runs are considerably longer than this, so a lower loss, better quality cable should be selected. The attenuation figure per 100 metres, at the usable frequency, is perhaps the most important parameter as far as the installer is concerned, although a lower attenuation figure coaxial cable is usually accompanied by an increase in diameter and subsequently cost. Clearly, for the domestic market, a compromise must be reached where an acceptable diameter cable at reasonable cost is selected.

Cable attenuation and line amplifiers

Cable attenuation is commonly measured in dB per 100 metre length and this increases with frequency. Figure 3.16 shows the frequency v attenuation/100 m relationship for a common cable type, CT100. Excessively long cable runs or poor choice of coaxial cable will result in sparklies appearing on the screen. These 'lower order' sparklies have a slightly different appearance to that associated with low carrier-to-noise (C/N) ratio at the head end. If cable runs are very long and a degraded picture results, the fitting of an inexpensive line amplifier will often improve things. It is usually unnecessary to fit one if the cable run is less than about 30 metres. A line amplifier is powered by the LNB supply which is fed up the coaxial cable by the receiver. The maximum cable attenuation that can be tolerated, without noticeable degradation

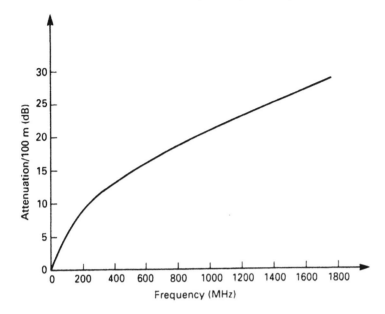

Figure 3.16   *Frequency v attenuation/100 m for cable type CT100*

of picture quality, is in the region of 12 dB to 15 dB. Thus a scan through the tables, presented in this chapter, will show which cable is most suited to a particular run.

For the average wall mounted single satellite installation, the most often used cable is CT100 (H109F). This allows a maximum cable run of about 45 m which is more than sufficient for the vast majority of cases, which in practice rarely exceed 20 m. For longer runs, fitting of inexpensive line amplifiers can compensate for cable attenuation but it is important to remember that it will be fruitless to fit one in order to reduce sparklies from a poor input. This is because the noise level injected into the cable, as well as the signal, are amplified which results in no overall improvement in the signal to noise ratio. Fitting a larger antenna or a lower noise LNB is the only way to cure this problem. The fitting of line amplifiers, every 30 m or so, can often be a cheaper solution, for occasional long runs, rather than stocking less-used, higher quality cables. Line amplifiers obtain their power from the LNB voltage feed, so extra wiring is avoided. Another important parameter, as far as the installation technician is concerned, is the minimum bending radius of the cable. As a general rule, the minimum bending radius of a coaxial cable should be about ten times its sheath diameter or undue attenuation may result. Cables are normally supplied in black as this colour provides maximum protection from ultra-violet radiation.

Cables suitable for direct burial

Manufacturers produce their cables in a choice of either PVC (polyvinylchloride) or PE (poly-ethylene) sheaths. PE sheathed cables may be directly buried, but it is inadvisable to use PVC sheathed cables in any underground situation. Volex Radex have further produced a patented form of bonded laminated sheath called RBS (Raydex Bonded Shield). The flexible, bonded outer jacket has several advantages over other cables for direct burial. Firstly, the jacket is impact and abrasion resistant and, secondly, good pull strength and slip characteristics make it ideal for use in underground ducts.

Cabling requirements for single satellite installations

There appears to be no standardization with regard to cabling requirements for the various manufacturers' fixed satellite packages. For example the first Amstrad/Fidelity and Ferguson Astra packages used single coaxial cable. The IF signals are fed down the cable from the dish and a d.c. voltage is fed back up the cable from the receiver to supply the LNB. This d.c. voltage can be varied to operate an inbuilt V/H polarization switch. The corresponding Grundig package used coax with an additional polarizer lead to alter the polarity of the polarizer. The LNB supply of 15 V is fed up the coax and a further 12 V supply cable is used to control the polarizer: zero volts for vertical polarization and 12 V for horizontal polarization. The earth return is via the coax braid. First launched Tatung and ITT models used coax with two separate polarizer leads. Rarely encountered motor driven polarizers will need three extra conductors. Combination cables incorporating these extra conductors in the same sheath are available from Volex Radex distributors, and broadly equivalent versions manufactured by Pope Cables, from Webro (Long Eaton) Limited. These special multiple cables are relatively costly, so a sometimes used, but far less professional solution, is to run a separate cheap multi-core cable for the polarizer requirements in conjunction with normal quality coaxial cable, such as CT100 or H109F.

Volex Radex range of cables for single satellite installations

Table 3.3 shows the parameters of cables, suitable for single satellite installations, from Volex Radex. Included in the table are 50 ohm types that may be occasionally needed to match certain LNBs and receivers. However, the 75 ohm type is by far the most commonly used.

The SAT 1000 series is a range of composite cables with single, twin or triple polarizer leads to suit any single satellite TV installation. They all employ a 'figure-eight' construction as shown in Figure 3.17, and

Table 3.3   *Coaxial cables manufactured by Volex Raydex*

Type	Number of polarizer conductors	Impedance (ohms)	Attenuation/ 100 m at 1750 MHz (dB)	Sheath diameter (mm) A	B	C
CT 100	0	75	28.3	–	6.45	–
CT 125	0	75	23.6	–	7.80	–
SAT 100	0	75	28.6	–	6.65	–
RA 519	0	50	17.5	–	10.30	–
RA 521	0	50	22.5	–	7.85	–
SAT 1001	1	75	28.3	11.00	6.85	4.15
SAT 1002	2	75	28.3	13.00	6.85	4.25
SAT 1003	3	75	28.3	13.00	6.85	4.50

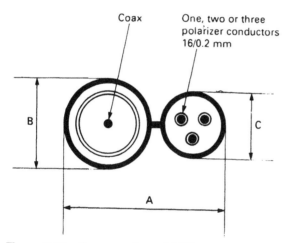

Figure 3.17   *Cross-section of SAT 1000 series cables from Volex Radex*

each retains its own sheath when separated for termination. Each polarizer conductor is again separately insulated.

Either CT100 or SAT 100 can be stocked for installations incorporating the V/H switch type of polarizer. In order to wire up all the other possible polarizer configurations, clearly SAT 1003 is a universal cable, since any unused conductors can be either snipped off or parallel connected. However, since motor driven polarizers are comparatively rare these days, perhaps SAT 1002 may be a reasonable alternative.

Pope range of cables for single satellite installations

A range of equivalent cables are manufactured by Pope Cables. The official distributors are Webro (Long Eaton) Ltd. Generally speaking

H109F is equivalent to CT100 and H47 is equivalent to CT125. A universal cable, shown in cross-section in Figure 3.18, capable of any polarizer wiring combination, is type H142A (all in a single PVC sheath) or H142B (figure-eight sheath). This cable incorporates type H125 coaxial cable, broadly equivalent to CT125 from Volex Radex, with three insulated 0.5 mm conductors for linear polarizer wiring. Again, any unused conductors can be snipped off or connected in parallel.

If your business is restricted to mainly fixed dish installations, then it is recommended that H109F and H142 are kept in stock to meet all present eventualities.

H142 A   H142 B

Figure 3.18   *Universal cable type H142 for additional polarizer wiring (Source: Webro (Long Eaton) Ltd)*

Multi-satellite motorized installations

The cabling requirements for multi-satellite motorized systems are slightly more complicated and usually comprise the following:

1   Normal quality, coaxial cable for feeding the block down-converted signals to the receiver. The 15 to 24 V LNB supply is also simultaneously fed via this cable.
2   Three polarizer conductors for mechanical polarizers or one/two for magnetic polarizers.
3   Actuator cable for the dish drive comprising two condcutors for the motor power plus three additional conductors for the position sensor.

These requirements can all be wired separately, but specially produced universal cables for such installations are available which enable an easy, neat and professional job to be performed.

Pope multi-satellite cable

The requirement may be met by using H142 cables for (1) and (2), and using H143 cable for (3). The cable H143 consists of two PVC insulated 1.4 mm^2 stranded copper conductors for the actuator motor power plus

H143

Figure 3.19 *Cable type H143 for actuator wiring*
(Source: Webro (Long Eaton) Ltd)

three 0.5 mm PVC insulated solid copper conductors, for the position sensor needs. The cable cross-section is shown in Figure 3.19.

For an even neater job, there is a combined cable, type H144, which consists of H142 and H143 integrated into the same PVC sheath, and available in two configurations, figure-eight or ribbon, as shown in Figure 3.20.

If all forms of installation are likely to be undertaken then it is recommended that H109F, H142 and H144 cables, or their equivalents, are stocked. In this way any job can be performed neatly and with the minimum of effort. The minimum reel size is usually 250 metres.

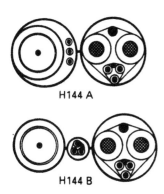

H144 A

H144 B

Figure 3.20 *Universal multi-satellite system cable type H144*
(Source: Webro (Long Eaton) Ltd)

Volex Radex multi-satellite cable

Volex Raydex produce a similar universal cable for multi-satellite installations, type K1005. This cable is available in a flat form, four component configuration and is shown in cross-section in Figure 3.21. The coaxial section of the cable is available in 75 ohm CT125 or 50 ohm RA521.

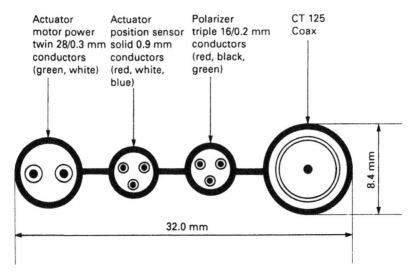

Figure 3.21   *Cross-section cable type K1005*

Connectors

The F connector is now established as the standard for coaxial terminations and is available in many forms. Some need to be crimped with a special tool or pliers and others are simply twisted onto the end of the stripped cable. I have found that the 'twist-on' type of connector is by far the most versatile and easiest to fit and can literally be fitted in less than thirty seconds. The connectors require a standard coaxial strip length and are available in a range of internal diameters to suit most cables. Due to the tapering of the entry guide a wide range of cable thicknesses can be used for each size. As a general rule, the internal diameter of the chosen connector should be in the region of 0.5 mm to 1 mm smaller than the outside cable diameter to ensure good strain relief without excessive distortion of the cable. The main advantage of this type over the crimped variety are outlined below.

1   Completely re-usable (no scrap or wastage).
2   Single piece construction.
3   Quick and easy to fit.
4   No special tools or soldering required.
5   Do not corrode (nickel plated brass).
6   Tapered entry guides ensure a good contact and high strain relief.

Some installers prefer the crimp type of connector but it is really a matter of personal choice. Waterproofing characteristics of F connectors are not really relevant to choice, since all outside connections should be either overwrapped with self-amalgamating tape or protected by a weatherproof rubber boot. Of the two, self-amalgamating tape is perhaps the better choice since rubber boots tend to crack and perish after a relatively short time. Other specialist compounds specifically formulated for sealing satellite connections are now available but avoid any compound that smells of the curing agent acetic acid (vinegar); this has been known to cause corrosion with LNB printed circuits. It is inadvisable to use any sealer in between waveguide connection flanges in case any seeps into the waveguide itself and absorbs energy. However, it is generally beneficial to overwrap flanges, on the outside, with self-amalgamating tape to prevent rain-water penetration.

Transmission lines

The place where a signal is *produced* is seldom the place where it is to be *used*. For example, a ground station transmitter on the ground will produce the signal but it must radiate from an aerial perched high above the building. So there is a transmission problem. Superficially, this doesn't seem much of a problem – just use a pair of wires to connect the transmitter output to the aerial. For moderate frequencies and moderate intervening lengths, a simple pair of wires can certainly be used. The trouble arises if either or both of these conditions are not met. There are three objections to the simple pair of wires:

1 Some of the energy is lost by direct radiation before it arrives at the destination. This loss increases with frequency.
2 A pair of wires, even if the ohmic resistance is negligible, still has *characteristic impedance* (Z_0). Unless this impedance happens to match the impedance of the load at the end (Z_L), maximum power will not be passed from the source to the load.
3 If the pair of wires are not correctly terminated by a matched load, a fraction of the signal voltage will be reflected back.

Characteristic impedance

Consider first a line of infinite length. It will have a certain inductance per unit length (L) and a certain capacitance per unit length (C). The impedance of the line (Z_0) is given by

$$Z_0 = \sqrt{L/C}$$

If the line is of finite length but terminated by a load (Z_L) equal to Z_0, it behaves as a line of infinite length. Under these conditions, the line is said to be feeding a matched load and, according to the maximum power law, maximum power is developed in the load.

Coefficient of reflection

When a line is mismatched, it is necessary to distinguish between the incident voltage wave (V_{inc}) and the reflected voltage wave (V_{ref}). Since the load is mismatched, it cannot absorb all the incident wave, so it reflects some of it back towards the source. The ratio of incident to reflected voltage is called the *coefficient of reflection* (*p*) or the *voltage reflection factor* and is defined as

$$p = V_{ref}/V_{inc} \tag{3.2}$$

In a matched line, there is no reflected voltage so $p = 0$. If, on the other hand, $V_{ref} = V_{inc}$, $p = 1$. So, *p* can vary between 0 and 1, depending on the degree of mismatch. The following is an alternative and more useful equation for *p*

$$p = \frac{Z_L - Z_0}{Z_L + Z_0} \tag{3.3}$$

Note: if $Z_0 = Z_L$ then $p = 0$; if $Z_L = 0$ then $p = -1$; if $Z_L = \infty$ (open circuit) then $p = 1$.

Similar equations apply to current incident and reflected waves except for a change of sign. Thus,

$$p = \frac{I_{ref}}{I_{inc}} \quad \text{and} \quad p = \frac{Z_0 - Z_L}{Z_0 + Z_L} \tag{3.4}$$

From the above equations, it follows that a well-behaved transmission line should have a zero coefficient of reflection.

Nodes and antinodes

If a line is feeding a matched load, the RMS voltages and currents along the line are propagated at constant amplitude from source to load, apart from slight attenuation due to resistive losses along the wires. If, however, the line is feeding a mismatched load, the reflections cause the voltage to increase at certain fixed points, called *antinodes*, and decrease at other fixed points, called *nodes*. Nodes occur at half-wavelength intervals from each other. Antinodes occur at half-wavelength intervals from each other. Therefore, nodes and antinodes are separated by one quarter-wavelength. Current nodes and antinodes also exist, but they are a quarter-wavelength apart from voltage nodes and antinodes. Thus, wherever there is a voltage node, there is a current antinode. Conversely, wherever there is a voltage antinode, there is a current node. Apart from the poor transmission capabilities of a mismatched line, there is a danger of voltage breakdown, either across the line or at some vulnerable component at either end. For example, if the line is designed to transmit 100 V and there is a severe mismatch at the far end, the nodal voltage points could approach 200 V. Taking an extreme case, if the line is terminated by a short circuit, there can be no voltage across the load, so a node exists there but the coefficient of reflection will be 1. One quarter-wavelength from the end, there will be a double voltage because the reflected wave and the incident wave will be additive. If, on the other hand, the load is terminated by an open circuit, an antinode will occur there and one quarter-wavelength from the end will be a voltage antinode.

Example
Assume a transmitter is sending a 300 V incident wave down a 75 Ω line to feed a load of 150 Ω.
From Equation 3.3, the coefficient of reflection is given by

$$p = \frac{150 - 75}{150 + 75} = 0.33$$

The antinodal voltage will be 300 + (0.33 × 300) = 300 + 99 = 399 V, and the nodal voltage will be 300 − (0.33 × 300) = 300 − 99 = 201 V.

Standing wave ratio

The presence of reflections on a mismatched line causes *standing waves* to appear because nodal and antinodal voltages appear at stationary points. The standing wave ratio (SWR) is the ratio of maximum voltage points to minimum voltage points and defined as

$$SWR = \frac{V_{max}}{V_{min}} = \frac{\text{antinode}}{\text{node}} \quad \text{or} \quad SWR = \frac{V_{inc} + V_{ref}}{V_{inc} - V_{ref}}$$

By rearranging, we get

$$SWR = \frac{V_{inc} + V_{inc}p}{V_{inc} - V_{inc}p} = \frac{V_{inc}(1 + p)}{V_{inc}(1 - p)}$$

Therefore,

$$SWR = \frac{1 + p}{1 - p} \tag{3.5}$$

In the above example, the antinode was 399 V and the node was 201 V, which gave an SWR of 399/201 = 1.99. If Equation 3.5 was used, SWR = (1 + 0.33) / (1 − 0.33) = 1.99.
Standing waves act as if they were 'stationary' waves because the nodes and antinodes appear to be locked to fixed points on the line. A simple experiment to grasp the idea of standing waves is to have a long length of rope with one end fixed to a wall. If the rope is now flicked up and down, standing waves will seem to appear on the rope when the flicking frequency is correct.

Coaxial cable

Figure 3.22 shows the cross-sectional view of a *coaxial* cable, consisting of two concentric conductors, an 'outer' and an 'inner'.

The outer conductor forms an effective shield against direct radiation, thus overcoming objection (1) above. The field patterns show that electric fields (*E* lines) are directed radially from inner to outer conductors while the magnetic lines (shown dotted) surround the central conductor. The characteristic impedance (Z_0) depends on the ratio of the diameters of the outer and inner conductors and the material between them. In the case of air-spaced conductors,

$$Z_0 = 138 \times \log_{10} \frac{d_2}{d_1}$$

where d_1 is internal diameter of outer and d_2 is external diameter of inner.

For minimum loss, the ratio of diameters has been found to be around 3.6, in which case the characteristic impedance is

$$Z_0 = 138 \times \log_{10} 3.6 = 75 \; \Omega \text{ approximately}$$

This is a fortuitous result since the input impedance at the centre of a dipole aerial happens to be 75 Ω.

The above equation is valid for air-spaced lines but, in general,

$$Z_0 = 138 \sqrt{\frac{\mu_r}{k_r}} \times \log_{10} \frac{d_2}{d_1}$$

where μ_r = relative permeabilty and k_r = relative permittivity.

In air, both of these constants equal 1 but, if the coaxial is filled with a solid dielectric, such as polythene, the impedance can fall to about 50 Ω. To bring the impedance back to 75 Ω, the ratio d_2 to d_1 can be

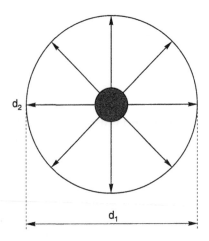

Figure 3.22 *Cross-sectional view of a coaxial cable*

increased. Coaxial lines are unbalanced with respect to earth which, in some situations, can be a disadvantage.

Parallel twin lines

Figure 3.23 shows an end view of a parallel twin transmission line.
The characteristic impedance is given by

$$Z_0 = 276 \times \log_{10} \frac{d}{r}$$

where d = spacing between conductor centres and r = radius of conductors.

With the average spacing employed, Z_0 is normally in the region of 600 Ω, so, in comparison with coaxial, twin lines are high impedance. The impedance does not vary a great deal with conductor spacing; for example, if the spacing is doubled, the impedance would only increase by about 80 Ω. Because of this, the value of 600 Ω is taken as the standard for parallel twin lines. Their advantage is their symmetry with respect to earth. A disadvantage is they suffer radiation loss more than coaxial lines.

Quarter-wave stubs

A length of line shorted at one end and exactly a quarter-wavelength long has the strange property of presenting an *open circuit* at the other end. This is because no voltage can possibly exist across the short-circuited end although there is a current antinode. A quarter-wavelength away is a voltage antinodal point and a current node. If this stub is connected across a transmission line, it will behave as a 'metallic insulator' because, at its point of connection, the stub can draw no current. Thus the stub can be used to support transmission lines without drawing any of the main current as shown in Figure 3.24.

Figure 3.23 *End view of a parallel twin transmission line*

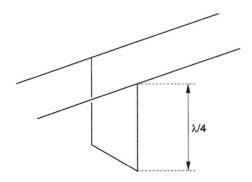

Figure 3.24   *Stub (quarter-wave)*

Although quarter-wave stubs can be efficient 'insulators' they are only so if the signal frequency is held reasonably constant. If the frequency wanders at all, the stub is no longer a quarter-wave and will start stealing power from the main line. We have said that a shorted-end stub behaves as an 'open circuit' at the other end. In point of fact, the stub behaves as a high-Q parallel resonant circuit having an enormous input impedance. Thus, for all practical purposes, the stub behaves as an open circuit at its point of connection to the main line.

Transmission lines as reactive elements

Depending on their length, and whether or not the ends are shorted or left open, sections of line can simulate inductive or capacitive reactance which, at centimetre wavelengths, are far superior to lumped conventional components. Considering first the case of shorted-end lines:

1   A stub less than $\lambda/4$ behaves as inductance, but at exactly $\lambda/4$ it behaves as open circuit.
2   A stub greater than $\lambda/4$ but less than $\lambda/2$ behaves as capacity, but at exactly $\lambda/2$ behaves as short circuit.

For longer lengths, the behaviour slides $\lambda/4$ along. For example, a line greater than $\lambda/2$ but less than $3\lambda/4$ acts as an inductance again. If the end of the line is left open-circuited, the behaviour is opposite to the above.

4 Satellite receivers

Introduction

The purpose of a satellite receiver is the selection of an individual channel, from the down-converted block, for viewing. The received signal must also be processed into a form suitable for interfacing to a conventional TV set and/or stereo equipment. Receivers may be incorporated as a subsection of a conventional TV set or built as a separate unit and contain, at least, the following blocks of circuitry.

1 Power supply.
2 Satellite tuner/demodulator unit.
3 Video processing circuits.
4 Audio processing circuits.
5 UHF modulator.

Jargon terms and abbreviations

Inevitably, a number of jargon terms and abbreviations have crept into telecommunications subjects which may be confusing to the newcomer. The commonly used terms appropriate to the understanding of satellite receiver schematic diagrams are defined and explained below.

AGC (automatic gain control) – This is a term used to describe a method of automatically backing off the strength of strong signals relative to weak ones thus protecting the signal circuitry from overloading effects, and maintaining a constant output level.
AFC (automatic frequency control) – This is a technique whereby the receiver tuned frequently is automatically locked on to the optimal tuning point. This reduces the risk of tuning drift with temperature and humidity changes.
Algorithm – An orderly set of programmed instructions designed to perform a particular task. For example, tuning algorithms are often used in voltage synthesized tuning systems.
Bandwidth – The total range of frequencies occupied by a particular signal.

Baseband – A term used to describe the unprocessed or raw video frequencies (unde-emphasized and unclamped) plus the audio sub-carriers.

De-emphasis – The reversal of pre-emphasis, a technique used to boost high frequencies prior to transmission. Since noise density increases with frequency, the subsequent de-emphasis reduces the signal level to normal and consequently attenuates the high frequency noise acquired during the transmission path.

Deviation – A measure of how much a carrier is deviated from its centre frequency by the modulating signal.

FM demodulator – A circuit which recreates the original signal from a frequency modulated carrier. This circuit is sometimes referred to as a discriminator.

IF (intermediate frequency) – A term used to describe the output frequency of signals from a down-conversion or mixer circuit (frequency changing circuit).

Local oscillator – A term used for a sinewave oscillator (sinusoidal signal generator) which is used in conjunction with a mixer stage.

MAC – Multiplexed analogue components. A TV system, to reduce cross-modulation effects and be compatible with future high definition TV developments.

Mixer – A circuit whose function is to generate sum and difference frequencies from an incoming RF signal and a local oscillator.

RF (radio frequency) – A term used to imply that the signal frequency is modulated onto a high frequency carrier.

uPc – An abbreviation for microprocessor or pre-programmed micro-computer chips. These are often used for system control and frequency synthesis tuning systems.

The tuner/demodulator

The satellite tuner demodulator is commonly manufactured as a single unit or 'can'. The down-converted range of frequencies (channels) is fed from the LNB via cable to the input of the tuner/demodulator can. This connection is normally via an F connector socket mounted directly on the can itself. Inside the tuner unit the following circuit blocks are frequently found.

1 Injection of d.c. LNB supply voltage.
2 d.c. isolation of the tuner input from 1.
3 An AGC controlled RF stage and filter.
4 A second down-conversion stage (2nd IF).
5 FM demodulator.
6 An AGC stage.
7 An AFC stage.

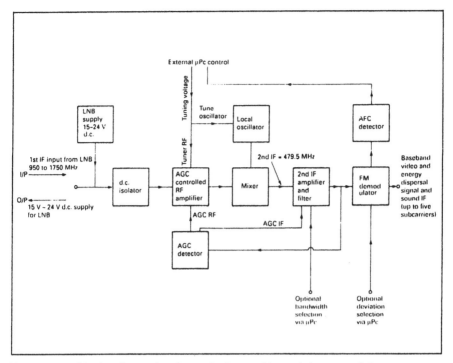

Figure 4.1 *Block schematic of a typical tuner/demodulator module*

Figure 4.1 shows a typical arrangement found in satellite tuner/demodulator modules.

LNB supply voltage injection

The d.c. voltage requirement of a typical LNB unit is in the region of 15 to 24 V. This voltage, derived from the power supply of the receiver, is injected into the input cable to power the LNB. The RF input stage of the receiver must be protected from this voltage so a d.c. isolation circuit is included. D.C. isolation can be effected by input capacitors or transformers. Some systems also have a voltage level shift method of switching the V/H polarizer. This is achieved by sending, say, 13 V d.c. up the cable to switch the polarizer for vertical polarized signals or 17 V to switch the polarizer for horizontal polarized signals.

The AGC controlled RF stage

The incoming block of channels is applied to the input of the RF stage. Here, a degree of tuning often takes place where the required channel is roughly tuned and RF gain is applied. The gain of the circuit is controlled by a voltage derived from the 2nd IF output because strong signals will require less amplification or gain than weak ones. Tuning is normally controlled by a microcomputer chip or tuning processor whereby a control voltage is made to vary the bias applied to vari-cap (variable capacitance) diode circuits. However, the majority of the tuning and filtering is performed by the 2nd down-conversion and IF amplifier stages.

Local oscillator and mixer stages

The local oscillator output and the incoming RF signal are fed into a mixer stage where both sum and difference signals of the required channel are generated. Only the difference signals are utilized so the sum frequency band is filtered out by the IF amplifier stage. The frequency of the local oscillator is varied by a control voltage provided by a microcomputer or tuning processor. Whatever channel is selected, the mixing of the two frequencies translates the incoming signal to a fixed band of frequencies centred around a nominal 460 MHz but this is not standardized and can vary depending on the particular model of receiver. The output of the mixer stage is referred to as the second down-conversion or 2nd IF frequency.

2nd IF amplifier and filter

The purpose of the 2nd IF amplifier is to shape the signal to the required band pass characteristic and provide the majority of the signal gain. Only the selected channel's range of frequencies are allowed to pass through the IF filter stage. The IF filter bandpass characteristic can vary between 16 MHz and 36 MHz but is commonly found to be around 27 MHz. For example, Astra satellite receivers have the IF filter characteristic shown in Figure 4.2. Receivers designed to receive

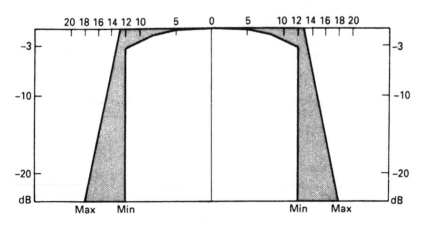

Figure 4.2   *IF filter characteristic for Astra receivers*

Eutelsat II series transmissions will need wider IF filter characteristics of 36 MHz. Overall, the IF amplifier stage can be considered as a channel band-pass filter with gain. In some high quality receivers it is possible to alter the bandwidth of the IF amplifier to suit transmissions from a wide range of satellites. This facility also has the advantage of reducing the noise content of signals during bad reception conditions by restricting the signal bandwidth to less than the normal value, although a loss of picture fidelity is the cost of the tradeoff. In the majority of designs, the 2nd IF amplifier is also AGC controlled in order to automatically boost weak signals relative to strong ones. The output of the second IF stage is used as the input to the AGC detector stage whose output is used to control the gain of the RF and IF stages.

The FM demodulator

The purpose of the FM demodulator is to reconstitute the original video band of frequencies and the accompanying audio subcarriers from the chosen channel bandwidth and strip off the carrier. The output from the FM demodulator is called the baseband signal.

Basic phase locked loop demodulator operation

Figure 4.3 shows the basic operation of a phased locked loop demodulator. The frequency of a voltage controlled oscillator (VCO) is made to follow the incoming changes in frequency. The control voltage input to the VCO is also the demodulated signal output. The frequency output of the VCO, with zero control voltage is called the natural frequency and its value is chosen to be equal to the 2nd IF frequency.

The signal from the VCO is compared with the incoming 2nd IF frequency. The phase detector generates an error signal which is related to the frequency or phase difference between the two inputs. The error signal from the phase detector is low-pass filtered, amplified and used

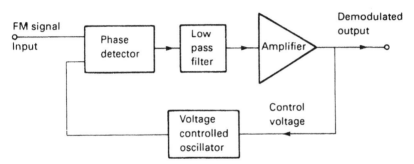

Figure 4.3 *Basic PLL demodulator*

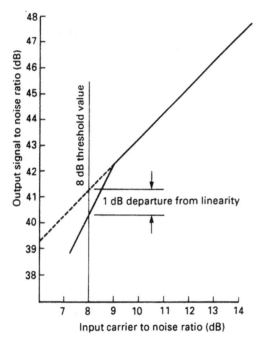

Figure 4.4 *Typical FM demodulator threshold performance*

to control the frequency of the VCO. If the incoming frequency is sufficiently close to the frequecy of the VCO then the feedback nature of the loop will cause the VCO to lock onto the incoming frequency. When in lock, any small phase errors detected are immediately corrected by the feedback loop so that the VCO, accurately and virtually instantaneously, tracks the deviations of the incoming signal. Providing the VCO is made to be sufficiently linear its control voltage is proportional to the required demodulated or baseband signal. The control voltage varies at a maximum rate equal to the maximum frequency of the modulated frequency. The low pass filter in the loop is thus set to this maximum frequency so that the effects of out of band interference and noise are suppressed. The principal advantage of phase locked loops over other forms of FM demodulation are:

1 It is easy to obtain a linear relationship between the incoming frequency deviations and the corresponding output voltages.
2 PLL circuits do not need any tuned circuits, which can be difficult to align.
3 PLL circuits exhibit better performance on signals affected by noise.

The output from the FM demodulator is the baseband signal which incorporates the video information, audio subcarriers and energy dispersal signal. The baseband signal has a typical bandwidth of 10 MHz.

Some up-market multi-satellite receivers incorporate selectable deviation values as well as bandwidths to match a wide range of transponder formats. These are usually controlled by microcomputer chips where the values can be stored and recalled from memory as part of the overall tuning process.

Extended threshold demodulators

One of the main reasons why frequency modulation (FM) is used is its typically 20 dB noise power improvement known as FM improvement (FMI) or FM modulation gain. To guarantee this the demodulator must operate in its linear region as shown in Figure 4.4. This corresponds to a slope of 1 dB per dB when the output signal/noise (S/N) ratio is plotted against input carrier-to-noise (C/N) ratio. This one-to-one linear relationship only exists providing C/N is above a certain threshold value known as the *demodulator threshold*. This point is generally agreed to be where the slope departs from the 1 dB per dB slope by 1 dB. In Figure 4.4 this is shown to occur at a rather unimpressive 8 dB. When the input level falls below demodulator threshold the advantages of FM rapidly fail. This results in impulse noise, which is seen on the screen as annoying comet shaped white spots (sparklies) or dark ones (darklies). On audio signals the same effect causes a series of clicking noises. If this symptom is encountered it is often worthwhile to check for optimal tuning and polarization before getting the ladders off, and re-aligning the dish or replacing the low noise block (LNB). Modern threshold extension techniques have steadily reduced typical threshold figures from a one time value of 11 dB to a little more exciting level of 5 or 6 dB. This allows other system parameters more tolerance and increases resistance to rain fades.

There are currently three basic types of extended threshold demodulator all sharing the principle of reducing the detection bandwidth of the demodulator. This leads to a significant reduction in noise power while leaving the signal power relatively unaffected. An improvement in the S/N ratio is the result.

Extended threshold PLL demodulator

This circuit shown in Figure 4.5 is similar to the basic version (Figure 4.3), but employs an additional secondary feedback loop network forming an active filter. The filter is designed to reduce the detection bandwidth and so effectively increase the S/N ratio. The secondary feedback loop can include a wide variety of specialist design tricks sometimes employing line or field delay networks. The net result of most designs is effectively to reduce the threshold from 11 dB to between 5 and 6 dB.

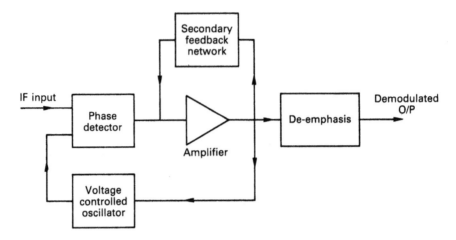

Figure 4.5 *A PLL demodulator with a secondary feedback loop*

Dynamic tracking filter (DTF) demodulator

This species of demodulator (Figure 4.6) employs an active tracking filter with a narrow bandwidth. The filter tracks the deviating FM carrier under feedback control. The detection noise bandwidth is equal to the reduced bandwidth of the tracking filter thus significantly improving the S/N ratio.

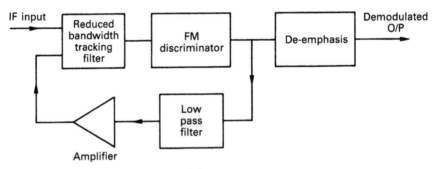

Figure 4.6 *Dynamic filter demodulator*

Frequency feedback loop (FFL) demodulator

Essentially the demodulator (Figure 4.7) manipulates the frequency difference signal at the mixer output of a frequency changer circuit such that the deviation of the carrier signal is reduced. This is decreased sufficiently to allow a reduced bandwidth IF filter to be used. In this way, the noise bandwidth is decreased thus improving the S/N ratio.

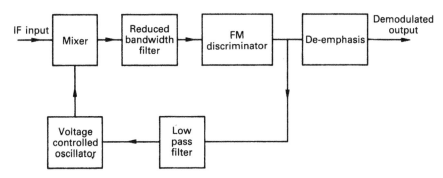

Figure 4.7 *Frequency feedback loop demodulator*

This type of demodulator is also known as a 'frequency modulated feed-back demodulator' (FMFB).

Energy dispersal signal

Under certain modulation conditions, frequency components in the FM spectrum can lead to interference with other users in the same frequency band. This is undesirable, so a 25 Hz triangular waveform, locked to the video signal field rate is produced, during uplink transmission, to cause energy dispersal. The corresponding carrier deviation of this signal varies from 600 kHz for D-MAC to 2 MHz for PAL.

Automatic frequency control (AFC)

An output derived from the FM demodulator is fed to an AFC detector. The purpose of this circuit is to provide an error signal of the correct polarity to achieve any necessary correction voltages required to trim the tuned input stages. This is sometimes controlled indirectly by a tuning processor or microcomputer chip. Without this facility, a tendency to drift off tune may occur due to changes in temperature or humidity levels.

Baseband processing

The baseband signal output from the tuner/demodulator module is typically 250 mV peak to peak and contains the video information, energy dispersal and audio subcarriers. A block schematic of a no-frills satellite receiver is shown in Figure 4.8. The baseband output is split three ways to the video processing stage, sound demodulator stage, and external baseband stage.

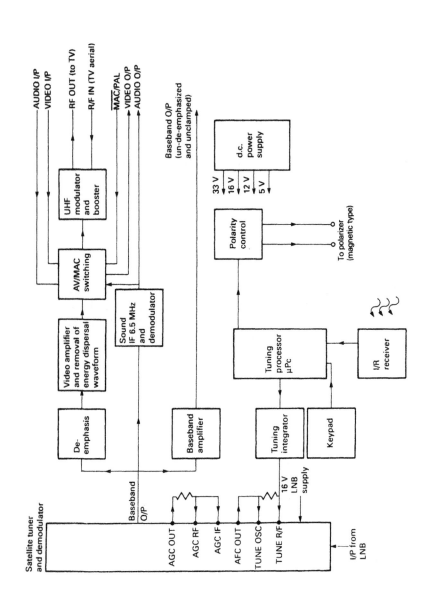

Figure 4.8 *Block schematic of a basic satellite receiver*

External baseband signal

The raw output from the FM tuner/demodulator is boosted to 1 V p-p into a 75 ohm load by the baseband amplifier. This signal, unde-emphasized and unclamped, is fed to external AV/MAC connectors, and is provided to enable the fitment of external decoders such as MAC. MAC decoders, D-MAC in particular, also require a baseband signal bandwidth of 25 Hz to 10.5 MHz. Later versions of receivers may incorporate MAC decoders internally.

Video processing

The raw baseband signal is first de-emphasized and fed into a video amplifier stage which not only makes good the losses incurred in de-emphasis but brings up the signal in level to 1 V p-p, the standardized composite video level. In some receivers the video de-emphasis can be switched between PAL or linear. Within this stage a clamping circuit removes the 25 Hz energy dispersal signal and a filter circuit removes all the audio subcarriers above the upper edge of the video band. The output of the video amplifier, called the baseband composite video signal, is fed to an AV/MAC switching stage which is explained later. Teletext signals are often transmitted in the vertical blanking period and up to fifteen lines of this period are commonly used. An oscilloscope trace of the baseband composite video signal is shown in Figure 4.9.

Sound processing

Again, the number and centre frequencies of sound subcarriers are not standardized. Furthermore, differing sound de-emphasis levels can be encountered, one is 62 µs and the other is known as the standard J17. Some receivers allow switching between the two. For purposes of simplification the following explanation will assume we have a dedicated Astra receiver; other satellite transmissions will have similar facilities but not necessarily using the same frequencies.

The baseband signal obtained from the Astra satellite is shown in Figure 4.10; the video band extends from zero to 5 MHz. A primary audio carrier centred on 6.5 MHz is provided for monophonic, single language transmissions and four separate subcarriers centred on 7.02, 7.20, 7.38 and 7.56 MHz are provided for stereo or multi-lingual transmissions. This later group of four sound carriers has smaller deviations and bandwidths.

In addition to the primary audio subcarrier, a stereo sound signal and two more audio channels can be transmitted using the four extra subcarriers. The extra two channels need not be related to the video service

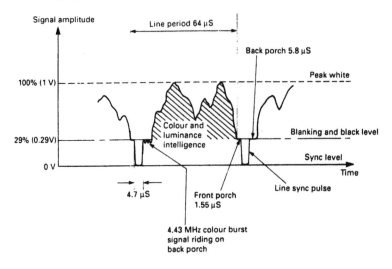

Figure 4.9 *Composite video signal (PAL)*

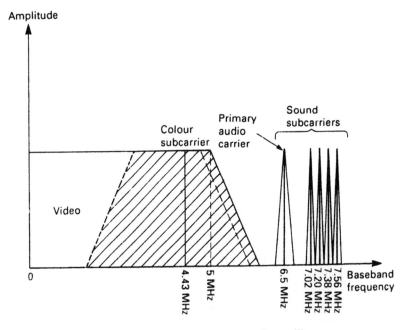

Figure 4.10 *Baseband signal from the Astra 1A satellite*

and may be radio transmissions, a second stereo language or two additional monophonic languages. Sometimes, all four channels can be utilized to transmit different languages in mono. This configuration is often found with multi-lingual commentaries on the sport channels. Some low cost receivers are totally monophonic and tend to rely on the primary audio subcarrier reception only and this is the simplest design. However, if the receiver is designed to demodulate all four narrow bandwidth sound subcarriers (stereo and multi-lingual) then it will not need to demodulate the primary subcarrier since this always carries a monophonic duplicate of the corresponding stereo channel or language. Figure 4.9 shows the PAL composite video signal.

Specification of primary audio signal

Subcarrier frequency	6.5 MHz
Audio bandwidth	20 Hz–15 kHz
Deviation	plus or minus 85 kHz
Pre-emphasis time constant	50 μs
Modulated subcarrier bandwidth	180 kHz
Modulation index	0.26

Specification of additional narrow bandwidth subcarriers

Subcarrier frequencies	7.02, 7.20, 7.38 and 7.56 MHz
Audio bandwidth	20 Hz–15 kHz
Deviation	plus or minus 50 kHz
Pre-emphasis	Wegener Panda 1 system (adaptive pre-emphasis)
Modulated subcarrier bandwidth	130 kHz
Modulation index	0.15

Audio combinations used in Astra 1A transmissions

Audio channel	A	B	C	D
Subcarrier freq. (MHz)	7.02	7.20	7.38	7.56
Mode 1	Lang 1 (L)	Lang 1 (R)	Lang 2 (L)	Lang 2 (R)
Mode 2	Lang 1 (L)	Lang 1 (R)	Lang 2 (M)	Lang 3 (M)
Mode 3	Lang 1 (M)	Lang 2 (M)	Lang 3 (M)	Lang 4 (M)

where Lang = language, L = left audio channel, R = right audio channel and M = monophonic.

In order to obtain a high S/N ratio the four stereo/multi-lingual sound channels use the licensed Wegener Panda 1 noise reduction system.

Decoder switching

Most satellite receivers have the facility for connecting external decoders and de-scramblers. In the simplest case, all that is required is a means of routing composite video and audio signals to the UHF modulator from either internal circuitry or externally from, say, an external decoder. An externally generated switching voltage from the decoder is routed via the connector to control the switching. Remember that the external decoder input is obtained from the raw baseband signal.

Modulator/booster

The UHF modulator generates a UHF carrier signal typically at 607.25 MHz (channel 38). The carrier frequency is usually adjustable between channels 24 and 40 in case there is interference patterning with local TV stations or VCRs operating at or near this channel frequency. Both the composite video and audio signals are AM modulated onto the carrier. The modulator normally provides a loop-through facility so that a terrestrial TV aerial can be connected at the rear of the receiver. A certain amount of gain is added, to compensate for the losses incurred in combining the two signal sources. Terrestrial TV signals and the RF signal output from the modulator are combined into one common RF output socket at the back of the receiver. Without the loop-through facility a passive combiner would have to be used leading possibly to a decrease in signal level. The UHF modulator is provided for connection to older TV sets not having SCART/PERITEL sockets which provide for direct composite video and audio connection.

Power supplies

A range of power supply voltages is required in satellite receivers. A stabilized 33 V line is typically required to bias the vari-cap diodes in the tuner/demodulator module. A regulated 12 V supply is required for all the analogue circuitry such as sound and video processing circuits as well as the main power to the satellite tuner/demodulator module. A regulated 5 V supply is often needed for digital circuits, remote control receivers and display drivers. Supplies for the head unit polarizer and LNB are typically 12 and/or 15 V, and regulated.

Polarizer control

The selection of the required linear polarization sense, either vertical or horizontal, is performed by the polarity control circuit. This is a fairly complex piece of circuitry which maintains a constant current through a solid state magnetic polarizer embedded in the feedhorn. Polarization sense is critically determined by the accuracy of the current flowing through the polarizer windings. In order to ensure this current is constant over a wide range of outside operating temperatures, highly sensitive regulatory circuits need to be employed to compensate. The direction of current flow (typically 35 mA) determines whether vertical or horizontal polarization is selected and this is commonly stored and controlled by the tuning processor or system microcomputer chip. A variation on this principle is that the polarization sense is selected by passing either zero current or 70 mA through the windings. Some head units, designed for single satellite reception, employ simple V/H switching of 90° spaced dual probes built into the head unit. In these cases, a simple d.c. voltage level shift in the LNB feed is all that is required from the receiver to control the polarizer.

On some up-market receivers employing magnetic polarizers, it is possible to program and store optimum polarization adjustments for each channel. The reason being that the amount of wave twisting that is needed for each polarization sense is not only governed by the current through the polarizer but also, to a lesser extent, by the channel frequency of the signal. The facility to program in the relevant trim, by finely adjusting the current flowing in the polarizer windings, ensures spot-on polarization for all channels in the band. The major disadvantages with this are that the initial setting up of the channel tuning can be tedious and the extra wiring can lead to increased installation time. A maximum wave twisting range slightly in excess of 90° is often provided.

A few polarizers are three-wire mechanical devices and work by feeding voltages pulses to a servo motor which physically rotates a polarizer probe through 90°. The mark to space ratio of the pulses determines the amount of rotation. The 3 wire connections to a polarizer of this type are +5 V line, ground, and a pulse line. Mechanical polarizers are not used much nowadays and the current trend leans more toward the use of the magnetic type in up-market models.

One of the main conclusions to draw from these possible polarizer/receiver variants is that extreme care must be exercised when putting together a system from different manufacturers' units. A sound knowledge of polarizer/receiver compatibility is necessary, otherwise costly and potentially embarrassing mistakes can be made.

Integrated receiver/decoders (IRDs)

The vast majority of receiver sales are designed to be operated using a small fixed mount dish directed to some particular satellite (or cluster of co-located satellites). These tend to be the more popular ones of interest to the wider population of the intended service area. Where a broadcaster has a sufficiently large customer base to implement subscription TV, manufacturers often integrate the necessary decoder technology into the same box as the receiver and rename it an IRD.

Satellite receivers built into TV receivers

In view of the general hatred of tangled wires, satellite receivers and decoders are increasingly being built into TV sets and possibly VCRs, having the obvious advantage over the 'sat-box' of allowing viewing of one satellite channel whilst recording another. Combined with the introduction of multi-media, 16:9 TV, and HDTV it can be envisaged that the concept of the 'three-man-lift entertainment centre' may ripen. Doctors may well notice an increase in the attendant 'glass-back' syndrome in the near future.

Decoders and AV sockets

Most satellite receivers have some method of connecting de-scramblers or decoders. In addition many have a method of direct connection of AV (audio/video) signals to a suitably equipped VCR or TV. This practice bypasses the UHF modulator section of the satellite receiver and will consequently lead to improved sound and picture quality. The vast majority of satellite receivers of European origin utilize either the 21 pin SCART/PERITEL type of socket, or the 15 pin subminiature 'D' type socket or sometimes both. Some models have one socket wired for direct AV to a VCR or TV and, in addition, another socket specifically for the connection of decoders. This latter socket is usually clearly labelled 'decoder only'. The reason is that the raw baseband signal, derived from the satellite tuner/demodulator, is required for input to decoders such as MAC, whereas ordinary VCR and TV sets require processed PAL composite video signals at a nominal 1 V p-p. Some low cost receivers dispense with this additional socket and use a single socket which can be used for either purpose. The provision of baseband or composite video is determined by a switching signal on a specific pin of the socket as explained later. Another method sometimes encountered is the provision of the baseband output on a non-standard or reserved pin, rarely used in AV connections to VCRs or TVs.

The standard SCART/PERITEL connector

The SCART/PERITEL connector, sometimes referred to as the Euroconnector, is a convenient 21 pin socket for the interconnection of various signals to/from domestic electronic equipment. The pin layout is shown in Figure 4.11. This type of wiring is the standard used for AV connection to VCRs and TVs and is often additionally used as a decoder interface connector on satellite receivers.

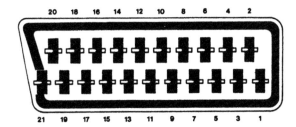

Figure 4.11 *SCART/PERITEL pin numbers (front view)*

Standard AV SCART/PERITEL pin connections:

PIN SIGNAL

1 Audio output (right)
2 Audio input (right)
3 Audio output (right)
4 Audio earth
5 Blue earth (RGB)
6 Audio input (left)
7 Blue video (RGB)
8 AV source switching voltage (see text)
9 Green earth (RGB)
10 Data line
11 Green video (RGB)
12 Data line
13 Red earth (RGB)
14 Data line earth
15 Red video (RGB)
16 Video blanking signal
17 Composite video earth
18 Video blanking signal earth
19 Composite video output
20 Composite video input
21 Overall shield earth

An extra SCART/PERITEL socket for external decoders

When a SCART/PERITEL socket is additionally fitted for the connection of external decoders the output available at pin 19 is the raw baseband signal and not the clamped and de-emphasized composite video as in the standard specification. Sockets of this type are normally labelled 'DECODER ONLY'. The source switching line on pin 8 is held high (+12 V) by the decoder to inform the satellite receiver to return, or switch in, audio and video signals from it. When this line is switched low (0 V, decoder switched off or disconnected) the receiver routes its own internally processed signals to the additional AV socket or UHF modulator. On most models pins 5 and 7 and pins 9 to 16 remain unused in the decoder socket, so these pins are normally linked through to the corresponding pins of the additional AV socket. This allows component RGB signals to be routed, from the decoder, directly to the TV or VCR. In this way the benefits of MAC decoded signals can be enjoyed providing the TV is a modern one which accepts RGB input and the MAC decoder derives the component RGB output.

A single SCART/PERITEL socket for AV or external decoder connection

Some satellite receivers have a single SCART socket that can be used for either connection to external decoders or AV direct connection to a VCR or TV. For AV mode, pin 8 is held high by the satellite receiver so that any connected VCR or TV is automatically switched to receive its internally processed signals. Thus de-emphasized and clamped composite video is available at pin 19 of the satellite receiver's SCART socket.

When an external decoder is connected, pin 8 must be pulled low by the decoder, or by human intervention, to enable the raw unde-emphasized and unclamped baseband signal to be available at pin 19. When pin 8 is low, in these designs, the returned audio and video signals from the decoder are switched to the UHF modulator. Internally processed signals from the receiver are switched out thus are no longer available.

Another method used to utilize the same SCART socket for AV or external decoder connection is to direct the baseband output signal to a pin normally reserved for other uses. Sometimes pin 8 is not used as the switching voltage. In these and other receivers, it is important to check the manufacturer's instructions or consult the model's service manual before connecting decoders. However, for AV use this is not particularly important, since the majority of SCART to SCART connector cables are simply composite video and audio connected. That is to say, most of the pins remain unconnected.

An extra 'D' type socket for external decoder connection

Some models of Astra satellite receivers employ the 15 pin, subminiature, 'D' type socket for connection to external decoders. The pin layout is shown in Figure 4.12.

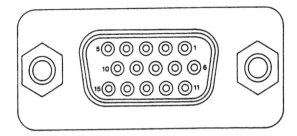

Figure 4.12 *Decoder 'D' type socket pin numbers (front view)*

Pin connections to D type socket:

Pin	Signal	Details
	Pin Signal	*Details*
1	Return audio input (left)	500–700 mV into 10 k load.
2	Return video input (PAL)	1 V p-p into 75 ohms.
3	Video switching voltage*	Switch receiver to return video from decoder if +12 V applied.
4	Baseband output	Unde-emphasized and unclamped video.
5	PAL video output (clamped)	1 V p-p into 75 ohms.
6	Return audio input (right)	500–700 mV into 10 k load.
7	Audio switching voltage*	Switch receiver to return audio from decoder if +12 V volts applied.
8	Earth	0 V.
9	Reserved	Future applications.
10	Reserved	Future applications.
11	Earth	0 V.
12	Audio output (left)	500–700 mV into 10 k load.
13	Audio output (right)	500–700 mV into 10 k load.
14	Reserved	Future applications.
15	Reserved	Future applications.

*Some receivers may have a single +12 V line, switching both audio and video at the same time.

The MAC system

MAC is short for multiplexed analogue components and was envisaged as the future for HDTV (high definition TV) in Europe. There are increasing rumours that MAC is dead and the future points to all digital TV systems. This may well be true, but MAC nevertheless still exists at present.

Advantages of MAC systems over PAL

MAC systems have considerable advantages over composite video systems like PAL or SECAM and these are briefly outlined below:

1   The absence of cross-colour and cross-luminance effects ('the checked jacket effect').
2   Improved FM signal to noise ratio performance.
3   The combined carrier of digital and analogue components leads to the availability of high quality sound and data services.
4   Absence of subcarriers for sound and colour signals allows an increase in video bandwidth and consequently in picture quality.
5   Picture aspect ratio can be standard 4:3 or 16:9 for future wide screen and HDTV developments.
6   The flexibility of conditional access of receivers using over-air addressing removes the need for external decoder boxes.

MAC transmission standards

There are several variants of the MAC system: B-MAC, C-MAC, D2-MAC, and D-MAC. The differences between the systems is not so important as what they have in common. Here the D-MAC variety is discussed, since D-MAC receivers are also capable of decoding D2-MAC signals. The entire D-MAC baseband signal is frequency modulated using a deviation of 13.5 MHz/V. A triangular energy dispersal waveform with a deviation of 600 kHz is added to this. A carrier to noise ratio of about 12 dB allows vision quality in excess of grade 4 on the CCIR scale (see Chapter 5). The basic signal consists of a baseband time division multiplex of time-compressed colour and video signals preceded by a 20.25 Mbit/s duobinary burst signal which carries audio, data and digital synchronization information. Duobinary is a term given to a logic system which is more economical on bandwidth for a given data rate. Extreme positive and negative transitions are 'logic level 1' and intermediate values (zero volts) are 'logic level 0'. Compatibility with existing PAL receivers is achieved with a 625 line, 50 Hz field rate with 2:1 interface. The layout of one active line of a D-MAC signal is shown in Figure 4.13. A 10 µs duration duobinary burst of 206 bits is transmitted at the start of each 64 µs line. This relatively high data rate can encode up to eight high quality sound channels or a combination of sound, data and 'over air' conditional access signals. This digital information is arranged by the broadcaster into packets or subsections related to a particular service. For example, two stereo sound channels, teletext, and conditional access signals may be chosen for a particular service.

The colour difference bandwidth is 0 Hz to 2.75 MHz and the corresponding luminance bandwidth is 0 Hz to 5.75 MHz. However, colour

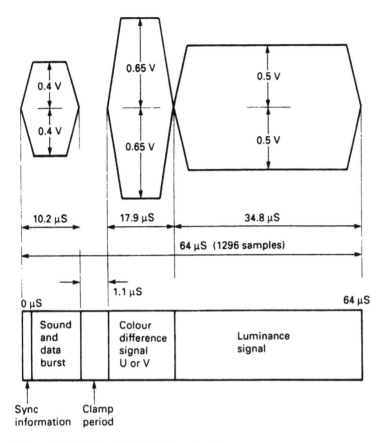

Figure 4.13 *Simplified MAC active line*

difference signals are time compressed, prior to transmission, by a factor of 3:1 and the luminance signals by 3:2. The result of this is to increase the bandwidth by the same proportion thus increasing the overall video bandwidth to 8.75 MHz. Within the D-MAC decoder the analogue signal is sampled digitally at 20.25 MHz and stored in memory for reading out at the normal (non-time compressed) rate. The choice of 20.25 MHz for the digital sampling rate is related through the corresponding time compression ratios to the CCIR recommended studio sampling rates of 6.75 MHz for colour difference signals and 13.5 MHz for luminance signals. During the encoding process the $E'U_m$ colour difference is transmitted on odd numbered lines whilst the $E'V_m$ colour difference signal is transmitted on even lines. In the receiver the missing component is recovered by averaging the two signals on adjacent lines. A one line time delay is used in the luminance signal to compensate for this. The first bit word of each sound/data burst signal contains line sync information. In addition line 625, a data only line, contains field sync information and other high priority data.

The sound/data burst multiplex

The 206 bit sound/data burst signal at the beginning of each active line (excluding line 625, which is the 1296 bit data only line) carry mainly sound and data. Most of the bits (198 per line) are subdivided into two groups of 99 bits per line. These are used over one frame period to construct two subframes of digital information as shown in Figure 4.14. Together these subframes convey 123 354 bits of information per television frame which corresponds to a data handling capacity of 3 083 850 bits/s. These subframes are further subdivided into discrete packets of 751 bits, each containing a 23 bit header for the receiver to select and identify those packets necessary for a particular broadcast while rejecting all others. The headers incorporate error protection codes so that erroneously received data can be detected and corrected.

Sound channels have built in flexibility, either high quality 15 kHz bandwidth (32 kHz sample rate) sound channels or 7 kHz bandwidth (16 kHz sample rate) commentary quality channels can be transmitted.

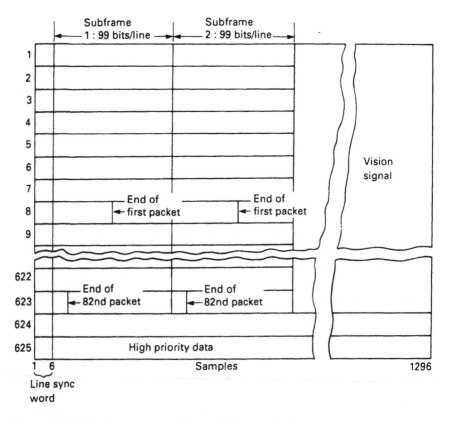

Figure 4.14 *Packet structure of sound/data multiplex*

High quality sound can be coded as 14 bit linear or companded NICAM formats. The D-MAC system also offers a much higher teletext/data capacity than conventional PAL systems. However, these signals need to be transcoded into the PAL vertical blanking period for decoding and display by a conventional teletext receiver. The D-MAC vertical flyback interval is reserved for future developments, such as HDTV, where control data, sent during the vertical flyback period, are to be used by specially adapted television receivers, to reconstruct a 1250 line picture. This technique is known as 'digitally assisted television' (DATV).

Receivers' data requirements

Since the D-MAC system is so flexible, certain signals and information need to be sent to the receiver in order that it can successfully process the correct picture information, data and sound channels. This information is sent as follows:

High-priority data – This information is contained in line 625 of each frame and contains two basic types of information. Firstly, static information which does not change significantly from frame to frame, such as satellite identification, date, and time. Secondly, dynamic information which contains details of the complete multiplex structure of a satellite channel signal so that the receiver can gain access to its transmissions. Included in this portion are the digital frame synchronization signals.

Service identification channel – This medium priority information is assigned packets in the sound/data multiplex and is used to convey information about the channel services and transmission characteristics. This allows receivers to be automatically configured to decode the correct service components of the channel. This information is repeated every so often so that a receiver that has changed channels or has just been switched on can quickly acquire the selected services.

Interpretation blocks – This low priority information is used to specify the correct audio channels corresponding to a particular transmission and prepare the receiver for any forthcoming changes in the selected sound service. This information is contained within packets of the sound/data multiplex.

MAC scrambling and encryption

The D-MAC sound and vision signals can be conveniently divided into a number of discrete samples or packets. The vision components can be subdivided into 1256 discrete time periods or samples per line, and blocks of these samples may be transmitted in any order. A technique known as 'double cut component rotation' is used to mix up the colour difference and luminance segments in a pseudorandom manner over each line period. The chosen positions from 256 equally spaced cut points vary from line to line thus rendering the picture unintelligible. Two 8 bit numbers, one for each waveform, are produced by a pseudorandom number generator to determine the cut points. Only if the receiver is made aware of these cut point sequences can it successfully reconstruct the signals, and this is done by generating an identical series of pseudorandom numbers in the receiver. If the receiver does not know the cut point sequence then the picture is said to be scrambled and the locking of such, with hidden keys, is known as encryption. The key signals to unlock this scrambling sequence clearly need to be sent to the receiver. For a non-subscription service this unlocking sequence is provided free to all receivers. For a subscription channel, the key information to unlock the sequence will only be sent to individual receivers on receipt of the appropriate payment. This concept is known as conditional access or over-air addressing of receivers. Due to a technique known as cross-fading the viewer will be unaware of any quality degradation due to the cutting actions used in the scrambling or descrambling process. The complexity of a D-MAC video decoder section can be appreciated from Figure 4.15, but thankfully the whole lot is buried within a handful of chips and a few discrete components.

Sound and data scrambling can be effected by juggling the relevant useful sound/data burst packets in a pseudorandom fashion. The reverse process in the receiver (descrambling) can also be controlled by conditional access techniques.

A typical PAL satellite receiver

This section briefly describes the complete block diagram of a versatile PAL satellite receiver, with stereo sound, incorporating facilities to control all three different methods of polarizer, electro-mechanical (pulse), electro-magnetic (ferrite) and V/H switched. Most PAL satellite receivers employ similar circuits, but only a few are capable, at the time of writing, of all three linear polarization control methods.

Power supply

The power supply section, the most likely section to fail in use, is shown in Figure 4.16. A.c. mains, 240 V in the UK, is connected to the power transformer with three secondary windings. Outputs are rectified and fed to a variety of different regulators. The first 5 volt supply (U401) supplies the electro-mechanical pulse type of polarizer

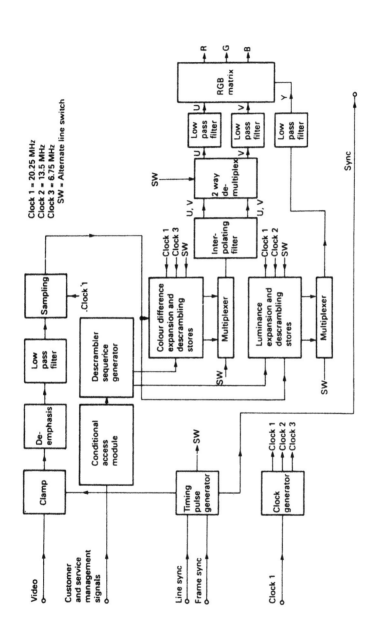

Clock 1 = 20.25 MHz
Clock 2 = 13.5 MHz
Clock 3 = 6.75 MHz
SW = Alternate line switch

Figure 4.15 *Video processing in a D-MAC decoder*

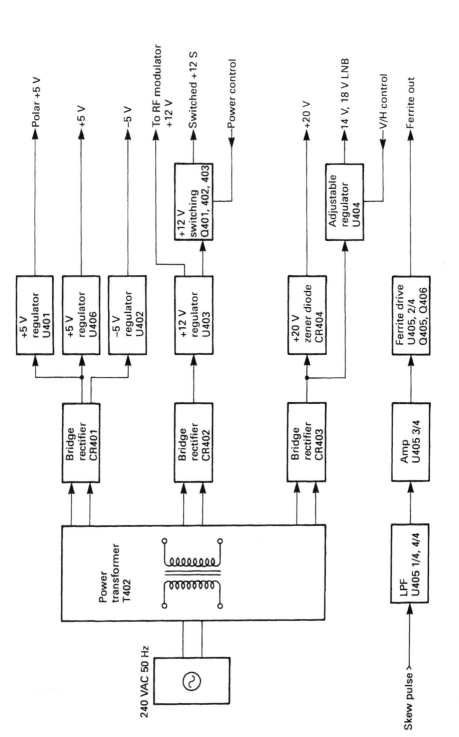

Figure 4.16 Power supply diagram (Samsung Electronic Components UK)

if fitted. Second and third 5 volt supplies (U406 and U402) supply power to internal chips and circuitry. The 12 V switched line (labelled Q401, Q402, and Q403) is enabled when the receiver is brought out of standby and powered up by the presence of the power control voltage from the microprocessor. The 20 volt line, stabilized by a zener diode, supplies tuning circuits of the RF tuner module. An adjustable regulator (U404) supplies the 14/18 V switching voltage for V/H switch type polarizers. Associated V/H selection control voltage comes from the microprocessor. A skew pulse for electromagnetic polarizers is low-pass filtered, amplified and fed to a ferrite drive circuit so the receiver can control antennae systems equipped with electro-magnetic (ferrite) polarizers.

RF, video and RF modulator stages

These stages are shown in Figure 4.17. The RF tuner, 950 MHz to 1750 MHz, is tuned using frequency synthesis technique similar to that used with most modern TV sets. The baseband output is split into two paths. The first path provides a de-emphasized but unclamped PAL output together with a buffered base band output, while the second path provides a PAL de-emphasized and clamped video signal for injection, along with mono audio, into the RF modulator as well as providing a separate buffered video output. A switching circuit (U103) is used to switch in video signals from an external decoder. The input required to most external decoders is the basic baseband signal (unde-emphasized and unclamped).

Audio stages

Referring to Figure 4.18, the baseband signal is bandpass filtered to extract audio sub-carriers then split to a pair of frequency changing mixer circuits and FM discriminators. This practice enables audio tuning to be continuously variable over the range 5.0 MHz to 8.8 MHz. Audio A and Audio B outputs thus allow stereo sound to be demodulated. Figure 4.19 follows on from Figure 4.18 and shows remaining audio processing, where both Audio A and Audio B signals are further split. One path is amplified, low-pass-filtered and de-emphasized, the other is fed to a noise reduction circuit for hi-fi stereo sound. All audio signal routes are fed to a mono/stereo switching circuit (U212) where switching is controlled by a line from the microprocessor. A further switching circuit (U103) is provided for routing audio signals from external decoders where scrambling of audio is present. Finally, both mono and stereo outputs are buffered to the outside world.

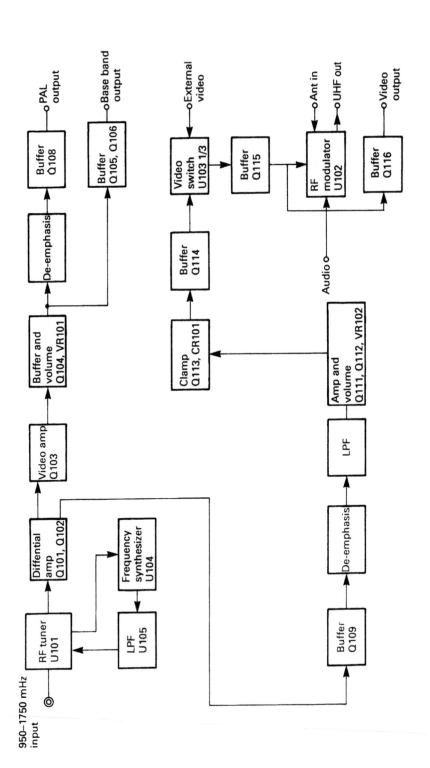

Figure 4.17 *Video and baseband processing (Samsung Electronic Components UK)*

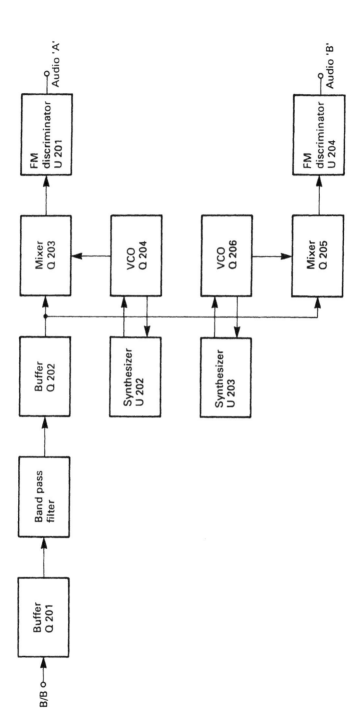

Figure 4.18 *Audio processing part-section (Samsung Electronic Components UK)*

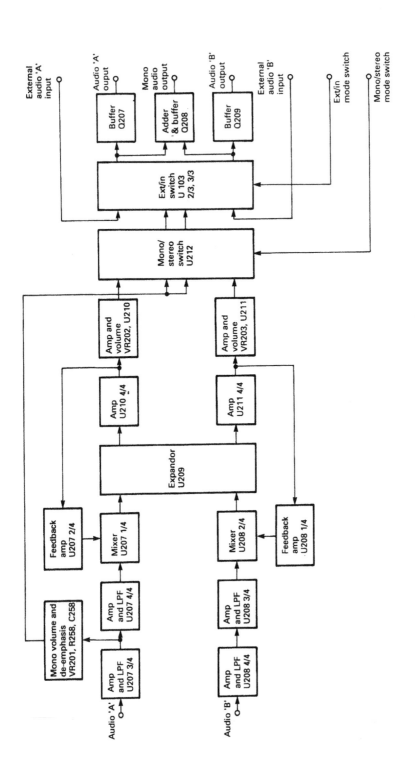

Figure 4.19 *Audio processing part-section (Samsung Electronic Components UK)*

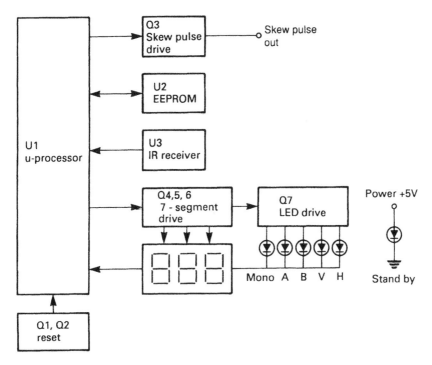

Figure 4.20 *Microprocessor control section (Samsung Electronic Components UK)*

Microprocessor circuit

Front panel display and overall control of receiver is performed by a microprocessor U1. As can be seen from Figure 4.20, Q3, is the skew pulse driver to control electro-mechanical or motorized linear polarizers. This output, as mentioned earlier, is also modified to provide current control for electro-magnetic (ferrite) polarizers. U2 stores relevant tuning and skew values in digital form in non-volatile memory, while U3 is the infra-red remote control receiver circuit. The microprocessor also controls the various multi-function seven-segment displays and mode status light emitting diodes (LEDs). The detailed method by which these functions are performed need not concern us unduly since internal software buried within the chip set controls all functions.

Video scrambling and encryption

To some the subject of video scrambling and encryption is nothing short of a total yawn while others may become obsessed with the finer points. It probably depends on your type of mind and whether or not you like cracking artificially set puzzles. To cover the subject in any detail would require allocation of much space and risk the author's sanity, so only a minimum 'black box' treatment is given here.

Distinguishing features

Scrambling and encryption are two quite distinct methods of rendering a video signal incomprehensible to viewers unless they have the necessary decoding equipment. Although either method produces the same effect, there is a fundamental difference in the way it is produced. The terms are distinguished as follows:

1 *Scrambling* – A *scrambled* video signal is one that is just a jumbled up version of the original. Without the correct descrambling or decoding equipment, the received pictures are without form and are meaningless.
2 *Encryption* – An *encrypted* video signal is one in which the video control signals have been altered according to some code. The information necessary for decoding the control signals is transmitted along with the video information.

Scrambling offers less protection against unauthorized code breakers (*hackers*) than encryption. Hackers are a determined breed and will willingly burn gallons of midnight oil just to beat a system which may be altered the next month. Their labours are unlikely to attract pecuniary benefits in the long term. Anyone with a smattering of electronics knowledge could beat some of the first generation scrambling systems which relied on rather crude modification to the video signal and sync pulses. For legitimate subscribers, the appropriate licensing authority supplied a descrambling black-box on the understanding that, should payments for the service fall behind, it had to be returned – a requirement not always easy to enforce! The modern tendency is to design composite protection systems which combine scrambling with encryption. Fortunately, digital transmissions earmarked for the future are eminently suited for advanced encryption algorithms.

Encryption, providing sufficiently involved algorithms are employed, offers greater resistance to the hacking fraternity. Indeed, the designers of one proposed encryption system boasted that their brain-child would take over a billion years to crack. Without wishing to throw cold water on their convictions, they might do well to remember that a similar claim was probably made by the designers of the *Enigma* machine during the disagreement with the Third Reich.

Need for protection

Programming authorities may need to encrypt or scramble their transmission for several reasons:

1 Services which attract no license fees or income from commercial advertising have to recover their operating and programme costs by imposing fees on a subscriber-only basis so the only way to prevent unauthorized viewing is by scrambling or encrypting the signals.
2 Some studio programmes or films are sold or rented subject to the proviso that transmission reception (the footprint) is geographically limited. An unrestricted viewing audience decreases the chance of future sales of a film or programme, so the proviso is understandable. However, it is not always easy to obey such a restriction to the letter because:
   (a) footprint outlines are not always too sharply defined and have a tendency to overlap; and
   (b) high performance receiver systems may still be able to receive acceptable pictures at a considerable distance outside the footprint area.
3 Certain programmes are intended for controlled audiences only. For example, channels which churn out programmes, euphemistically classified as fit for 'adult only' viewing may offend public taste or, in some cases, offend state laws unless the channel is available on a subscription-only basis to the 'dirty mac brigade'.

Encryption basics

A message signal may be anything from text, through an analogue video signal, to a stream of binary digits. In its simplest form a written text message is about the easiest for us to visualize. To encrypt a written message is to mess it about in some way such that only the intended receiver of the information can understand it. This can only work if the alterations have been made according to some pre-arranged plan known only by the sender and the receiver. There are two ways of doing this:

1 *Substitution* – This is where each data element is *replaced by another* according to a fixed plan. For example, in the simplest case, a letter could be replaced by the one, say, three alphabetical places ahead. Thus HECTOR would be sent as KHFVRQ, and BUNS as EXQV.
2 *Permutation* – this is where the original data are used but the *order* has been mixed up.

Substitution codes, used on their own, give little protection against code breakers. Going back to the simple text analogy, an 'average' 12-year-old could, given sufficient time, break the code without too much difficulty because the occurrence of each letter in most languages, particularly in English, is subject to statistical laws. The most commonly employed letters are E, N, A and T, and the most common paired letters are TH, EE and ST. So our bright 12-year-old would start by scanning through the encrypted message for the presence of these popular letters. Once these have been substituted, it will lead the way to further substitutions by combining trial and error with inspired guesswork. Modern methods of encrypting a satellite channel usually employ a mixture of substitution and permutation because the resultant combination greatly enhances the protection.

VideoCrypt

A reasonably secure encryption system, called VideoCrypt, is used by British Sky Broadcasting and others. It employs a composite mixture of scrambling and encryption and, for extra protection, requires a smart card at the receiving end which is normally changed every few months. Although it is about the size of a normal credit card, it contains the following electronics, all on a *single chip*:

1 A microprocessor.
2 A masked ROM (read only memory) for general housekeeping operations.
3 An EPROM (erasable programmable random access memory) for holding the service data, authorization and subscription details.
3 A RAM (random access memory) used for general storage and for holding the decryption algorithm.

Hackers with access to university or medical equipment have been known to ferret out the secrets of microchips by examining them under the electron microscope, but they are unlikely to succeed in cracking the smart card because the radiation falling on the EEPROM will destroy its contents.

System overview

At the transmission end, an 8-bit random number generator cuts the line video at any one of 256 points which is then rotated around this point. At the receiving end the picture is descrambled by reverse rotation. The starting position of the random number generator, called the *seed*, is sent out within the vertical blanking interval. For extra protection the seed is scrambled by a complex algorithm before it is sent out. When the seed arrives it is combined with the smart card information and used for reversing the line video rotation back to its original position. Each invididual card can be addressed and 'switched off' at any time by the transmission authority. On the face of it, the procedure appears rather simple and not too difficult to crack but the security rests firmly and squarely on the secrets buried within the smart card. Figure 4.21 shows the VideoCrypt information flow during transmission and reception, beginning with the studio programme and ending at the customer's television screen.

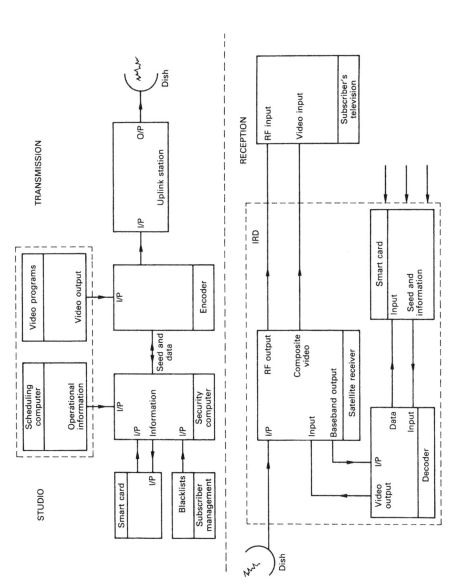

Figure 4.21 *VideoCrypt system*

It is, of course, still possible to use a 'mini-hack' in that a small number of households in a tight local community need only subscribe to one smart card in order to distribute a single decoded channel via a simple RF video sender or clandestine cable network. However, all viewers would be limited to the same decoded channel at any one time.

Transmission system details

Subscriber management system – This may be considered the starting block of the system because it handles all subscriber information, including new orders for the service, payment details, the occasional message and cancellation reminders or, where necessary, final demands accompanied by the usual thinly disguised threats of cut-off.

The security computer – This is where the random number generator is housed and where the seeded output is scrambled. It is also responsible for preparing the information received from the subscriber management system in the correct form for passing to the encoder.

The Encoder – The scrambled seed received from the security computer together with the video programme generated at the studio are encoded and passed to the uplink station.

Receiver system details

The satellite receiver – This is of conventional design, but if not an IRD, data from the baseband output may be sent to an external decoder.

The decoder – The data and the control elements, received from the baseband output, are sent to the smart card.

The smart card – The received data are reformed by the chip algorithms and the result sent back to the decoder section for the final descrambling of the video picture. As mentioned above, the card is periodically changed so that in the unlikely event that determined hackers have cracked its contents they are obliged to start all over again – not a very profitable occupation!

VideoCrypt features

1   Security is extremely high in the long term because it rests almost entirely in the smart card which can be changed periodically.
2   It is easy to blacklist illegal cards.
3   Only a few office and administration staff are needed because most work is handled automatically by the subscriber and management computer.
4   Studio controllers can send *individual* messages to subscribers.
5   VideoCrypt is compatible with most analogue transmission standards and can easily be adapted to suit all future digital transmissions.

DVB receivers

Figure 4.22 shows a block schematic of a typical DVB/MPEG-2 receiver. The first generation of receivers will be the set-top integrated receiver/decoder (IRD). IRDs have the familiar combination of RF connections and SCART sockets plus a personal computer interface for downloading internet services via satellite. Data channels are used to receive technical parameters of the transmission and EPG (Electronic Program Guides). A smart card slot for pay-TV access control will also be present. MPEG-2 is an audio/video digital compression standard based on the discrete cosine transform with a maximum bit rate of 15 Mbit/s.

Technical parameters received include broadcast frequency, modulation parameters and details of services available within the transmitted multiplex. The receivers can automatically configure themselves, using in-built software, to changes in broadcast parameters or even rearrangement of satellite transponders without retuning. Refer to Chapter 11 for more details of DVB.

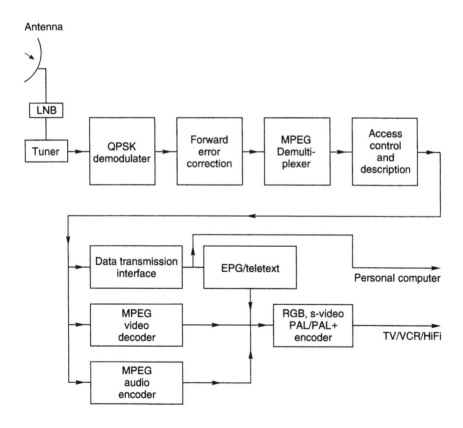

Figure 4.22 *Block schematic of a DVB receiver*

EPG information received includes:

a Distribution network (e.g. Astra, Eutelsat).
b Channel (e.g. Bravo, Sky One, UK Gold).
c Program ID (e.g. Monty Python, Fawlty Towers).
d Program content (e.g. film, news, erotic).
e Service provider (e.g. Sky Television).
f Bouquet (e.g. Sky multi-channels package).
g Time (e.g. 05:00 to 06:00).
h Description (e.g. Film: '101 Ways to Slaughter a TV Cop', cast, director, etc.).
i Future program details.

5 Link budgets

Introduction

Strictly speaking, the entire link, including both the uplink and down-link, should be analysed for dish-sizing. However, with broadcast W, the use of uplink power control and certain other traditional assumptions allow us to use an isolated downlink budget approach. In practice the simplified calculation outlined in this chapter is sufficiently accurate for the specification of TVRO equipment.

The basic purpose of a link budget calculation is to specify or check the suitability of equipment to achieve a predictable reception from a chosen satellite at a given location. This may be pictured as an accountant's profit-and-loss report but, rather than involving funds, various noise contributions, power losses and power gains are involved. In order to start, an opening balance is required, by way of a few basic operating parameters of the chosen satellite/s. These are the equivalent isotropic radiated power (EIRP) 'frequency of operation' and channel bandwidth. Calculations essentially follow the same format, bar a few parameters, in whatever band the satellite is active. Currently, the S, C, Ku and Ka bands are most commonly allocated for TV broadcasts. Table 5.1 shows the approximate frequency ranges of the various microwave bands.

Table 5.1 *Microwave bands*

Band	Frequency range (GHz)
P	0.2–1.0
L	1.0–2.0
S	2.0–3.0
C	3.0–8.0
X	8.0–10.0
Ku	10.0–15.0
Ka	17.0–22.0
K	26.0–40.0

It is entirely possible to rely either on general link budgets provided by satellite operators or concentrate on off-the-shelf packaged systems for popular satellites. There is nothing wrong with this, but it does restrict you in many ways. For example, general link budgets are necessarily a compromise and dictate a certain overall standard of signal quality and availability. This may not fit your customer's requirements. What would you do if a customer, perhaps a foreign national living in your country, wants to receive signals from his or her own country. What antenna size would you need? In short, there are a multitude of reasons why you might need to perform such a calculation reasonably accurately.

The calculation may be performed for a range of satellite transponders to ensure that good reception is obtained for each channel. It is more important where a system is put together from a variety of manufacturers' component parts since, at one extreme, poor results may be experienced and, at the other, 'over-engineering' may unnecessarily add to equipment cost and look less aesthetically pleasing.

The final result of a link budget calculation is a signal-to-noise (S/N) ratio value which may be compared with the International Radio Consultative Committee (CCIR) five-point scale of impairment as shown in Table 5.2. These grades, obtained from a series of subjective tests, are most often used as the criteria for overall system performance. Grade 4 is the accepted standard for domestic systems and this translates to a weighted S/N ratio greater than 42.3 dB.

Table 5.2   *CCIR five-point scale of impairment*

Quality	Grade	Impairment	Weighted S/N ratio (dB)
Excellent	5	Imperceptible	>46.6
Good	4	Perceptible but not annoying	>42.3
Fair	3	Slightly annoying	>38.0
Poor	2	Annoying	>33.2
Bad	1	Very annoying	>29.2

This chapter necessarily includes some rather tedious looking equations in order to provide a detailed analysis of any downlink path. If you find the maths difficult to follow, the chapter ends with a simplified 'bare bones' link budget calculation, which broadly approximates that of clear-sky conditions. The latter method relies on the addition of an arbitrary fixed carrier-to-noise (C/N) ratio margin over the receiver demodulator threshold to allow for atmospheric propagation effects and other operational losses. This is particularly suited to simple S and C band calculations where rain attenuation and atmospheric absorption effects are negligible.

Calculation aids

On first sight, link budget calculations may appear rather complex. Fortunately, a knowledge of elementary trigonometry, decibel notation, and logarithms to the base 10, and perhaps the ability to read graphs are the only mathematical skills required.

A scientific calculator with trigonometrical functions is the basic calculation tool and some models may be programmed to perform chain calculations. A common problem with trigonometric and logarithmic function buttons on calculators is a failure to realize the synonymous terms listed below. Press the calculator key on the right-hand side of the equals sign to perform the functions on the left-hand side:

$$\text{arc sin} = \sin^{-1}$$
$$\text{arc cos} = \cos^{-1}$$
$$\text{arctan} = \tan^{-1}$$
$$\text{antilog} = 10^x$$

Example using the 10^x button

$$F_{LNB} = 10^{(1.5/10)}$$

Divide 1.5 by 10 and press the equals button in the normal manner, then press the 10^x button on the calculator. This should give the approximate value of 1.41.

As with most fields of technology, once the learning phase is over, the computer is the ideal tool for performing such repetitive calculations in practice. Those with programming knowledge may like to code the calculation process themselves; however, this is rather like re-inventing the wheel since there are many well tried and tested commercial software packages available to run on mainstream IBM PC compatibles (see Appendix 2). Even if a computer is used, it is a worthwhile exercise to see how the various parameters interact so that the relevance of each parameter is basically understood.

A detailed link budget

The reception of S and C band signals are relatively immune from such factors as rain attenuation and atmospheric absorption but for Ku/Ka band reception it is very important to allow for these extra losses. There are many routes and alternate formulae, some more accurate than others, that may be used to arrive at a detailed link budget. The particular method adopted here is fairly comprehensive and takes into account attenuation due to precipitation, increase in noise due to precipitation, misalignment losses, coupling (waveguide) losses, and the parameters' 'nominal G/T' and 'usable G/T', where G/T is the ratio of the net antenna gain to the total system noise temperature.

As each equation of the link budget is introduced a practical example on its use will be presented. The arbitrary chosen receive site is Chester, England (53.20°N, 2.90°W). We will evaluate a well known packaged system intended to receive Ku band signals from the Astra 1A satellite located at 19.2°E. Each of these channels is frequency modulated using a 26 MHz bandwidth with a deviation of 16 MHz/V. The video bandwidth is 5 MHz.

Factors affecting satellite reception

The performance of a satellite TV receive only (TVRO) system is affected by a number of physical factors. Some of these are outlined below.

1 The equivalent isotropic radiated power (EIRP) of the satellite in question.
2 The effective antenna diameter.
3 The low noise block (LNB) noise figure or noise temperature.
4 Coupling losses by waveguides and polarizers.
5 Antenna pointing losses:
    (a) initial pointing error (degrees),
    (b) antenna stability in wind or other environmental conditions (degrees), and
    (c) satellite station keeping accuracy.
6 Polarization losses.
7 Transponder ageing.
8 Rain attenuation for a given signal availability (typically 99.5% of average year).
9 For Ku and Ka band, noise increase due to precipitation (rain, snow or hail).
10 Atmospheric absorption by oxygen and water vapour (depends on humidity).
11 Temperature variations.
12 The receiver (demodulator threshold) figure.
13 The signal modulation characteristics.
14 Scattering of signals due to blockages such as trees, buildings, birds and aircraft.
15 Spreading loss through the atmosphere.

Transient effects such as passing birds and aircraft are largely unpredictable so can be neglected from the calculation. The others can all have a significant long-term effect, although factors 8, 9 and 10 can be neglected for S and C band reception.

Receive site relative to satellite position

Each geostationary satellite occupies a unique position or orbital slot at 35 784 km (22 235 miles) directly above the equator. The actual position is specified by the longitude of the sub-satellite point (the point directly below the satellite on the equator). Within the intended footprint area an antenna will need to be precisely positioned in both *azimuth* and *elevation* in order to capture its signals.

Elevation

Elevation (EL) is the angle of upward tilt of the dish with respect to the ground. This can be calculated using:

$$EL = \tan^{-1}\left[\frac{m\cos(A)\cos(B) - 1}{m\sqrt{1 - \cos^2(A)\cos^2(B)}}\right] \text{(degrees)} \tag{5.1}$$

where: A = the latitude of the earth station (positive for northern hemisphere, negative for southern hemisphere)

B = the longitude east of the Earth station minus the longitude east of the satellite

m = 6.61, the ratio of the geostationary orbital radius to that of the equatorial radius of the Earth.

For low elevations less than about 30°, the geometric elevation may be slightly modified to take into account average refraction due to the atmosphere by using the approximate Equation 5.2. The true elevation angle calculated must always be greater than the geometric value to be valid.

$$\text{True EL} \approx \frac{EL + \sqrt{(EL)^2 + 4.132}}{2} \text{(degrees)} \tag{5.2}$$

where EL is the result of Equation 5.1.

Latitudes and longitudes are obtained from atlases in degrees and minutes. These must be converted to degrees (decimal) for use in calculations. To do this simply divide the number of minutes by 60 and multiply the result by 100 and add to the whole degree portion. For example, converting 53°15′ N to degrees:

53 + [(15/60) × 100] = 53.25°N

Longitude values west must be converted to their corresponding east value and are measured from 0°E (the Greenwich meridian) through 180°E to 360°E which is the same as 0°E again. Thus for longitudes west of Greenwich, subtract the °W term from 360° to arrive at the equivalent longitude east. For example, −3°W would evaluate to:

360° − (3°W) = 357°E

It is worth remembering that latitudes above about 81° cannot 'see' any part of the geo-arc of satellites. Similarly, the longitude difference between the earth station and the wanted satellite cannot exceed this value.

Example

Earth station: Chester, England (53.20°N, 2.90°W)

Satellite: Astra 1A (19.2°E)

Plugging values based on the above co-ordinates into Equation 5.1 and remembering that $\cos^2(A)$ is shorthand for $\cos(A) \times \cos(A)$ gives:

$$EL = \tan^{-1}\left[\frac{6.61\cos(53.20)\cos(337.90) - 1}{6.61\sqrt{1 - \cos^2(53.20)\cos^2(337.90)}}\right] \text{ degrees}$$

$$= \tan^{-1}\left[\frac{2.667}{5.499}\right] \text{ degrees}$$

$$= \tan^{-1}(0.485) = 25.87°$$

Azimuth

True azimuth (AZ) (swing of dish) is the angle from true north that targets the chosen satellite. Compass bearings are measured in degrees from zero to 360°. North, east, south and west have bearings 0°, 90°, 180° and 360°, respectively. The geo-arc of satellites is targeted by azimuth bearings between 90° and 270° in the northern hemisphere or 270° to 90° in the southern hemisphere. The equation for calculating true azimuth is:

$$AZ = 180 + \tan^{-1}\left[\frac{\tan(B)}{\sin(A)}\right] \text{ (degrees)} \qquad (5.3)$$

The term 180 is deleted for the southern hemisphere.

Example

Earth station: Chester, England (53.20°N, 2.90°W)

Satellite: Astra 1A (19.2°E)

$$AZ = 180 + \tan^{-1}\left[\frac{\tan(337.90)}{\sin(53.20)}\right]$$

$$= 180 + \tan^{-1}\left(\frac{-0.406}{0.801}\right)$$

$$= 180 + \tan^{-1}(-0.507)$$

$$= 153.11°$$

Compass bearing

Once the true azimuth angle has been calculated, a compass bearing can be easily calculated by simply adding or subtracting the magnetic variation appropriate to the receive site. For all regions of Europe the westerly magnetic variation value is added to the true azimuth value. The value of magnetic variation will vary depending on where the ground station is situated and can be obtained from local survey maps. In some cases, rather than use a compass the position of the sun at various times of day can be used for azimuth setting, but this is only practical where a large installation is envisaged. You are unlikely to install many systems in a day using this method!

Example
The local magnetic variation for Chester, England, is 6.5°W so the compass bearing is:

$$153.33° + 6.5° = 159.83°$$

Downlink path distance

The path distance, sometimes called the 'slant range', is the distance between the ground station and the satellite of interest. Clearly the further away from the equator this is, the longer the path distance. An equation used to calculate this is:

$$\text{Path distance } (D) = 6378.16\sqrt{m^2 + 1 - 2m[\cos(A)\cos(B)]} \text{ (km)}$$

$$(5.4)$$

Example

$$D = 6378.16\sqrt{43.69 + 1 - 13.22[\cos(53.20)\cos(337.90)]}$$

$$= 6378.16\sqrt{44.69 - [13.22(1 - 0.555)]}$$

$$= 38\,982\,\text{km}$$

Wavelength

In many equations, including those that follow, a wavelength (λ) value rather than frequency is required for simplification. Conversion from frequency to wavelength can be done using:

$$\lambda = c/f \qquad (5.5)$$

where: c = the speed of light (2.998×10^8 m/s)
f = frequency (Hz).

Example
To convert a channel frequency of 11.332225 GHz to its corresponding wavelength:

$$\lambda = 2.998 \times 10^8 / 11.332225 \times 10^9$$
$$= 0.0265 \text{ m}(2.65 \text{ cm})$$

Free space loss

The free space loss (L_{FS}), or path loss, expresses the attenuation of microwave signals on their Earth-bound journey and occurs due to the spreading out of the beam. A good analogy is visualized by the intensity fall-off of a car headlight beam with distance. The path loss increases with frequency and is greatest for low antenna elevation angles. A suitable equation for calculating its value is:

$$L_{FS} = 20 \; \log[(4000\pi D)/\lambda] \; \text{(dB)} \tag{5.6}$$

where: $\pi = 3.14159$
          D = path distance (km)
          λ = wavelength (m).

Example

$$L_{FS} = 20 \; \log \; [(4000 \times 3.14159 \times 38982)/0.0265]$$
$$= 205.34 \text{ dB}$$

Antenna gain

The antenna gain (G_a) increases with the effective antenna size which takes into account the efficiency (p) of the antenna. The gain can be expressed as:

$$G_a = 10 \; \log \left[\frac{(\pi d)^2 p}{100\lambda^2} \right] \; \text{(dB)} \tag{5.7}$$

where: d = the antenna diameter (m)
          p = the percentage antenna efficiency (60–80% typically)
          λ = wavelength (m).

Note: the antenna efficiency may be specified as a normalized value less than 1 (e.g. 0.67 or 0.80) rather than as a percentage. In such cases delete the term 100 in the denominator and substitute the normalized factor for p.

Example
The antenna size is 0.65 m (65 cm) with an efficiency of 67% at 11.332225 GHz (wavelength = 0.265 m):

$$G_a = 10 \log \left[\frac{(3.14159 \times 0.65)^2 \times 67}{100 \times 0.0265^2} \right]$$

$$= 36.00 \text{ dBi}$$

Total system noise temperature

For a ground receiving station, the total system noise temperature (T_{SYS}) is the combination of noise temperatures from all contributing components in the receiving system and includes noise introduced by the LNB, the waveguide components and the effective or modified antenna noise.

Figure 5.1 shows the main components affecting the system noise temperature. The plane PQ represents the point to which the total system noise is referenced. This is usually taken to be a point immediately before the LNB or the connecting flange between the waveguide components and the LNB. The effective antenna noise temperature (T_A) is made up of all the noise contributions incident on the antenna but reduced by the feed *fractional transmissivity*, σ.

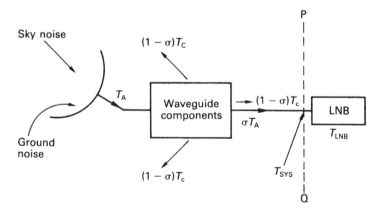

Figure 5.1 *Determining system noise temperature*

Fractional transmissivity

Fractional transmissivity (σ) is defined as the fraction of incident energy between zero and 1 that passes through a medium and emerges from the other side. A value of zero indicates total absorption by the medium and a value of 1 indicates that the medium is not absorbent or transparent.

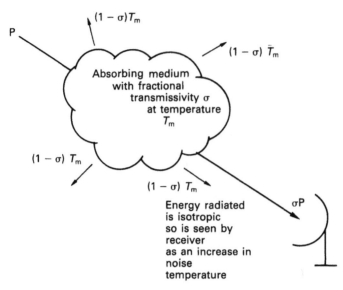

Figure 5.2 *An absorbing medium with a fractional transmissivity, σ*

When an absorbing medium is in equilibrium with its surroundings, it will isotopically radiate as much energy as it absorbs. Referring to Figure 5.2, if the absorbing medium is raised to a temperature T_m by, say, absorbing energy from the ground, the efficiency with which it is absorbing and re-radiating energy can be quantified by its fractional transmissivity, σ. A signal, of power P, passing through the absorbing medium will emerge at the other side at a reduced power level of σP. Since the radiated energy is isotropic, any receiver detecting this signal will also detect an increase in noise temperature of $(1-σ)T_m$. The principles are the same whether we are referring to a rain cell or an absorbing or lossy feed. It is more common to substitute more convenient attenuation values rather than to use the feed transmissivity parameter directly. The transmissivity of a medium (σ) is related to the attenuation (*A*) by the simple relationship:

$$A = 10 \log\left(\frac{1}{σ}\right) \text{ (dB)} \tag{5.8}$$

or, rearranging,

$$σ = \frac{1}{10^{0.1A}} = 10^{-0.1A} \tag{5.9}$$

Two points can immediately be seen from this. Firstly, an intervening rain cell in the earth space path will not only attenuate a signal but the receiver will also detect an increase in noise temperature. Secondly, the incident antenna noise as well as the signal are absorbed by the insertion of waveguide components such as feedhorns, polarizers, and othomodal transducers (OMTs).

Quantifying the system noise temperature

The general equation used to quantify the total system noise temperature is:

$$T_{SYS} = T_{LNB} + (1 - \sigma)T_C + \sigma T_A \text{ (K)} \tag{5.10}$$

or its equivalent expression using attenuation values:

$$T_{SYS} = T_{LNB} + (1 - 10^{-0.1A_{feed}})T_C + 10^{-0.1A_{feed}} T_A \text{ (K)} \tag{5.11}$$

where: T_{SYS} = total system noise temperature (K)
T_A = effective antenna noise temperature either for clear sky or for a given percentage of time (K)
T_{LNB} = the equivalent noise temperature of the LNB (K)
T_C = physical temperature of coupling (waveguide) components (K)
α = coupling losses (dB)
σ = feed transmissivity
A_{feed} = the feed attenuation or insertion loss figure.

The equivalent LNB noise temperature

The first term (T_{LNB}) in Equations 5.10 and 5.11 is the overall LNB noise factor expressed as an equivalent noise temperature and is the major contributor to the overall system noise temperature. A noise factor when expressed as a power ratio in decibels becomes a noise figure. The noise performance of a LNB can be expressed as an equivalent noise temperature in kelvin or, more commonly, as a noise figure in decibels. In the latter case we need to convert the noise figure to an equivalent noise temperature in order to calculate the total system noise:

$$T_{LNB} = 290(10^{(NF/10)} - 1) \text{ (K)} \tag{5.12}$$

where: T_{LNB} = noise temperature (K)
NF = noise figure (dB)

Example
You intend to fit a 1.5 dB LNB in your system. What is its equivalent noise temperature?

$$\begin{aligned} T_{LNB} &= 290(10^{(1.5/1.0)} - 1) \\ &= 290(0.412) \\ &= 119.6 \text{ K} \end{aligned}$$

It is usually more difficult and expensive to achieve low noise figures the higher the frequency. For the Ku band, low-cost LNBs are in the range 1.2–1.5 dB. Lower noise figures can be achieved with the use of high electron mobility transistors (HEMT) devices. Typical noise figures for the Ku band are 0.8–1 dB.

Coupling noise

The second term $((1 - \sigma)T_C)$, in Equations 5.10 and 5.11 is the noise isotropically radiated by the feed components. These will absorb energy principally from the ground and thus have a fractional transmissivity value or inherent loss. This isotropically re-radiated portion $(1-\sigma)T_C$ will be detected by the LNB. Insertion losses or attenuation experienced by waveguide components are normally quoted by manufacturers as a decibel power ratio so Equation 5.11 is the one to use in practice. The total feed attenuation figure is the sum of the attenuation contributions of waveguide components such as feedhorns, OMTs, and polarizers, etc. T_C is the physical temperature of the feed, and so is normally taken as 290 K.

Example
You might fit feed components with a total insertion attenuation of 0.3 dB in your head unit, what additional noise temperature will be detected by the LNB?

$$\begin{aligned}
(1 - \sigma)T_C &= (1 - 10^{-0.1 A_{feed}})290 \\
&= (1 - 10^{-0.03})290 \\
&= (1 - 0.933)290 \\
&= 19.43 \text{ K}
\end{aligned}$$

Modified antenna noise temperature

The third term (σT_A) in Equations 5.10 and 5.11 is the modified antenna noise temperature which is the effective noise temperature of the antenna, T_A (i.e. comprising all the noise components incident on the antenna), reduced by the feed transmissivity.

Example
Suppose the effective antenna temperature is 68 K and the feed has an insertion loss of 0.3 dB. What is the modified antenna temperature seen at the LNB input?

$$\begin{aligned}
\sigma T_A &= 10^{-0.1 A_{feed}} T_A \\
&= (10^{-0.03})68 \\
&= 0.933 \times 68 \\
&= 63.46 \text{ K}
\end{aligned}$$

Effective antenna noise temperature

The effective antenna noise temperature (T_A) defined above is now discussed in a little more detail. The effective antenna noise temperature is determined by many factors, such as antenna size, elevation angle, external noise sources and atmospheric propagation effects. During clear-sky conditions, the principal noise component of the effective antenna noise temperature is ground noise pick-up. This is easy to see since, neglecting atmospheric propagation effects (rain, etc.), this is virtually all the noise entering the antenna. This is the 'antenna noise' parameter that manufacturers often tabulate for a range of elevation angles; it may also include a relatively small contribution by galactic background noise. There are three main contributions to the overall antenna noise.

1 *Antenna noise temperature due to ground noise* (T_{ANT}) – The smaller the antenna the wider and more spread out are the side lobes intersecting the warm earth and, consequently, the more ground noise is picked up by the antenna. It can also be seen that these side lobes, principally the first side lobe, would intersect the ground at a higher elevation angle than that of a larger antenna and so would be a noisier device when set at a given elevation. Ground noise pick-up may be reduced, at the expense of gain, by under-illuminating the dish; thus this factor essentially determines the efficiency of the dish. Size being equal, a prime focus antenna would detect increased ground noise over an offset design since the head unit, directly mounted in the signal path, would be 'seen' at the same temperature as the Earth.

   Since the antenna noise temperature has so many variable factors, it is apparent that in the absence of a manufacturer-supplied figure, an estimate is perhaps the best we can hope for. Equation 5.13, which takes into account the elevation and the diameter, may be used to calculate a reasonable approximation for the antenna noise under clear-sky conditions.

$$T_{ANT} \approx 15 + 30/D + 180/EL(K) \tag{5.13}$$

where: D = antenna diameter (m)
      EL = dish elevation angle (degrees)

Example
You hope to use a 0.65 m dish. Estimate the worst-case antenna noise temperature at an elevation of 25°.

$$T_{ANT} = 15 + 30/0.65 + 180/25$$
$$= 68 \text{ K}$$

2 *Cosmic or galactic noise component* – This is background cosmic noise, principally the residual noise of the 'big bang'. It has a small noise temperature of about 2.7 K. This component is relatively small in relation to the error in estimating the ground noise component, and may be omitted from practical calculations. In any case, depending on how 'antenna noise' is defined in manufacturers' specifications, this may be incorporated.

3 *Atmospheric propagation components* – There are two main propagation effects experienced on the downlink. Firstly, atmospheric gaseous absorption by water vapour and oxygen; this is basically a
   clear-sky effect. Its value depends on the absolute humidity or vapour density measured in grams per square metre, the antenna elevation and the frequency involved. It is a relatively minor contributor
   below about 7.5 GHz. Typical values for atmospheric absorption for
   Europe are given in Figure 5.3. The details of how to calculate a
   specific value for any given slant path and frequency in the world
   is beyond the scope of this book, but it may be easily calculated
   using computer software such as that described in Appendix 2.
   The second propagation effect is attenuation due to precipitation.
   Considering the uplink situation, a receiver on board a satellite will
   'see' a fairly constant but high noise temperature emitted from the
   warm Earth of around 290 K, so further thermal energy emission by
   rain will have a negligible effect. In the downlink situation, the receiver is directed toward a relatively cool sky so, in a relative sense, the
   additional thermal noise contribution by rain is by no means a negligible component of the total system noise, especially if the receiver
   (LNB) is a low noise device operating in the Ku or Ka band. The
   effects of rain and atmospheric absorption are negligible in the S
   and C bands.
   Precipitation will not only directly attenuate the signal (known as a
   'rain fade'), but the system noise temperature will also increase since
   the temperature of the intervening medium approaches that of the
   Earth. It is important that the increase in system noise is taken into
   account and not just the attenuation experienced by a rain fade. The
   combination of the two is known as the *downlink degradation* (DND).

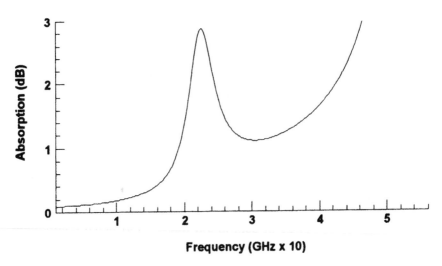

Figure 5.3 *Typical atmospheric gaseous absorption values for Europe*

The effects of precipitation become significant above about 8 GHz. Rain, or to a lesser extent snow, fog, or cloud, attenuate and scatter microwave signals. The magnitude depends more on the size of the water droplets (in cubic wavelengths) rather than the precipitation rate itself. Heavier rain tends to comprise larger droplets so the two are normally related. As a general rule, the physical-medium temperature, of all forms of precipitation, is taken as 260 K. For clouds and clear-sky use 280 K. Again, to calculate specific values for any specific Earth–space path and signal availability use computer software such as that described in Appendix 2 (see Figure 5.4 for typical European values).

Signal availability and operational margins

An attenuation figure for rain has to be predicted from long-term rainfall statistics for the receive site of interest. Rather than allow a massive operational margin over threshold for the worst ever rain storm likely, we are normally content with specifying a signal availability figure for an average year, which potential customers find acceptable. In other words for a percentage of time the signal will not fall below some predetermined C/N (or S/N) ratio. For example, when we say a CCIR grade 4 (good) signal is available for 99.7% of an average year we mean that the S/N ratio is not expected to fall below 42.3 dB for 99.7% of the time (or 99% of the worst month). However, it will be expected to occasionally fall below this for 0.3% of the time during severe storms. The higher

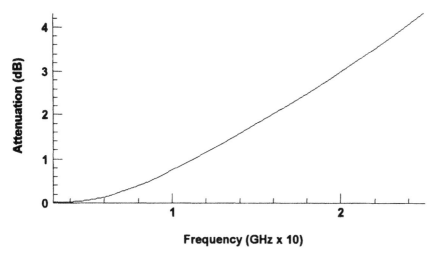

Figure 5.4   *Typical rain attenuation values for Europe for a signal availability of 99.7% of an average year (99% worst month)*

the signal availability designed into a system, the better will be the protection against the effects of rain attenuation. The dish size needed also grows alarmingly as the designed signal availability increases. Rain attenuation, or more specifically the downlink degradation, is the major component of the overall loss margin for Ku and Ka band systems. For typical direct-to-home (DTH) systems, a figure of 99.5% availability is normally considered acceptable. In fact most packaged fixed dish systems for popular satellites are designed around this figure. For satellite master antenna TV (SMATV) you may require a higher figure of 99.9%, and for cable head even higher. The law of diminishing returns eventually intervenes since 100% availability is impractical.

Noise increase due to precipitation and atmospheric absorption

During clear-sky conditions the only attenuation experienced between the satellite and the ground station will be that due to atmospheric absorption, A_{atm}, by oxygen and water vapour. During rain there will be a combined atmospheric gaseous absorption, A_{atm}, and attenuation due to the rain, A_{rain} (dB). The overall consequence is to increase the effective antenna noise temperature, T_A, above operational frequencies of about 8 GHz. For the S and C bands this calculation is not considered necessary since the contributions are negligible, but for the Ku and Ka bands it becomes increasingly significant, particularly at the low system noise temperatures achieved today. Even during clear-sky conditions an allowance for the temperature increase due to atmospheric absorption should be added to the effective antenna temperature, T_A. Equation 5.14 may be used to calculate this increase. During rain an additional noise temperature increase can be calculated using Equation 5.15 to allow for statistical rainfall effects. To obtain values for A_{atm} and A_{rain} for Europe use Figures 5.3 and 5.4, respectively; or, more accurately, use computer software such as that described in Appendix 2.

$$T_{clear\ sky} = (1 - 10^{-0.1A_{atm}})T_m + 10^{-0.1A_{atm}}T_g\ (K) \tag{5.14}$$

$$T_{rain} = (1 - 10^{-0.1(A_{atm}+A_{rain})})T_m + 10^{-0.1(A_{atm}+A_{rain})}T_g\ (K) \tag{5.15}$$

where: T_m = the physical temperature of the medium (260 K rain, 280 K clear sky or cloud)

T_g = cosmic or galactic noise temperature (2.7 K typical at frequencies of $\geq$4 GHz)

A_{atm} = gaseous attenuation due to atmospheric absorption (dB)

A_{rain} = rain attenuation for a given percentage of the time (dB).

Adding either $T_{clear\ sky}$ or T_{rain}, as appropriate, to the effective antenna temperature, T_A, and recalculating using Equation 5.11 will yield a modified value of the total system noise temperature, T_{SYS}, taking into account gaseous and/or rain attenuation.

If you wish to calculate the increase in noise due to a given rain fade expressed as a power ratio in decibels, use the equation:

$$\text{Noise increase (rain)} = 10 \, \log\left(\frac{T_{SYS_{rain}}}{T_{SYS_{clear \, sky}}}\right) \, (dB) \tag{5.16}$$

where: $T_{SYS_{rain}}$ = system noise temperature calculated during rain for a specific percentage of the time of an average year (K).

$T_{SYS_{clear \, sky}}$ = system noise temperature calculated during clear sky conditions including atmospheric gaseous absorption (K)

The downlink degradation (DND) experienced during a given rain fade is given by:

$$DND = A_{rain} + 10 \, \log\left(\frac{T_{SYS_{rain}}}{T_{SYS_{clear \, sky}}}\right) \, (dB) \tag{5.17}$$

Example
Calculate the DND for the example link budget where the atmospheric gaseous absorption is 0.17 dB and the rain attenuation for 99.5% of an average year does not exceed 0.83 dB.

$$\begin{aligned}
T_{clear \, sky} &= (1 - 10^{-0.1 \times 0.17})280 + (10^{-0.1 \times 0.17})2.7 \\
&= (1 - 0.96)280 + (0.96)2.7 \\
&= 11.2 + 2.6 \\
&= 13.8 \text{ K}
\end{aligned}$$

$$\begin{aligned}
T_{rain} &= (1 - 10^{-0.1 \times (0.17+0.83)})260 + (10^{-0.1 \times (0.17+0.83)})2.7 \\
&= (1 - 0.79)260 + (0.79)2.7 \\
&= 54.6 + 2.13 \\
&= 56.7 \text{ K}
\end{aligned}$$

$$\begin{aligned}
T_{SYS_{clear \, sky}} &= T_{LNB} + (1 - 10^{-0.1A_{feed}})T_C + 10^{-0.1A_{feed}} T_A \\
&= 119.6 + 19.43 + 0.933(68 + 13.8) \\
&= 215.3 \, K
\end{aligned}$$

$$\begin{aligned}
T_{SYS_{rain}} &= T_{LNB} + (1 - 10^{-0.1A_{feed}})T_C + 10^{-0.1A_{feed}} T_A \\
&= 119.6 + 19.43 + 0.933(68 + 56.7) \\
&= 255.4 \, K
\end{aligned}$$

$$DND = 0.83 + 10 \, \log\left(\frac{255.4}{215.3}\right)$$

$$= 1.57 \text{ dB}$$

The noise increase due to the 0.83 dB rain fade is given by Equation 5.16 or the second term of Equation 5.17 and evaluates to 0.74 dB. From this, it can be seen that, although the rain fade is 0.83 dB, the corresponding degradation in the downlink is significantly higher due to increased noise detection. This is an important point to note.

Nominal figure of merit

G/T is the ratio of the net antenna gain and total system noise temperature. The 'nominal figure of merit' (G/T$_{nom}$) is the maximum obtainable figure for a given elevation angle and comprises the net antenna gain (antenna gain − coupling loss) divided by a noise temperature factor made up from contributions of the equivalent receiver noise temperature (i.e. LNB), the coupling noise of inserted polarizers and waveguide components such as orthomodal transducers (OMTs) and the 'clear sky' modified antenna noise temperature. This is expressed mathematically by Equation 5.18. No operational margins are included such as antenna misalignment losses, ageing, or the increase in antenna noise for a given percentage of time due to rain. This is the highest value of the G/T ratio allowing qualitative comparison between different outdoor units. The higher the ratio, the better the system will perform. G/T, in general, is the figure which has the greatest effect on the final C/N ratio. All other contributory factors are relatively constant, as will be seen later.

$$\text{G/T}_{nom} = 10\ \log\left[\frac{10^{0.1(G+\alpha)}}{T_{SYS}}\right]\ (\text{dB/K}) \tag{5.18}$$

where: G = antenna gain (dB)
α = coupling loss (dB) by waveguide components (loss = negative gain)
T_{SYS} = clear-sky system noise temperature excluding propagation effects.

Example
Substituting the previously calculated values for antenna gain and the 'clear-sky' system noise temperature into Equation 5.18 gives:

$$\text{G/T}_{nom} = 10\ \log\left[\frac{10^{0.1(36.00+(-3))}}{215.3}\right]$$

$$= 10\ \log\left[\frac{3715.35}{215.3}\right]$$

$$= 12.37\ \text{dB/K}$$

Usable figure of merit

The required G/T parameter needed in a detailed link budget is the 'usable (degraded or minimum) figure of merit' (G/T$_{usable}$) this allows for further operational losses due to antenna pointing errors, polarization effects, ageing, and the increase in system noise due to precipitation for a given percentage of time and comprises the net antenna gain (antenna gain − coupling loss − operational losses) divided by the total system noise temperature. This G/T thus characterizes the 'in service' performance and is the one used in detailed link budgets. An additional noise temperature contribution is added to T_{SYS} to allow for the increase in system noise due to precipitation for a certain specified percentage of the time. This is expressed mathematically by:

$$G/T_{usable} = 10 \, \log \left[\frac{10^{0.1(G+\alpha+\beta)}}{T_{SYS_{rain}}} \right] \, (dB/K) \tag{5.19}$$

where: G = antenna gain (dB)
  α = coupling loss (dB) by waveguide components (loss = negative gain)
  β = losses due to antenna pointing errors, polarization errors and ageing (dB) (loss = negative gain)
  $T_{SYS_{rain}}$ = modified total system noise temperature which includes the increase in noise temperature due to precipitation for a given percentage of the time (K).

Example
Calculate the in-service 'usable' G/T for the example link budget using a value of 0.5 dB to allow for pointing errors, polarization errors and ageing using the appropriate parameters calculated previously.

$$G/T_{usable} = 10 \, \log \left[\frac{10^{0.1(36.00+(-0.3)+(-0.5))}}{255.4} \right]$$

$$= 10 \, \log \left(\frac{3311.31}{255.4} \right)$$

$$= 11.13 \, dB/K$$

Calculating antenna pointing loss

The antenna pointing loss, P, may be calculated as:

$$P = 12 \left[\frac{\theta_1^2 + \theta_2^2 + \theta_3^2}{\theta_0^2} \right] \, (dB) \tag{5.20}$$

where: θ_1 = the initial pointing accuracy of the fixed mount antenna to the satellite (degrees); this is typically around 10–20% of the half-power beamwidth

θ_2 = the pointing stability of the installation due to environmental factors such as wind and ageing (degrees)

θ_3 = the station keeping accuracy of the satellite (degrees) ($\pm 0.16°$ typical)

θ_0 = the half-power beamwidth of the receiving antenna (degrees).

Example

The half-power (-3 dB) beamwidth of a 0.65 m antenna is typically $3°$ (depending on feed or illumination method adopted). Calculate the antenna pointing loss if the initial pointing accuracy is $0.3°$ and the pointing stability is $0.5°$.

$$P = 12 \left[\frac{0.3^2 + 0.5^2 + 0.16^2}{3^2} \right]$$

$$= 12 \left[\frac{0.09 + 0.25 + 0.026}{9} \right]$$

$$= 0.49 \, \text{dB}$$

The larger the antenna diameter the greater is the pointing error due to the effects of wind pressure, so large antennas greater than 1 m have a significant disadvantage in this respect. The pointing stability may be as high as $1°$ for solid large antennas in windy conditions. The use of mesh dishes can reduce this effect considerably.

Equivalent isotropic radiated power (EIRP)

An isotropic radiator is defined as one which radiates uniformly in all directions. This is not obtainable in reality but is easy to visualize. By using a reflector an isotropic radiator can concentrate all its energy into a narrow beam which appears to some distant observer, at the other end of the beam, as an isotropic source of several magnitudes greater power output. Thus the term equivalent isotropic radiated power is used as a measure of signal strength that a satellite transmits to earth. EIRP is measured in dB relative to one watt (dBW) and is highest at the beam centre. This value decreases logarithmically at distances away from the beam centre. The EIRP of any satellite can be obtained from the appropriate footprint map, as contours of equal magnitude. Modern satellites can shape their EIRP contours to a certain extent to fit the desired service area although the methods used need not concern us here. A typical value of EIRP for medium power semi-DBS satellites such as Astra is 52 dBW. High power DBS satellites have EIRP values in excess of 60 dBW.

Carrier-to-noise ratio

For the Ku and Ka bands the system carrier-to-noise (C/N) ratio is given by:

$$C/N = EIRP - L_{FS} + G/T_{usable} - 10 \log (kB) - A_{rain} - A_{atm} \text{ (dB)} \quad (5.21)$$

where: $EIRP$ = the equivalent isotropic radiated power from the satellite at the site location (dBW)

L_{FS} = free space path loss on the earth to satellite path (dB)

G/T_{usable} = minimum degraded value of the system figure of merit (dB/K)

k = Boltsmann's constant (1.38×10^{-23} J/K)

B = receiver's pre-detection intermediate frequency (IF) bandwidth (Hz)

A_{atm} = gaseous attenuation due to atmospheric absorption (dB)

A_{rain} = rain attenuation for a given percentage of the time (dB).

Note: (a) A_{atm} and A_{rain} can be omitted for operation frequencies of <8 GHz; and (b) for a 'clear-sky' calculation omit the A_{rain} term and substitute the nominal figure of merit, $G/T_{(nominal)}$, for $G/T_{(usable)}$.

Example 1
Calculate the clear-sky carrier-to-noise ratio, for the example thus far, if the worst case EIRP is 51 dBW at the receive site.

$$\begin{aligned}
C/N_{clear sky} &= 51 - 205.34 + 12.39 - 10 \log (1.38 \times 10^{-23} \times 26 \times 10^6) \\
&\quad -0.17 \\
&= 51 - 205.34 + 12.39 - (-154.45) - 0.17 \\
&= 12.33 \text{ dB}
\end{aligned}$$

Example 2
Calculate the degraded sky carrier-to-noise ratio for the example thus far, if the worst case EIRP is 51 dBW at the receive site.

$$\begin{aligned}
C/N_{degraded sky} &= 51 - 205.34 + 11.13 - 10 \log(1.38 \times 10^{-23} \\
&\quad \times 26 \times 10^6) - 0.83 - 0.17 \\
&= 51 - 205.34 + 11.13 - (-154.45) - 0.83 - 0.17 \\
&= 10.24 \text{ dB}
\end{aligned}$$

Signal-to-noise ratio

Providing the individual deviations of a small number of audio channels are small in relation to the video deviation, it is assumed for practical purposes that the overall peak-to-peak deviation of a baseband signal (including the multiple sound carriers) approximates that of the video signal alone.

For frequency modulated (FM) television signals, the signal-to-noise (S/N) ratio on demodulation can be calculated as:

$$S/N = C/N + 10 \log\left[3\left(\frac{f_{(p-p)}}{f_v}\right)^2\right] + 10 \log\left(\frac{b}{2f_v}\right) + k_w \text{ (dB)} \qquad (5.22)$$

where: S/N = the peak-to-peak luminance amplitude to weighted r.m.s. noise ratio (dB)

C/N = carrier-to-noise ratio (dB)

$f_{(p-p)}$ = peak-to-peak deviation by the video signal including the sync pulses (Hz)

f_v = highest video frequency present (Hz) (see Table 5.3 for details of analogue colour TV systems in use)

b = radio frequency bandwidth (usually taken as $f_{(p-p)} + 2f_v$ (Hz)

k_w = combined de-emphasis and weighting improvement factor in FM systems (dB).

Table 5.3 *Analogue TV systems including the noise weighting improvement factor (triangular noise), k_w (dB), for FM systems*

System	Primary sound carrier (MHz)	Video bandwidth (MHz)	Baseband channel width (MHz)	Weighting (dB)	Combined noise weighting and de-emphasis improvement, k_w (dB)
PAL B	5.5	5	7	16.3	16.3
PAL D1	6.5	6	8	17.8	18.1
PAL G	5.5	5	8	16.3	16.3
PAL H	5.5	5	8	16.3	16.3
PAL I	6	5.5	8	12.3	12.9
PAL I 1	6	5.5	8	12.3	12.9
PAL M	4.5	4.2	6	16.3	16.3
PAL N	4.5	4.2	6	–	–
SECAM B	5.5	5	7	16.3	16.3
SECAM D	6.5	6	8	17.8	18.1
SECAM G	5.5	5	8	16.3	16.3
SECAM H	5.5	5	8	16.3	16.3
SECAM K	6.5	6	8	17.8	18.1
SECAM K1	6.5	6	8	17.8	18.1
SECAM L	6.5	6	8	17.8	18.1
NTSC M	4.5	4.2	6	10.2	13.8

Line frequencies: 15625 Hz (625 lines) except NTSC M and PAL M 15734 Hz (525 lines).
Field frequencies: 50 Hz except NTSC M and PAL M (59.94 Hz colour, 60 Hz field).
All systems negative video modulation except SECAM L.
All systems FM sound except SECAM L.
Note: when using CCIR recommendation 405, the pre-emphasis, k_w, is approximately equal to the weighting factor alone.

Note: (a) Equation 5.22 is only valid for systems operating above the demodulator threshold. (b) The effect of the additional deviation for multiple sound sub-carriers located above the video baseband tends to improve the video S/N ratio slightly (by a fraction of a decibel) over that calculated using Equation 5.22. However, depending on your point of view, a slightly pessimistic link budget need not be a bad thing. There are other formulae claiming to be slightly more accurate, but these are also based on approximations. For practical purposes the overall peak-to-peak deviation may be taken as the overall peak-to-peak deviation by the video signal, provided the individual deviations of the audio channels is small in comparison. (c) The combination of the second and third terms of Equation 5.22 is sometimes called the 'FM modulation gain' or 'FM improvement'.

Example
Calculate the degraded sky video S/N ratio after demodulation, using Equation 5.22, for our example Astra 1A system where: $k_w = 13.2$ dB (11.2 dB weighting plus 2 dB de-emphasis); the radio frequency bandwidth is 26 MHz; the peak-to-peak frequency deviation is 16 MHz; the highest video frequency is 5 MHz.

$$S/N = 10.24 + 10 \ \log\left[3\left(\frac{16 \times 10^6}{5 \times 10^6}\right)^2\right] + 10 \ \log\left(\frac{26 \times 10^6}{2(5 \times 10^6)}\right) + 13.2$$

$$= 10.24 + 14.87 + 4.15 + 13.2$$

$$= 42.46 \text{ dB}$$

Comparing this S/N value with the values listed in Table 5.2 renders a CCIR grade of just above grade 4 (good) for 99.5% of the time with the equipment and parameters specified in the example. Raising the dish size further in effect increases the signal availability by raising C/N, and reduces the risk of sparklies degrading the picture during rain storms.

Analogue TV systems

Table 5.3 shows the world analogue TV systems currently in use; the video bandwidth can be taken as the highest video frequency, f_v, for use in Equation 5.22.

An analogue television signal is a composite of two basic signals: a black-and-white (monochrome) component called 'luminance', and a colour component called 'chrominance'. Systems are classified as follows:

1 The luminance signal type is described by a CCIR system letter, A, B, C, D, D1, E, F, G, H, I, K, L, M, N. (Systems A (405 lines), C (625 lines) and E (819 lines) are monochrome only).
2 The chrominance signal can be NTSC, PAL or SECAM.

Example
In the UK, system I is used for the monochrome signal, and PAL for the colour system. We simply label the system PAL I.

Bandwidth

The bandwidth of a microwave signal is relatively large compared with its terrestrial AM counterpart, and is normally in the range 24–36 MHz. For medium power FSS and DBS satellites, a transponder of around 27 MHz is commonly used, although a few (the Eutelsat II series for example) have bandwidths of 36 and some of 72 MHz. With 72 MHz channels it is possible to transmit two 36 MHz bandwidth channels using the same transponder (so-called half-transponder format). Since the frequency spectrum of a FM signal is infinite (produces an infinite range of sideband frequency components) an infinite bandwidth would be needed to transmit it. Clearly some form of compromise or band-limiting is necessary in practice which must be related to the deviation value used. From subjective tests, it has been found that picture quality derived from 27 MHz channels is indistinguishable from that of 36 MHz or more, and that bandwidths as low as 16 MHz still produce reasonable picture quality. In fact some receivers allow the user to reduce the band-width of the IF filter to 15 or 16 MHz to reduce noise, thus increasing the pre-detection C/N ratio. The trade-off with wide bandwidths is a correspondingly lower number of channels that may be fitted into a given frequency allocation. Higher 36 MHz bandwidth signals produce a better improvement in the S/N ratio on demodulation than do 27 MHz signals (FM improvement), so a particular value of S/N can be achieved with a lower C/N ratio.

Deviation

With frequency modulation, the instantaneous frequency of the carrier signal is varied in response to the instantaneous voltage of the video signal (including sync tips). This modulation method produces an infinite number of frequency components as sidebands. The amplitude of these components decreases with the distance from the carrier frequency. For practical purposes, only a limited number of these components need be sent without affecting the perceived picture quality. Band-limiting these smaller components produces very little distortion and a minimum bandwidth, somewhat larger than the maximum deviation, is normally sufficient. (See Carson's rule, defined below). The maximum frequency deviation of the modulated signal is the frequency difference between the maximum modulated frequency and the unmodulated frequency and corresponds to the maximum and minimum amplitudes of the message signal, respectively. The ratio of peak deviation and the highest video frequency is called the *frequency modulation index*. This depends on the sensitivity of the modulator, and increasing this has the effect of spreading out the signal spectrum. Increasing the deviation of the transmitted signal results in a higher S/N ratio (less noise). FM deviation is a measure of the modulator sensitivity (in units of MHz/V but is often quoted in MHz). This assumes that the peak-to-peak value of the video signal is 1 V, including synchronization pulses.

In a link budget calculation, we need the *peak-to-peak deviation* value, $d_{(p-p)}$, of the video signal (in Hz) in order to calculate the S/N ratio after demodulation in the receiver. Figure 5.5 illustrates Carson's rule and the difference between the peak and the peak-to-peak deviation. If a peak deviation value is quoted, remember to double it to obtain the peak-to-peak value (sync tips to peak white). Watch out for this one, or a 6 dB error may be introduced into your S/N calculations.

With satellites operating on the half-transponder format the FM deviation value may be reduced (halved) to simulate the effect of reduced S/N since signals from two channels are modulated onto the same carrier. The half-transponder format is where two channels are simultaneously modulated onto a single, say, 72 MHz bandwidth transponder.

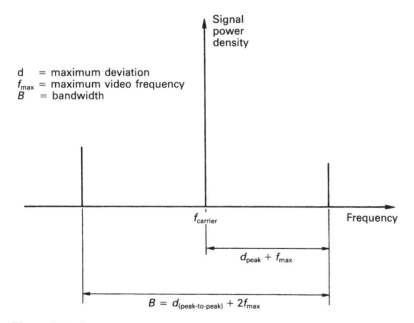

Figure 5.5 *Carson's rule*

Estimating FM deviation

If the FM deviation, or video deviation is not known, but you know the bandwidth of a required channel you can use Carson's rule to arrive at a reasonable estimate of the peak-to-peak frequency deviation. FM deviation is defined above.

Deviation (peak-to-peak) = RF bandwidth − 2 (maximum video frequency) (Hz)

Example 1
Astra 1a (Europe) uses 26 MHz bandwidth channels for 5 MHz video:

Video deviation (peak-to-peak) = 26 − 2(5) = (26 − 10) = 16 MHz

The quoted figure is 16 MHz/V (standardized video signals are typically 1 V (p–p) amplitude including sync pulses).

Example 2
The Eutelsat II series uses 36 MHz bandwidth transponders for 5 MHz video:

Video deviation (peak-to-peak) = 36 − 2(5) = (36 − 10) = 26 MHz

The quoted figure is 25 MHz/V so, as you can see, a reasonable approximation is obtained by using Carson's rule.

C/N, S/N and threshold

The carrier-to-noise ratio (C/N) is relevant before demodulation in the receiver. The signal-to-noise ratio (S/N) is that relevant after demodulation. The S/N ratio is thus dependent on both the C/N ratio and the modulation characteristics.

Another important link parameter is the receiver's demodulator 'threshold' figure. Threshold is the point where the linear relationship between demodulator C/N input and S/N output begin to break down. The demodulator threshold is the point at which the demodulator in the receiver loses its linear relationship between input C/N and output S/N. Thus if a system is operating near or below threshold a small temporary reduction in C/N caused by rain, etc., can result in a non-linear reduction in S/N. If the C/N sinks below threshold then the calculated S/N value is invalid. At the time of writing, typical values obtained using extended threshold demodulators are in the range 5–6 dB.

For continuous good reception, two criteria must be satisfied for a specified signal availability:

1 The antenna system should provide a degraded sky C/N value exceeding the receiver demodulator threshold figure.
2 The degraded sky S/N should exceed 42.3 dB for CCIR grade 4.

It should be noted that 27 MHz bandwidth Ku-band reception, all else being equal, depends more on satisfying (2), whilst 36 MHz bandwidth reception depends more on meeting (1). Also S/N is dependent on both C/N and the signal modulation characteristics, unless receiver IF band-limiting is used. Some upmarket receivers allow optional band-pass filtering in the IF circuit prior to demodulation. This practice can improve the C/N for systems operating near demodulator threshold and may eliminate sparklies at the expense of picture distortion. A bandwidth reduction to about 50% is regarded as the absolute limit. Commonly used bandwith values are 27 and 36 MHz for FM systems.

Pre-emphasis (de-emphasis) improvement

Since the noise power density of a receiver demodulator output increases with frequency, high frequencies are boosted or pre-emphasized prior to transmission. When the signal is subsequently demodulated in the receiver the signal and its acquired noise is de-emphasized or reduced by an equal amount. The overall effect is to reduce the noise component and leads to a typical improvement in S/N of 2 dB for PAL I signals or 2.5 dB for NTSC M signals.

Noise weighting factor

When a high bandwidth signal is transformed to a lower baseband value, an increase in the S/N ratio is to be expected. Although the FM improvement value (also called FM modulation gain) may be calculated, viewers vary in their perception of differing spectra noise accompanying the video signal. As a result of many subjective tests, standardized noise weighting figures have been introduced for various TV systems to correct for this effect. Table 5.3 shows the values for triangular noise for various analogue TV systems. Values are typically around 11.2 dB for PAL I, 10.2 dB for NTSC M, and 13 dB for MAC.

Digital link budgets

This section provides information on extending the calculation from an FM modulated carrier to a digital or phase modulated one.

Information theory is classically divided into two separately defined areas:

1   Source coding.
2   Channel coding.

Television picture signals are sampled at a rate of, at least, twice the highest frequency present and converted to a digital stream of bits known as the information source. The output of the information source is input to the source encoder whose function is to reduce the average number of data bits per second which must be transmitted to the user over the channel. Source coding, basically another subject area, involves the study of data compression techniques such as that used in the MPEG-2 standard. This need not concern us here since we are only interested in the final transmitted data stream in order to calculate a link budget. It is convenient in such cases to ignore the details of source encoding and refer to the combined output of the information source and the source encoder as the information source.

A transmitted information-bearing signal may not be correctly interpreted at the receiver due to distortion of the signal over a noisy channel, so the output of the information source is fed to the channel encoder where redundancy (extra bits inserted) is introduced to reduce the bit error probability. This practice is known as forward error correction (FEC) and is the only known method of providing error correction without calling for retransmission. The bit error probability and the bit error rate (BER) of the receiver's decoder are numerically equal. It seems strange at this stage that we go to all the trouble of using digital compression techniques only to add extra bits again before transmission over the channel. However, there are sound reasons for this, as will become apparent.

Shannon's capacity theorem

Forward error correction is achieved by incorporating redundancy into the channel coding system. Extra bits are added in a predictable and predefined way so as to aid the decoder in its task of interpreting the transmitted bits correctly. The details of the actual codes are extremely tedious to follow, but fortunately we can treat it as a 'black box' area.

In the late 1940s a US-based engineer, Claude E. Shannon, placed the whole subject of information theory on a scientific basis. Essentially he showed that the capacity, C, of a channel is the number of information bits per second which can theoretically be transmitted over a channel with an arbitrary low bit error rate. Capacity is a function of bandwidth and the signal to noise ratio.

The equivalent parameter to S/N in digital systems is E_b/N_0, which is defined as the energy per information bit to noise density ratio. For a given digital modulation and encoding scheme there is a specific value of E_b/N_0 corresponding to a given BER expected on decoding. Experiments have shown that a BER better than 10^{-10} is roughly equivalent to Grade 5 reception.

Provided that the output of the source encoder is less than the capacity of the channel then it is possible to reduce the BER to any desired level, using FEC, without increasing the transmitter power above the value for which the capacity was calculated. In other words, there is an upper limit on the rate of error-free communication that can be achieved over any given channel. However, there is a trade-off. The complexity of the channel encoding system grows alarmingly near capacity and the bandwidth is also increased. A graph showing the capacity of a channel in relation to E_b/N_0 and bandwidth is shown in Figure 5.6.

Coding gain

Shannon did not specify codes that would enable operation near capacity. Much work has gone on since in an effort to achieve this theoretical limit. It follows that the use of FEC results in a coding gain (or decoding gain) on demodulation of the transmitted signal. Coding gain is defined as the difference between the E_b/N_0 required to achieve a particular BER without coding and the E_b/N_0 required to achieve the same BER with coding. Obviously the more efficient the coding, the higher will be the coding gain on demodulation, but the higher will be the complexity and cost. A fixed outer code used with DVB is a (188, 204) Reed Solomon code concatenated with an inner convolutional code chosen to suit the broadcaster's needs. Together this channel coding can result in coding

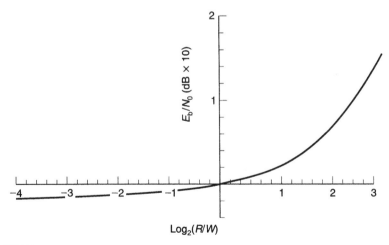

Notes:
(1) R = Data Rate (bit/s), W = Transmission Bandwidth (Hz).
(2) Graph denotes a capacity boundary for error-free communication.
(3) Operating points for error-free communication lie above curve.
(4) Infinite bandwidth is required for $E_b/N_0 \leqslant -1.6$ dB.
(5) Power limited operating lie on left of E_b/N_0 axis ($R < W$).
(6) Bandwidth limited points lie on right of E_b/N_0 axis ($R > W$).

Figure 5.6   *Shannon's capacity boundary*

gains in excess of 7 dB. Do not confuse channel coding with source coding (digital compression) here.

Forward error correction (FEC)

In satellite communications digital carriers in the great majority of cases use either QPSK, OQPSK (offset QPSK) or BPSK modulation with FEC. QPSK with coherent demodulation in conjunction with an inner code rate 0.5 or 0.75 is common, as is coherent demodulated BPSK with an inner code rate of 0.5.

Decoder implementation losses

The receiver's demodulator will cause a certain amount of loss in the overall channel, through non-linearity of filters, etc. These losses are normally small in relation to the decoding gain but still significant and should be taken into account in the link budget. Values in the order of 1 to 1.5 dB are commonly found in practice.

Digital modulation

Digital modulation, also known as phase modulation, is very similar to FM in many respects. As with FM, the spectral analysis is notoriously difficult and the two spectra would appear similar. The most suitable digital modulation methods for digital TV via satellite is BPSK (Binary Phase **Shift** Keying), QPSK (Quadrature Phase **Shift** Keying), 8-PSK, and possibly 16-QAM (Quadrature Amplitude Modulation). Of the four, QPSK is the most common. QPSK has the advantage that it can operate with transponder power close to saturation so is energy efficient. Table 5.4 shows the theoretical values of E_b/N_0 required to achieve a BER of 10^{-10} without channel coding for a range of modulation schemes. It is also apparent why 64-QAM is often chosen for cable distribution: it has a good bandwidth efficiency with currently available bandwidths of 6 to 8 MHz.

Modifications for DVB

DVB uses phase modulation, which is closely related to FM in character. Thus the parameters that are relevant to analogue FM signals are similarly valid for digital system link budgets, but with one exception. Just as S/N is indicative of received signal quality in analogue FM systems, the ratio E_b/N_0 to achieve a specified BER is the S/N equivalent for digital systems. The relationship between C/N and E_b/N_0, expressed in dB form, is

$$E_b/N_0 = C/N + 10 \log(1/R) + 10 \log(B) \quad \text{dB} \qquad (5.23)$$

Table 5.4 *Comparison of digital modulation schemes for a BER of 10⁻¹⁰*

Modulation scheme	E_b/N_0 uncoded (dB)	Bandwidth efficiency (bits/s/Hz)
BPSK	13.06	1.0
QPSK	13.06	2.0
8-PSK	16.55	3.0
16-PSK	21.09	4.0
4-QAM	13.06	2.0
16-QAM	16.98	4.0
64-QAM	21.40	6.0

where

E_b/N_0 (dB) = ratio of energy per bit, E_b (J), to noise power density, N_0 (W/Hz)
R = the data rate (bit/s)
B = transmission bandwidth (Hz)
C/N = carrier to noise ratio within bandwidth B (dB).

A feature of practical digital systems is that, for a given bit-rate-to-bandwidth ratio, a signal-to-noise ratio (E_b/N_0) exists above which error-free reception is possible and below which it is not. Unlike the gradual deterioration of analogue signals affected by noise, digital systems tend to be relatively unaffected up to a point where the error correction system fails to operate effectively. This results in a rapid deterioration or a 'crash'. It is this robust nature of digital systems that precludes the need for grading of received picture quality. Provided that the overall degraded value of E_b/N_0 is greater than some required value corresponding to a suitable 'designed in' bit error probability, P_e (or specific BER), then picture quality will be relatively unimpaired. The BER is the ratio of the number of bits received erroneously to the total number of bits transmitted per second. The relationship between P_e and E_b/N_0 depends on the particular species of digital modulation adopted, so satellite operators usually specify the minimum level of E_b/N_0 required. Values around 8 dB are typical for most DVB broadcasts.

A short form 'clear-sky' link budget (minimum calculation)

A frequently used method for calculating link budgets is given below and relies on a prior knowledge of operating parameters such as statistical rain fade margins and atmospheric absorption. The method takes no account of waveguide losses or antenna pointing errors. However, it is well suited to simple S- and C-band calculations which are relatively unaffected by rain. For Ku, and especially, Ka band, it is necessary to allow a conservative margin over the demodulator threshold to allow for all the above effects. This would be typically 2–3 dB for the Ku band in Europe.

Step 1: Calculate the elevation (Equation 5.1 and, optionally, Equation 5.2 for low latitudes).

Step 2: Calculate the azimuth (Equation 5.3).

Step 3: Calculate the path distance to satellite (Equation 5.4).

Step 4: Calculate the wavelength (Equation 5.5).

Step 5: Calculate the free space loss (Equation 5.6).

Step 6: Calculate the noise factor, F, of the LNB using Equation 5.24. By convention, LNB the noise figure (in dB) is quoted for Ku and Ka bands while equivalent noise temperatures are normally given for S and C bands. If the latter is the case, then Steps 6 and 7 may be skipped.

$$F_{LNB} = 10^{NF/10} \tag{5.24}$$

where NF is the noise figure of the LNB (dB).

Step 7: Convert the noise factor to its equivalent noise temperature at an ambient temperature of 290 K using:

$$T_{LNB} = 290(F_{LNB} - 1) \text{ (K)} \tag{5.25}$$

Step 8: Calculate the total system temperature using:

$$T_{TOT} = T_{LNB} + T_{ANT} \text{ (K)} \tag{5.26}$$

where: T_{ANT} = the antenna noise temperature (K) as quoted by the antenna manufacturer (or estimated using Equation 5.13)
T_{LNB} = the LNB noise temperature (dB), as calculated above.

Step 9: Calculate the noise bandwidth (NB) using:

$$NB = 10 \log (BW) \text{ (dB HZ)} \tag{5.27}$$

where BW (Hz) is the receiver IF bandwidth.

Step 10: Calculate the figure of merit, G/T, neglecting atmospheric absorption, rain attenuation, pointing errors and waveguide losses, using:

$$G/T = 10 \log \frac{10^{(G_a/10)}}{T_{TOT}} \text{ (dB K)} \tag{5.28}$$

Step 11: Calculate the C/N ratio using:

$$C/N = EIRP + G/T - BC - NB - L_{FS} \text{ (dB)} \tag{5.29}$$

where: EIRP = effective isotropic radiated power of satellite at receive
site (dB W)
G/T = figure of merit (dB/K)
BC = 10 log (Boltzman's constant) = (−228.6 dB J/K)
NB = noise bandwidth (dB Hz)
L_{FS} = free space loss (dB)

Step 12: Calculate the S/N ratio using Equation 5.22.

Working backwards

By rearranging the above formulae it is possible to work backwards
and specify the dish size and LNB noise figure required to achieve a
particular level of 'clear sky' C/N. Suitable rearrangements are given
below:

$$G/T = C/N - EIRP - BC - NB + L_{FS} \qquad (5.30)$$
$$G_a = 10 \log (10^{GT/10} \times T_{TOT}) \text{ (dB)} \qquad (5.31)$$

$$D = \frac{\sqrt{(100\lambda^2 \times 10^{0.1G_a})/p}}{\pi} \qquad (5.32)$$

where *p* is the percentage efficiency.

EIRP level v antenna diameter

Using the foregoing equations a table can be constructed showing the
relationship between *EIRP* level and recommended minimum antenna
diameter for a range of *C/N* values. Such a computer generated version
is given in Table 5.5. The *C/N* values chosen consist of a worst-case
receiver threshold at 8 dB and then further increments of 3 dB giving
11 dB, 14 dB and 17 dB. Other uses such as SMATV would require the
higher figures. As with all tables and graphs of this type, certain assumptions need to be made, so their accuracy cannot be taken too literally.
However, they do give a general indication of antenna size requirements
to receive any present or future satellite service. The main trends that
can be deduced from Table 5.5 and the corresponding graph (Figure 5.7)
is that the antenna size increases with *C/N* value but decreases with *EIRP*
level. In compiling the table, the following worst case assumptions are
made:

1 The free space loss is assumed to be 205.7 dB worst case.
2 The frequency chosen is 11 GHz; this is toward the lower end of the
allocated European fixed satellite service, again worst case.
3 Total noise temperature is assumed 240 K, this is rather higher than
the majority of systems inherit.

Table 5.5   *EIRP level v antenna diameter for various C/N ratios*

Based on following fixed values:

Free space loss = 205.7 dB
Antenna efficiency = 60 per cent
Total noise temperature = 240 K
Frequency = 11 GHz
Bandwidth = 27 MHz

	Dish diameter (metres)			
EIRP level (dBW)	C/N = 8 dB	C/N = 11 dB	C/N =14 dB	C/N =17 dB
30.00	5.13	7.25	10.24	14.46
31.00	4.57	6.46	9.13	12.89
32.00	4.08	5.76	8.13	11.49
33.00	3.63	5.13	7.25	10.24
34.00	3.24	4.57	6.46	9.13
35.00	2.89	4.08	5.76	8.13
36.00	2.57	3.63	5.13	7.25
37.00	2.29	3.24	4.57	6.46
38.00	2.04	2.89	4.08	5.76
39.00	1.82	2.57	3.63	5.13
40.00	1.62	2.29	3.24	4.57
41.00	1.45	2.04	2.89	4.08
42.00	1.29	1.82	2.57	3.63
43.00	1.15	1.62	2.29	3.24
44.00	1.02	1.45	2.04	2.89
45.00	0.91	1.29	1.82	2.57
46.00	0.81	1.15	1.62	2.29
47.00	0.72	1.02	1.45	2.04
48.00	0.65	0.91	1.29	1.82
49.00	0.58	0.81	1.15	1.62
50.00	0.51	0.72	1.02	1.45
51.00	0.46	0.65	0.91	1.29
52.00	0.41	0.58	0.81	1.15
53.00	0.36	0.51	0.72	1.02
54.00	0.32	0.46	0.65	0.91
55.00	0.29	0.41	0.58	0.81
56.00	0.26	0.36	0.51	0.72
57.00	0.23	0.32	0.46	0.65
58.00	0.20	0.29	0.41	0.58
59.00	0.18	0.26	0.36	0.51
60.00	0.16	0.23	0.32	0.46
61.00	0.14	0.20	0.29	0.41

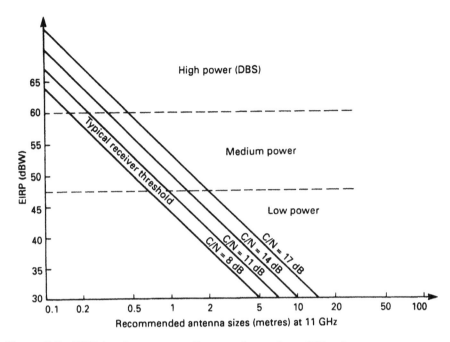

Figure 5.7 *EIRP level v antenna diameter for various C/N values*

4　Antenna efficiency is taken as 60 per cent, an easily obtainable figure these days.
5　Finally, the signal bandwidth is assumed to be 27 MHz.

Importance of beamwidth

As the geo-arc becomes increasingly crowded, the theoretical minimum antenna size may not always be the ideal choice. With continual improvements in demodulator threshold-extension techniques coupled with more efficient feeds and lower noise LNBs, smaller antennas for use with a given satellite may at first seem tempting. However, this may cause problems if due regard is not given to antenna bandwidth treated in Chapter 2. Pushing antenna size reductions too far may cause interference problems from adjacent satellites already in place or planned for the future. As a general rule, assume medium power Ku-band satellites are spaced about 3° apart. Remember, the smaller the antenna the wider the beamwidth, so be careful to perform beamwidth calculations and leave an adequate safety margin if you are putting together a package from different manufacturers' component parts. However, as we will see later, satellite spacings are often slightly further apart than it would first appear. If follows that there must be a limit to the number of satellites accommodated within the geo-arc, thus much work is going on throughout the world on various techniques to cram more channels into the available frequency space.

Satellite spacing

When a geo-stationary satellite position is quoted, given by a longitude, east or west of the Greenwich meridian (0° longitude) we find satellite position varies in azimuth depending on the receiving site location. This is because the position quoted is that from which a hypothetical observer, at the centre of the earth, would see it. If the spacing of two adjacent satellites is, say, 3° we find that on the surface of the earth at the equator they appear to be spaced further apart than this as shown, greatly exaggerated, in Figure 5.8. In fact, from any position on the Earth's surface, satellite spacing is often greater, but never smaller than the quoted

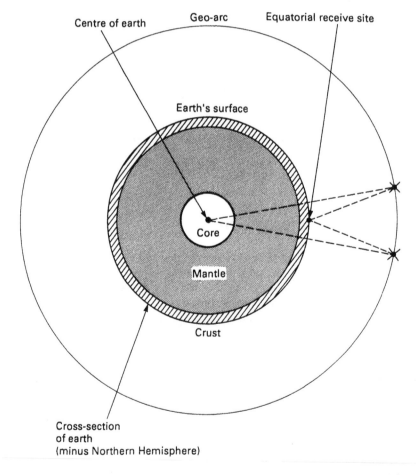

Figure 5.8 *Satellite spacing from Earth's centre and surface, viewed from over north pole*

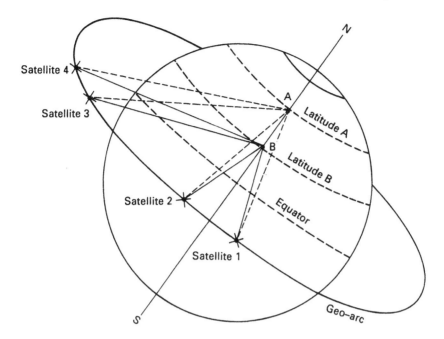

Figure 5.9   *Satellite spacing from various latitudes*

spacing. Figure 5.9, not drawn to scale, exaggerates the general idea. An adjacent pair of satellites appear closer together at latitude A than at latitude B. So, generally, as we move toward the equator satellite spacing appears to increase to a maximum, greater than the difference in quoted positions. Another point worth a mention is, for any given receive site latitude, satellite spacing appears smaller at the horizon or extremity of the geo-arc than at the due south position but at all times spacing is either equal to, or greater than, quoted positions. Try calculating, using Equation 5.3, azimuthal difference between a pair of southerly adjacent satellites for your latitude to see this effect more clearly. Repeat the calculations for a pair closer to the horizon. When deducing the possibility of interference from adjacent satellites operating on similar frequency and polarization this effect should be taken into account when analyzing the appropriate beamwidth diagram for a particular antenna.

Epilogue

This chapter has, unfortunately, had to cover some fairly complicated concepts that to some may appear irrelevant to the practice of satellite TVRO installation. However, those that have persevered will have acquired a sound basic knowledge of the underlying principles and the realization of how various parameters interact. This should help the installation technician to deal with most potential technical problems that may arise in the field and to be reasonably competent in the specification of equipment to fit any need.

6 Installation: site surveys

Introduction

The purpose of the rather pretentiously named 'site survey' is to check out the lie of the land and offer the customer the maximum amount of choice as to the location of the dish, bearing in mind that microwave signals will not pass through buildings, trees, etc. Some customers like their dishes to be unobtrusive whilst others like to advertise the fact that they have one. If properly conducted, a site survey can eliminate wasted journeys for tools and equipment not loaded onto the service vehicle. The need for a formal survey differs from region to region and listed below are the main types of terrain likely to be encountered, together with advice on the course of action.

1 *Flat open countryside* – This is the ideal terrain and presents no problems even to DIY installation. Providing you know the rough direction required, it may be possible to dispense with the formal site survey altogether since no obstructions of the geo-arc are likely to be encountered.
2 *City areas and large towns* – Locations in towns or cities with high rise flats, large industrial complexes, or high density housing may present problems when choosing an antenna mounting location. For example, many towns in northern England have rows of terraced houses built on hillsides. Because of the elevated nature of the adjacent row of houses it may not be possible to achieve the necessary clearance of the next rooftop using a standard wall mounting bracket. A similar difficulty can occur when attempting to find a gap between adjacent 'tower blocks' in city areas. Here, an accurate site survey is essential since inaccuracies of even a degree or two in siting the antenna may result in failure, leaving an unsightly array of mounting holes as evidence. It is not rare for so-called 'professionals' to do this, and in some cases they have even informed customers that reception is not possible when an accurate azimuth/ elevation calculation and precision survey has often proved them wrong. The installation can then proceed with confidence reducing the risk of the aforementioned 'Laurel and Hardy' performance which does little to instil customer confidence in your company. Where tall

buildings or trees are blocking, or partially blocking, the line of sight to the required satellite the customer should be informed that reception is not possible or likely to be degraded.

3 *Leafy suburbs and villages* – The main problem here is trees, requiring a varying degree of accuracy in site survey work. It is important to remember that trees grow in girth as well as height and the customer should be informed if future reception is likely to be degraded as a result of tree growth. If the offending trees are on the customer's land they may agree to lop off certain branches from time to time.

4 *Highland regions* – In mountainous terrain such as that encountered in the north of Scotland, siting an antenna can be critical. Other than obvious blocking problems by mountains, the following additional complications arise:

(a) EIRP levels often fall off in these northern areas and the need for a larger antenna leads to correspondingly more accurate alignment requirements (remember beamwidth in Chapter 2).

(b) Elevation angles to the geo-arc are low in northern regions increasing the risk of blocking of the signal by mountains.

Again, a formal link calculation and site survey are recommended in the majority of cases.

Finding satellite coordinates

Each geo-stationary satellite is located above the equator at a height of some 36 000 km and its position is given as a simple longitudinal position either west or east of the Greenwich meridian (0°). From this information we need to look up from records, or calculate using simple trigonometry, the required azimuth and elevation angles to capture a required satellite's signals from any particular location within the service area. The receiving site is described by a unique set of coordinates called latitude and longitude. For accurate information on these coordinates refer to an ordnance survey map of your local area.

Whether multi-satellite or fixed satellite reception is the object of the survey the *AZ/EL* angles for each required satellite must be checked for obstructions such as buildings or trees. There are four items of information, 1 to 4 shown below, needed to perform an accurate receiving site survey. Two of these need to be calculated, or looked up from records, for each required satellite before arriving at the customer's address. The keeping of detailed records of pointing angles for each satellite will alleviate the need for repeat calculations since angles will not vary significantly across a small local service area. A programmable scientific calculator with continuous memory or a computer will help here. If you find difficulty with the calculations there are many mathematically inclined individuals who will be delighted to help you.

1 *The geostationary position of the satellite* – For example, the Astra cluster of satellites are positioned above the equator over Zaire, 19.2° east of south of the Greenwich meridian. Appendix 3 gives this information for a range of the more popular satellites.
2 *Elevation* – The angle of inclination from the horizontal to the satellite line of sight and given by Equation 5.1 in Chapter 5.
3 *Azimuth* – The bearing of the satellite depending on the location of the receiving site and given by Equation 5.3.
4 *The local magnetic variation* – True south can be offset by a few degrees from magnetic south depending on the location of the receiving site, and varies by a small amount each year.

Magnetic variation

The indicated north on a compass can vary considerably from true north depending on the geographical location. The effect is known as *magnetic variation* in nautical terms or *magnetic declination* in scientific terms. Variation is said to be easterly if the direction of magnetic north lies to the east of the true meridian, and westerly if it lies to the west. Points of equal variation on the globe are contoured with *isogonal lines*. Where true north and magnetic north are the same (i.e. variation = 0), the contour is called the *Agonic line*. Magnetic variation is subject to three types of change:

1 *Secular change* – a continuous alteration decreasing by about nine minutes of arc annually in the UK.
2 *Annual change* – small seasonal fluctuations.
3 *Diurnal change* – a daily fluctuation which increases with latitude.

Items 2 and 3 can be neglected for our purposes since they are relatively insignificant and contribute a worst case error of about a quarter of a degree. Angles are small as this cannot be accurately resolved on most hand held compasses. Thus item 1, the secular change, is the most important.

If the variation is westerly, the variation is negative, therefore the correction to be applied to the compass is positive. The value is added to the indicated compass bearing to obtain the true bearing. Likewise if the variation is easterly, the variation is positive therefore the correction to be applied to the compass is negative. The value is subtracted from the indicated compass bearing to find the true bearing. Figure 6.1 clearly illustrates this point. Most good sighting compasses have a rotating dial which can be set to compensate for local variation and thus indicate true bearings. In the UK true north/south is between 4° and 10° clockwise from magnetic north/south, depending on location and this further decreases by about nine minutes clockwise each year. Current information can be obtained from the latest edition Ordnance Survey maps (or similar) for your area.

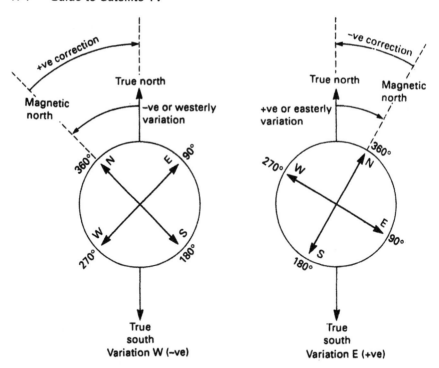

Figure 6.1 *Magnetic variation and compass readings*

Ordnance Survey maps

Local area Ordnance Survey maps, or their overseas counterparts, are singly the most useful source of information regarding accurate site latitude and longitude coordinates and can be consulted at your local library. Latitudes are given at the sides of the map and longitudes are given at the top or bottom of the maps. The current magnetic variation for your local area is obtained from the top centre of the map, although this variation is in relation to grid north and not true north. An additional correction factor is given on the maps, typically about one degree, and this should be added to the variation between grid north and magnetic north to find the total magnetic variation. Figure 6.2 illustrates this point.

If an Ordnance Survey map is not conveniently to hand the appropriate magnetic variation can be estimated to a reasonable degree of accuracy in the UK from Figure 6.3. The magnetic variation for any location in Europe may be estimated from Figure 6.4.

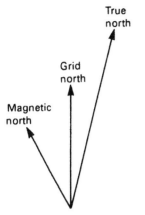

Figure 6.2   *Relationship between grid north and true north*

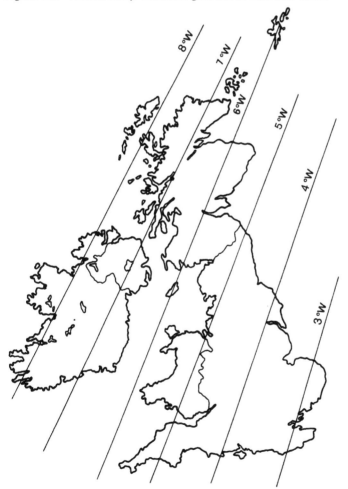

Figure 6.3   *Isogonal map of UK and Eire*

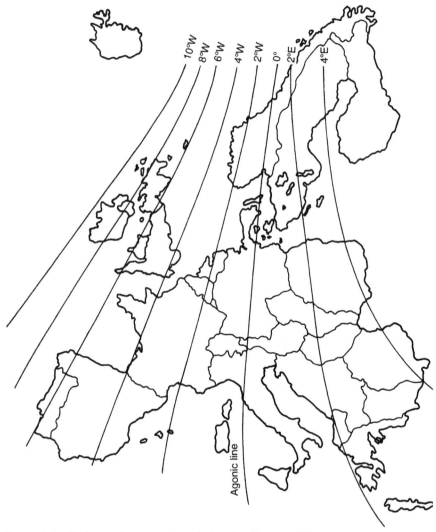

Figure 6.4   *Estimated magnetic variation for Europe 1998*

Azimuth survey

Once the magnetically corrected angle is known, the only piece of equipment required to perform an accurate azimuth survey is a compass. The
cheap Christmas cracker variety, although adequate in some open aspect areas, is a poor substitute for a quality sighting compass. These are
usually equipped with a movable graduated dial, mirror and sighting
line or notch to enable some distant object to be used as a reference to
the required bearing. The magnetically corrected bearing is set on the
compass so that the sighting line corresponds with the satellites
azimuthal position. Correctly orientating the compass so that the needle

lines up with the N/S markings will then give the required bearing at the sighting line as shown in the simplified diagram, Figure 6.5. When conducting an azimuth survey with a compass stay well clear of metal objects which affect compass readings. Stray influences can be discovered by moving about in the general area to see if the compass needle varies considerably.

Elevation survey

The items of equipment needed to perform an elevation survey are a device called an inclinometer and a sighting bar (normally a large spirit level). Inclinometers vary in design from a glorified protractor and

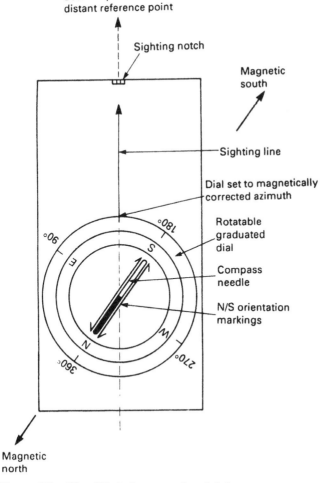

Figure 6.5 *Simplified diagram of a sighting compass*

plumb line to sophisticated moiré pattern types which can be accurate to a fraction of a degree. The choice is yours but the latter type is well worth the extra money in view of its further use in the setting of antenna elevation angles.

The inclinometer is set, according to the manufacturer's instructions, to the required elevation angle and placed on a sighting bar. The bar is inclined until either the moiré patterns are parallel or, with the cheaper type, the required elevation is indicated on the dial. By looking along the bar, as shown in Figure 6.6, any obstructions can be noted. Another method used by some installers for ground erected polar mounts is to attach a piece of pipe or a spent biro casing to an old camera tripod. The elevation angle is set for each required satellite using the inclinometer and if clear sky is seen when viewing through the tube then reception will be satisfactory. For DIY surveying, a makeshift inclinometer can be made from a piece of wood and a plastic protractor mounted at right angles to it, as shown in Figure 6.7. The reading coinciding with the weighted string is the elevation angle of the piece of wood relative to ground.

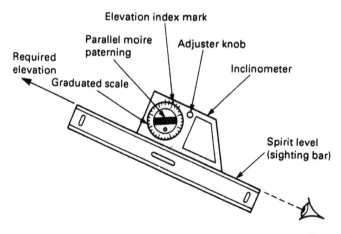

Figure 6.6   *Using an inclinometer for elevation surveying*

Purpose designed site survey instruments

There are many types of purpose designed instruments designed to help with site surveying. A typical low cost example, which works rather well, is the 'Standard Satfinder' from Satfinder (UK) Ltd, which is shown in Figure 6.8. The unit is manufactured under licence from the Independent Broadcasting Authority in the UK. It combines a modified Ebbco sextant, well known in the marine industry, with a high accuracy Sisteco compass and magnifier. A novel idea is the combination of mercury

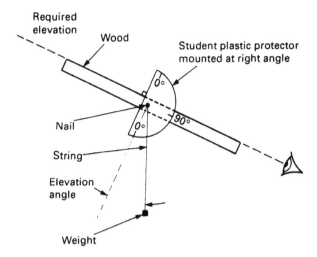

Figure 6.7 *Makeshift inclinometer for DIY surveying*

Figure 6.8 *The Standard Satfinder (Satfinder (UK) Ltd)*

switches and battery operated LEDs which indicate that the sextant is level. It is used in the following way. When the sextant is set to the desired elevation angle, look through the eyepiece and scan the horizon until the required azimuth angle appears on the compass magnifier. Ensuring that the LED lights indicate the unit is level, look for obstructions appearing in the sextant mirrors. There is also a special version available (Figure 6.9) which has the following added features: liquid-filled remote reading compass to signal the correct azimuth through a fibre optic cable to the LEDS, and the facility to be used with a dish mounting kit for installation work. The use of professional equipment does a lot for your company's image.

Figure 6.9   *The Special Satfinder (Satfinder (UK) Ltd)*

Single satellite reception

A fixed wall mounted antenna is the one most commonly encountered and is often part of a manufacturer's package for single satellite reception. A site survey typically involves running up and down ladders at various points of the dwelling checking the line of sight. Since wall mounted dishes can be rotated through 180° a south facing wall is not essential. Complications can commonly arise with some dwellings and the solution is to dispense with the wall mounting bracket and specify pole mounting. Common scenarios are listed below with the appropriate solution.

Problem:
   The awkward customer who does not want the antenna mounted on the front or sides of the house even though this is the most convenient position.

Solution:
   Check if the antenna can be mounted on an outbuilding or pole in the rear garden. Finally if this is not feasible check the minimum height pole required to capture signals from over the roof apex. This of course will depend on the roof pitch which can be easily measured with an inclinometer.

Problem:

The most suitable mounting position, with a wall bracket, is the front of the house but the dwelling is of the dormer bungalow type with tiled upper front walls or wooden planks. A secure mounting is not possible on such a surface.

Solution:

The antenna may be positioned, using a pole and T & K bracket to the side of the house (usually brick). The length of the pole must be such that the roof pitch, where necessary, is taken into account. In some cases a wide chimney construction emerges from a flat roof section below roof apex height. This is sometimes adequate for fixing a standard wall bracket.

Problem:

The wall mounting bracket cannot be placed on a suitable wall because part of the dish aperture is likely to be obstructed by the eaves. This is a common problem encountered with terraced housing where the line of the housing row is the same or close to the required azimuth bearing.

Solution:

The antenna will need to be mounted using a T & K bracket and pole to lift the antenna out and above the eaves. If possible always site at the rear of the house in preference to the front. It is not normally advisable to pole mount dishes on chimneys with aerial lashing kits as done with terrestrial TV aerials. The wind loading effect, on even a relatively small dish, can be such that the whole structure may become unstable and dangerous. Especially so if further compounded by poor mortar condition on older chimneys. Planning permission is also required for mounting above roof apex height.

Problem:

Adequate line of sight clearance above an adjacent structure prevents the standard wall mounting bracket from being used.

Solution:

Use a T & K wall bracket and pole to achieve the extra height or mount the pole in the attic. A special weatherproof tile is available for the pole exit from the roof space.

When specifying these alternative forms of mounting it is important to inform the customer that an extra charge, over and above the standard installation charge, is required to cover the extra work and materials used. It is also important to remember that some manufacturers' packaged systems are unsuitable for pole mounting and the customer should be advised accordingly.

Live tests

More often than not, with single satellite systems, installation follows immediately after a successful site survey – in order to keep costs down. Although a compass provides a fairly accurate guide to antenna location in a site survey, in certain critical cases, such as possible blocking by a tower block or a situation where the antenna needs to be clamped at or near right angles to a wall, even a compass may not be sufficiently accurate. A quick live hook-up can be performed using a battery-powered signal strength meter slung round the neck and the antenna hand-held at the selected site. Reception can then be physically tested. With practice, live tests such as these can be made quickly, even up a ladder, and mounting holes subsequently drilled with confidence.

Multi-satellite reception

Large antenna multi-satellite systems designed to include reception from the present crop of low power satellites are normally ground or patio mounted but later generations of polar mount dishes designed to receive only medium to high power satellites are much smaller with the possibility of wall or pole mounting. For multi-satellite reception the surveying process would need to be repeated for each required satellite. It is surprising how few domestic locations are suited for multi-satellite reception, due to obstruction of part of the geo-arc. However in many cases 'windows' can be found to receive signals from a reduced number of the more popular satellites. It is important to discuss this with customers at the time of the site survey so they are in no doubt as to the blocked satellites. For this type of survey there are various cardboard cutout creations that can be used to simultaneously check the line of sight to each satellite. Unfortunately these are rarely up to date and are short lived. It is advisable to use strong reliable instruments in conjunction with a checklist of the type outlined in Figure 6.10. As the coordinates of each required satellite are checked the form can be filled in. The results will clearly indicate which of the required satellites in the geo-arc will be received satisfactorily. The forms can be either printed or photocopied from a partially completed master. Multi-satellite coordinate surveying is not as tedious as it seems since it will be obvious when part of the geo-arc is unobstructed thereby increasing the rate of box ticking without using instruments.

Site plans and survey maps

The need for a site plan varies, but in most cases it is not considered necessary. One case where it may be necessary to spend some considerable time drawing a detailed plan is where the installation is

Required satellite name	Azimuth angle (magnetically corrected) (degrees)	Elevation angle (degrees)	Tick box as appropriate			
			Azimuth OK?	Elevation OK?	Receive	
					Yes	No

Figure 6.10   *Checklist for multi-satellite reception*

to be carried out by persons other than the surveyor. However, once the antenna position has been decided the area can be simply marked with chalk or a stick. A list of required tools and materials not always carried on the service vehicle is normally sufficient in practice.

Distribution amplifiers

On the site survey always ask if a multi-point distribution amplifier is installed for terrestrial TV. The likelihood is that the customer will want the RF output of the satellite receiver to be distributed along with his normal TV channels for viewing in a number of rooms. The extra materials and installation time will need to be charged for in the initial quote.

Planning permission

In the UK, it may not be necessary to apply for planning permission for a single antenna mounted below roof apex height unless it exceeds 90 cm in diameter. However, if the receiving site is in a conservation area or the dwelling is a listed building of architectural interest the customer should be advised to check with the local council planning department. Many councils have their own guidelines with regard to satellite installations and these are reputed to be up to seven A4-sized pages in length!

Antennas over 90 cm normally require planning permission if within public view or in sight of neighbours. This also applies if more than two dishes, in total, are mounted on a single building or blocks of flats. This section can only be a guide, there are many exceptions to those briefly indicated here. In the UK, a pamphlet is available from the Department of the Environment or your local HMSO which gives the latest details.

Vandalism

When surveying it is also necessary to judge the likelihood of vandalism. Obviously, if a ground mounted the antenna is located in the front garden of a dwelling close to a licensed premises, the likelihood of vandalism is high. Always try to mount the antenna out of open sight or at such a height that one person standing on another's shoulders cannot reach it. That is to say mount over 'two drunks high' (12 ft). It is really surprising the lengths vandals will go to perfect their 'art' and satellite dishes potentially attract a lot of attention.

Environmental considerations

Besides selection of environmentally friendly outdoor units such as glass or mesh dishes, consideration should also be given to the overall visual impact of an installation siting. Where there is a large choice of possible sitings it is often tempting to maximize profit by choosing a location requiring minimum effort or use of materials. Although there are sound technical reasons for using a minimum cable run, sometimes it is necessary to balance this with aesthetics. For example, mounting a dish in the centre front of a semi-detached house may look ridiculous whereas mounting closer to the eaves appears more pleasing. Sometimes it is possible to shield or partially shield a dish from public view using a building's own projections; siting it low down, say, on an adjacent wall above a flat roof extension. Always try to avoid the antenna being seen against the sky and never mount it above roof ridge level like terrestrial television aerials. Some installation companies are now operating a 'code of practice' whereby mounting antennas on the front of domestic premises is likely to suspend the managing director's monacle-wearing capabilities. This is likely to be popular with customers and increased business may offset extra costs of fitting elevated brackets and poles for rear-mounting. Personally, I think elevated pole-mounting of satellite antennas, either at the back or the front of a house, looks far more ugly than a simple thoughtfully placed wall bracket. I have always maintained that the antenna itself it not particularly ugly, it is the bracketry and *its* long-term appearance which requires attention of designers.

Brick composition

Where a wall-mounted antenna is specified, check a building's brick-work composition. Some bricks can be incredibly difficult to drill and may burn out or blunt several masonry drill bits during installation. Such bricks are high-density glazed engineering types used in some older houses and some bricks used on modern housing estates – iden-tified, sometimes, by the presence of bluish-tinged shiny areas or patches. These bricks are soon recognized with experience and it is important to warn a customer that an extra charge may be levied for drilling time and bits used. The cost of prematurely replacing masonry bits can eat significantly into your profit margin. An electro-pneumatic rotary hammer drill is best suited for drilling such materials with ease, but they are expensive.

7 Installation: antenna mounting and cabling

Introduction

This chapter is basically a practical 'nuts and bolts' guide to the assembly and physical mounting of a variety of satellite brackets and poles and discusses the principal tools, equipment, cabling and fixing methods needed to execute a professional installation that will last for years. It is not really necessary to understand the finer details of satellite TV to effect a perfectly adequate installation.

Insurance

Before starting any satellite installation work for the general public ensure that you are adequately insured. Public liability insurance cover should be taken out, along with employee liability if someone is working for you. The latter is a legal requirement. At the time of writing, cover to the value of one million pounds is the recommended level. Further insurance against personal injury is also wise, due to the potentially dangerous nature of the ladder work.

Installation staff

Ideally, an installation team should consist of a technician and a labourer skilled in ladder agility. Unfortunately, many companies with an eye on profits rather than safety insist on single person installation. Although it is perfectly possible for the job to be performed adequately by one person it is not to be encouraged. In any case, a skilled and coordinated two-person team can complete the job in approximately half the time.

Customer relations

Most customers are interested in the background technical details of satellite TV, particularly in such matters as the satellite's height, geographical position, etc. This exchange of 'chit-chat' creates a friendly atmosphere in which to work and is often punctuated by regular cups of tea or coffee. Cowboy practices, such as the operation of 'ghetto blasters' or raucous singing irritates many customers and should be avoided. Installers should work as tidily as possible, using dust sheets for all interior drilling and should discuss with the customer the dish siting, cable route and method of cable entry.

Two points worth a mention; firstly, never conduct an installation if the householder or spouse is not present. It is not uncommon to install a system, after consultation with a teenage son or daughter, only to be asked to move the dish elsewhere on the householder's return. Secondly, avoid mounting an antenna above a doorway. Rain drips can saturate customers and their callers and may lead to a request for it to be moved at a later date.

Guarantees

It is customary to guarantee your work for a period of at least a year against wind altering the alignment of the dish or water penetrating the coax etc. It should be pointed out to the customer that this does not cover vandalism or freak weather conditions such as the force ten hurricanes experienced in the South of England in 1987. It is the customer's responsibility to insure against these risks.

Basic tools and equipment

It is never worthwhile skimping on tool prices, stick to established brand names associated with quality. Cheap 'noddy' tools are rarely adequate, except for the occasional domestic use.

Ladders

The most important requirements are a roof rack and a set of ladders. A versatile size is a 28 ft double or an equivalent triple section ladder, which extends a few feet above gutter height of the average house. Aluminium ones are light and easy to handle but ensure that the rungs are flat topped and inclined so that they are level when the ladder is at its recommended working angle. Round section rungs, such as the type often fitted to wooden ladders are liable to make your feet ache after only a short time. A 25 ft roof ladder is a necessity if working on sloping roofs, although in my experience such working is rarely needed. In any case at the time of writing, planning permission must be sought if a dish is to be mounted above roof apex height, so is best avoided. Many modern houses have flat garage roofs adjoining their walls so an additional single ladder or step ladder is required. I have found that a two way single/step ladder is the most versatile in these cases and can be used where the 28 ft double is either too high, when not extended, or the resulting slope is excessively large. Alternatively, a short single ladder and a dedicated step ladder can be acquired.

Safety precautions when using ladders

All ladders used should be equipped with non-slip feet and when erected should be perfectly vertical, looking from the front. The recommended ladder slope is 4:1 (vertical:horizontal). If it is necessary to use a steeper slope, due to say a restricted width alleyway then ensure that another person is holding the ladder or lash the top to prevent the ladder slipping sideways. In the other case where the slope is less than recommended ensure the feet are firmly wedged with blocks of stone or concrete. Often the base for the ladder is not level; to ensure that the ladder is not inclined sideways, pack or firmly wedge one of the feet as appropriate. When erecting wall mounted brackets, the top of the ladder can be securely lashed to the bracket using strong multi-wrapped luggage straps. This is particularly important in windy conditions. Never over-reach when working on ladders, be sure of your footing and balance at all times. If snow or frost is present then clear the area where the feet are to be positioned and as a further precaution a mixture of sand and salt should be scattered over the area. Never work on sloping roofs unless a roof ladder is used fitted with a bracket designed to securely overhang the ridge tiles.

Occasionally, where the dish is to be mounted on a high wall, such as the third storey of a maisonette, it is possible to hire large rope operated ladders which extend to 44 ft or more. These ladders are very heavy and two persons, at least, are needed to manipulate them into position! The first use of such ladders can be a fairly 'hair raising' experience due to the whip witnessed as they are ascended. The ladders seem to exhibit a life of their own in this respect, and the feeling is rather akin to that experienced whilst walking on a trampoline, albeit an elevated one! As experience is gained the initial terror wears off, but you should still treat them with the greatest respect. It's a long way down, and falls of over 40 ft are nearly always fatal. If you've not got a head for heights then subcontract the high altitude work. When using ladders be careful of overhead power lines: these can be low in some rural areas and there is a risk of electrocution. Take care and treat *all* overhead cables as if they are exposed power lines.

Electric drills

An electric drill is used a lot in installation work for boring dish bracket and cable entry holes. When drilling wall mounting bracket holes it is vital that fixings be made into brick and not the mortar course, therefore a powerful hammer drill (650 W or greater) with at least a 13 mm ($\frac{1}{2}$ in) chuck is needed. There are a large variety of electric drills on the market to choose from, and below are some of the features provided:

1 Drilling/hammer drilling changeover facility.
2 Two-speed mechanical gearbox.
3 Continuously variable speed control.
4 Forward and reverse changeover switch.
5 Electronic torque control for electric screwdriving.
6 Soft start.
7 Safety handle.
8 Scaled drill depth stop.

For maximum versatility, a drill with all the above functions combined with a power rating of 650 W or greater is a good choice. The slow speed electric screwdriver facility can be useful for mounting small dishes with plastic plugs and screws. The drill is best connected to the mains via a quick release plug/socket and 30 m extension lead drum. Although most drills are double insulated, 'belt and braces' safety can be effected with the incorporation of a residual current circuit breaker plug or an isolation transformer.

Electro-pneumatic rotary and combi-hammers

Although standard hammer drills of the high street DIY variety are functional, they do lack performance when drilling engineering brick or other hard surfaces. The best machines for this sort of work are professional electro-pneumatic rotary hammer or combi-hammer varieties. These tend to be expensive but if you are to do a lot of installations the investment is well rewarded in speed and ease of drilling. These machines provide high performance, low vibration and low contact pressure and usually have safety clutches to protect both operator and machine. My initial investment in such a tool was soon recouped simply because drill bits can last at least ten times longer. Most of these models employ SDS drill bits which have the added advantage of quick bit changing without the eternal hunt for a chuck key! The bit runs cool, even when drilling the hardest of materials, resulting in much extended life. Models such as the Hilti TE14 or equivalent are ideal for this trade although they can be a bit over-zealous when drilling through-holes for cable entry – a large chunk of brick can be shattered on the far side as the bit emerges, looking unsightly and unprofessional. The solution is to use the combi-hammer for, say, the first three quarters of the way through, then resort to the more stately conventional hammer drill for the last quarter.

Safety precautions when drilling

The most common injuries sustained when drilling are eye injuries, so always wear safety goggles. Also, if you value your long term health, use ear defenders and a dust mask. There has been a lot of interest lately in a condition nicknamed *vibration white finger* which, in its mild form, leads to numbness and whitening of the fingers. Although risk in this trade is small there are a few precautions to reduce risk still further:

1 Hold machine in a comfortable and balanced manner.
2 Grip tool tightly but not so tight as your knuckles turn white.
3 The effects of vibration are worse in cold weather so wear dry gloves.

Drill bits

Set out below in Table 7.1 is a list of commonly used drill bit sizes and length included in the average tool kit. The lengths specified are the minimum length, it does not matter if they are longer. Usually 380 mm length is sufficient to drill a standard cavity wall. If possible obtain the double spiral type (usually matt black finish), such as Rawlbor manu-factured by the Rawlplug company. This design doubles the removal of spoil, resulting in faster, easier drilling and subsequently less wear on the drill and bit.

Table 7.1 *Commonly used drill bit sizes*

Diameter (mm)	Min. length (mm)	Type	Principal use
8	120 mm	Masonry	Mounting holes
10	120 mm	Masonry	Mounting holes
14	120 mm	Masonry	Mounting holes
16	120 mm	Masonry	Mounting holes
8	150 mm	Auger	Cable entry
10	380 mm	Masonry	Cable entry
14	380 mm	Masonry	Multi-cable entry
18	380 mm	Masonry	Multi-cable entry

Extra tools and materials for installation work

Other useful tools and materials are:

1 An assortment of screwdrivers including cross point types.
2 Claw hammer.
3 Spirit level.
4 Craft knife or coax stripper.
5 Wire cutters.

6   A set of metric ring or combination spanners (6 mm to 22 mm).
7   A pair of mole grips or pump pliers.
8   A metric socket set (optional).
9   A tool belt (saves a lot of footwork when working up ladders).
10 Solder gun (some receivers need special plugs soldering onto cable).
11 Rolls of self-amalgamating tape (weatherproofing outdoor connections).
12 Weatherproof rubber 'boots' (optional, to cover F connectors).
13 7 mm round section coax clips (CT100/H109F cables).
14 Assortment of larger clips for tacking up combination cables.
15 F connectors (twist-on type).
16 F to F connector line sockets.
17 Standard coax plugs.
18 Coax plug to coax plug line sockets.
19 Coax combiners (for receivers without loop-through facilities and for RF inputs to distribution amplifiers).
20 Multimeter.
21 Inclinometer.
22 Compass.
23 Signal strength meter.
24 Digging tools (where appropriate).
25 Torch or inspection lamp (for working in attic spaces).
26 Assortment of fixings to suit popular sized dish mountings.

Wall mounting

The most common mounting medium is brickwork, in which the compressive strength can vary from 7 N/mm^2 to 70 N/mm^2. In addition to this, variations in mortar compositions and strengths make it very difficult to provide accurate fixing recommendations so all information provided in this chapter is given as a guide only.

   The term 'static load' is used where a constant force is applied to a fixing or where known peak low frequency fluctuations are superimposed, such as in wind loading. Static loading can further be resolved into two components as shown in Figure 7.1.

1 *Tensile load* – This is the load applied along the axis of the fixing and has the tendency to pull the fixing directly out from the wall. This load is sometimes referred to as the axial or pull-out force of a fixing.
2 *Shear load* – This is the force acting at right angles to the axis of the fixing. Most bolt on structures involve loads predominantly in shear although some tensile loading is present.

   Although the total weight of the antenna and mounting assembly plays some part in the forces acting, the major consideration must be

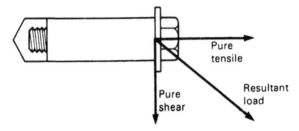

Figure 7.1 *Forces acting on wall fixings*

given to wind loading. The effects of wind loading can considerably increase the forces acting on fixings so an additional safety margin should be included to take this into account. Calculation of wind loading effects are complicated, but generally, for small antennas under 90 cm, a wind loading margin of (× 10) per fixing is normally sufficient to estimate a peak working load.

Screws and plastic wallplugs

Most wall mounting brackets, for small antennas under 90 cm, have four or five fixing points. There are many different fixing methods that can be used, but the cheapest involves screws and plastic wallplugs. This method should always be used in preference to expansion type anchors in low density blocks such as breeze. Use good quality plugs that are either made of nylon, polypropylene, or polyamide as these do not deteriorate with age. The associated screws and washers should be corrosion resistant, either plated or stainless steel. Do not use ordinary wood screws as these will rapidly corrode in exterior environments.

Pull-out loads (plastic plugs)

Fixings of this type predominantly rely on friction between the plug and mounting material for their strength. The tendency for a screw and plastic plug to be pulled directly out of a material is often termed the 'axial' or 'pull out' load. Assuming zero wind speed an estimate of the axial load, per fixing, is given by the following equation:

$$\text{Axial load} = 9.81 \times \frac{\text{Weight of dish and mount}}{\text{No. of fixings}} \text{ Newtons(N)}$$

(9.81 is the acceleration due to gravity (9.81 m/s^{-2}))

Allowing a wind load margin of (× 10) per fixing, we arrive at a rule of thumb estimate for the peak working load:

$$\text{Peak working load} = 100 \times \frac{\text{Weight of dish and mount}}{\text{No. of fixings}} \text{ N}$$

This is still not the end of the story, a further safety factor of 3 to 5 is needed to allow a general safety margin between working load and failure load; thus the equation for minimum pull-out load is approximately given by:

$$\text{Minimum pull-out load} = 500 \times \frac{\text{Weight of dish and mount}}{\text{No. of fixings}} \text{ N}$$

Practical example

Suppose the total weight of a dish, head unit and mounting bracket is 14 kg and the wall bracket has four fixing points. Using the above equation we can estimate the minimum pull-out force per fixing needed to safety secure the assembly.

$$\text{Minimum pull-out force} = 500 \times \frac{14}{4} \text{ N}$$
$$= 1750 \text{ N or 1.75 kN}$$

Therefore, a fixing with a pull-out value of at least 1.75 kN per fixing is needed. If you choose a higher value, all the better. Some manufacturers give a guide to the pull-out values of their plastic plug fixings. Table 7.2 shows the details for the 'Rawlplug' series of screw fixings. From the performance data, a correct combination of plug, screw and drill bit size can be found.

Shear loads and plastic plug fixings

Some manufacturers do not supply pull-out loads, which are essentially failure loads, for their plastic plugs. Instead, they give recommended working loads in tension and shear. The failure load is usually taken to be five times greater than this by applying a general safety factor. As mentioned above most bracket fixings have loads predominantly in shear, although some tensile loading is present particularly on the upper fixings. An estimate of shear load, assuming no wind loading is given by:

$$\text{Shear load} = 9.81 \times \frac{\text{Weight of antenna} + \text{bracket}}{\text{Number of fixings}} \text{ N}$$

Again, if we allow a wind load margin of ($\times 10$) per fixing for small antennas, we arrive at the following rule of thumb equation for the peak working load in shear:

$$\text{Shear working load} = 100 \times \frac{\text{Weight of antenna} + \text{bracket}}{\text{Number of fixings}} \text{ N}$$

Table 7.2   *Rawlplug data*

Types and sizes: boxes of 1000 fixings

Product description	Presentation	Pack contents	Screw sizes	Plug length mm	Drill size mm	Cat. no
Yellow 'Hundreds'	uncarded	10 clips of 100	Nos. 4,6 8, 10	25 (1")	5	**67–125**
	carded					**67–126**
'Twenties'	uncarded	50 panels of 20	3–5 mm			**67–128**
Red 'Hundreds'	uncarded	10 clips of 100	Nos. 6, 8 10, 12	35 (1$\frac{3}{8}$")	6 or 6.5*	**67–130**
	carded					**67–134**
'Twenties'	uncarded	50 panels of 20	3.3–5.5 mm			**67–138**
'Hundreds'	uncarded	10 clips of 100	Nos. 10, 12, 14	45 (1^3/$_4$")	7	**67–231**
Brown	carded					**67–233**
'Twenties'	uncarded	50 panels of 20	5–6 mm			**67–237**

*6.5 mm drill for No. 12 (5.5 mm) screw

Trade packs

Product description	Presentation	Pack contents	Screw sizes	Plug length mm	Drill size mm	Cat. no
Yellow Trade pack 300 Rawlplugs	Shrink film pack	3 clips of 100	Nos. 4, 6 8, 10 3–5 mm	25 (1")	5	**67–900**
Red Trade pack 300 Rawlplugs	Shrink film pack	3 clips of 100	Nos. 6, 8 10, 12 3.3–5.5 mm	35 (1^3/$_4$")	6 or 6.5*	**67–902**
Brown Trade pack 300 Rawlplugs	Shrink film pack	3 clips of 100	Nos 10, 12 14 5–6 mm	45 (1^3/$_4$")	7	**67–904**

*6.5 mm drill for No. 12 (5.5 mm) screw

Performance data Rawlplug

Rawlplug type	Screw size (no.)	Pull-out loads kN					
		4	*6*	*8*	*10*	*12*	*14*
Yellow		0.75	1.25	3.25	3.25	–	–
Red		–	1.5	2.5	4	4	–
Brown		–	–	–	3	4.75	4.75

Results obtained in common brick (1760 kg/m³) after fixing a standard wood screw to full penetration depth
Source: The Rawlplug Company

Practical example

Using the previous example of a 14 kg antenna and mounting bracket and four fixing points, the shear working load would be:

$$\text{Shear working load} = 100 \times \frac{14}{4}$$

$$= 350 \text{ N or } 0.35 \text{ kN}$$

To provide a sound fixing, a working shear value of at least 0.35 kN is needed. Table 7.3 shows the drill size, fixing and working loads for the Hilti HUD series of universal anchors.

Purpose designed fixings

If you do not wish to use the cheapest method of fixing, a far easier and quicker method at little extra cost can be to use the purpose designed fixing, type HRD-HS 10/10 supplied by Hilti (GB) Ltd. This fixing, for small 60/70 cm antennas, includes a polyamide anchor suitable for a variety of masonry materials and comes complete with a galvanized and yellow-chromated screw. After drilling and clearing a 10 mm diameter hole, the fixing is simply hammered in as far as it will go, and finally tightened with a 13 mm spanner. A suitable, corrosion resistant washer is needed for each fixing. The recommended maximum working load for these fixings is 0.5 kN and the failure load is 3.5 kN. Figure 7.2 illustrates the method of installation and construction of the fixing.

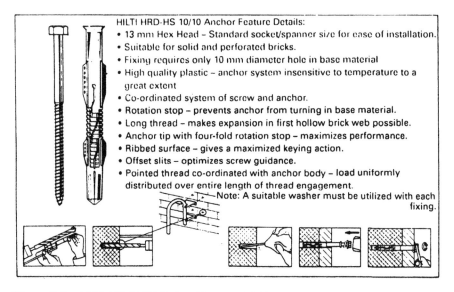

Figure 7.2 *HRD-HS 10/10 fixing (Source: HLTI (GB) Ltd)*

Expansion anchors

For larger wall mounted dishes, patio stands or pole mounted dishes using T & K wall brackets, 'Rawlbolt' expansion anchors are the ideal fixing (see Figure 7.3). M8 and M10 are typical sizes used in brickwork and sizes of M12 or above are used for patio stands bolted to concrete paving slabs. These fixings are plated to withstand corrosion and provide one of the most secure fixings into brickwork. As a general rule, expansion anchors above 16 mm hole size (M10 Rawlbolts) are not normally recommended for brickwork although 20 mm hole size (M12 Rawlbolts) may be used in engineering brick used in some older Victorian houses. Always try to keep as much brick round the fixing as possible and do not fix higher than four brick courses from the top of an unrestrained wall or on a corner brick. Expansion anchors should not be used with breeze or other low density blocks since these materials have low compressive strengths. Only plastic screw fixings or special resin bonded fixings work well with this mounting medium since the fixing method does not rely on compressive forces. A word of warning when using Rawlbolts, never overtighten because the compressive forces involved can easily crack brickwork. A torque wrench should be used and Rawlbolts should be tightened to the recommended torque shown in Table 7.4. The Rawlbolt installation method is as follows:

1 Drill holes to the recommended diameter and depth and clear the debris.
2 Insert the expansion sleeves into the holes and position the bracket.
3 Pass the bolt through the mounting bracket and into the expansion sleeves and tighten to the recommended torque.

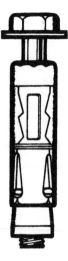

Figure 7.3 *Expanding Rawlbolt anchor*

Table 7.3 *Hilti HUD anchor data*

Advantages: No turning in hole
  Smooth surface maximizes friction grip
  Can be used for in-place fastenings
Material: Polyamide PA6
  Working temperature range: $-40°C$ to $+80°C$
  Setting temperature range: $-10°C$ to $+40°C$
  Screw: UTS = 400 N/mm^2
  YS = 240 N/mm^2

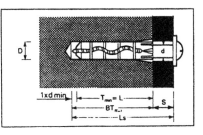

Setting details

Details		HUD 5	HUD 6	HUD 8	HUD 10	HUD 12	HUD 14
D hole diameter	(mm)	5	6	8	10	12	14
BT minimum hole depth	(mm)	35	40	55	65	80	90
T_{min} minimum depth of embedment in structural material	(mm)	25	30	40	50	60	70
L anchor length	(mm)	25	30	40	50	60	70
Ls required screw length	(mm)	29 + S	35 + S	46 + S	58 + S	70 + S	82 + S
d required screw shank diameter	(mm)	2.5–4	4.5–5	5–6	7–8	8–10	10–12
d required screw number		4–7	9–10	10–14	16–18	18–24	24–32

Working loads in tension (Zwl) and shear (Qwl) (factor of safety = 5)

Base material			HUD 5	HUD 6	HUD 8	HUD 10	HUD 12	HUD 14
Concrete	$\beta w = 35$ N/mm^2	Zwl(kN)	0.30	0.55	0.85	1.40	2.00	3.00
		Qwl (kN)	0.40	0.90	1.25	2.20	3.00	5.60
Sand-lime block*	$\beta w = 45$ N/mm^2	Zwl (kN)	0.20	0.40	0.70	1.00	–	–
		Qwl (kN)	0.25	0.56	0.74	1.32	–	–
Solid fired brick*	$\beta w = 47$ N/mm^2	Zwl (kN)	0.17	0.35	0.55	–	–	–
		Qwl (kN)	0.24	0.30	0.44	–	–	–
Aerated concrete*	$\beta w = 7.5$ N/mm^2	Zwl (kN)	0.09	0.15	0.20	–	–	–
		Qwl (kN)	0.13	0.18	0.30	–	–	–
Aerated concrete*	$\beta = 2.5$ N/mm^2	Zwl (kN)	0.02	0.03	0.06	–	–	–
		Qwl (kN)	0.04	0.05	0.08	–	–	–

* For brickwork and masonry, specific performance data cannot be assured owing to the wide diversity and variety of these materials. The loads given above must therefore be regarded as guide values only. For critical applications, load tests should be carried out. Please contact our Technical Advisory Service for further details. *Source: Hilti (GB) Ltd.*

Table 7.4   Rawlbolt data

Bolt size		M6	M8	M10	M12	M16	M20	M24
Shield length	mm	45	50	60	75	115	130	150

Fixing thickness loose bolt

		M6	M8	M10	M12	M16	M20	M24
Max.	mm	10 25 40	10 25 40 50	10 25 50 75	10 25 40 50	15 30 60	60	100 100 150
Min.	mm	0 0 0	0 0 0 0	0 0 0 0	0 0 0 0	0 10 30	25	60 25 100

Material

	M6	M8	M10	M12	M16	M20	M24
Electro plate / Aluminium	• • • • • •	• • •	• • • •	• • •	• • • •	• •	• • •
Bronze	• •		• •	• • • •	•		•

Fixing thickness bold projecting

		M6	M8	M10	M12	M16	M20	M24
Max.	mm	10 25 60 10	25 60	15 30	60 15 30	75 15 35 75	15 30	100 75 120
Min.	mm	0 0 0 0	0 0	0 0	0 0 0	0 0 10 35	0 10	30 0 75

Material

	M6	M8	M10	M12	M16	M20	M24
Electro plate	• • • •	• • • •	• •	• •	• • • • •	• •	• •

		M6	M8	M10	M12	M16	M20	M24
Hole diameter in structure	mm	12	14	16	20	25	32	38
Min. hole depth in structure	mm	50	55	65	85	125	140	160
Hole diameter in fixture	mm	6.5	9	11	13	17	22	26
Bolt tightening torque in concrete	Nm	6.5	15	27	50	120	230	400
Bolt tightening torque in 20.5 N/mm² brick	Nm	5	7.5	13.0	23.0	–	–	–

Loads for plated steel Rawlbolt with 5.8 grade bolt

Size		M6	M8	M10	M12	M16	M20	M24
Safe static load for 30 N/mm² concrete								
Tension	kN	3.3	4.5	5.8	9.1	20.0	29.4	35.8
Shear	kN	2.1	4.4	6.1	12.4	27.6	36.4	50.0
Ultimate load 30 N/mm² concrete								
Tension	kN	12.2	15.4	21.2	30.9	73.1	100.5	125.5
Shear	kN	7.6	16.4	24.5	40.4	86.0	131.4	165.2
Normal edge distance	mm	90	105	120	150	190	240	285
Normal spacing between bolts	mm	130	155	175	220	275	350	420

Suggested design loads for 20.5 N/mm² class 3 brick

		M6	M8	M10	M12
Tension	kN	1.8	2.3	2.9	4.3
Shear	kN	1.8	2.3	2.9	4.3

Bolts over M12 size are not generally recommended for use in brickwork

Ultimate load

		M6	M8	M10	M12
Tension	kN	9.2	11.35	14.35	21.5
Normal edge distance and spacing	mm	300	300	300	300

Loads for bolt projecting and aluminium bronze Rawlbolts vary slightly from those specified. Please consult Rawlplug Technical Advisory Department for further details.
Source: The Rawlplug Company.

Sleeve anchors

A cheaper alternative to expansion bolts is to use sleeve anchors such as 'Rawloks' manufactured by the Rawlplug company (see Figure 7.4). These can be used to secure wall mounting brackets in brickwork and can be fitted very quickly. Table 7.5 shows the product range and performance data.

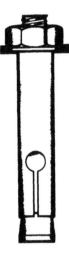

Figure 7.4   *Rawlok sleeve anchor*

Fitting of wall mounted antennas

Some manufacturers supply a template, on the back of the flatpack, in order to drill the dish mounting holes. A slight breeze up a ladder with a large flapping piece of cardboard in one hand and a hammer drill in the other is not a pleasant scenario. Throw it away. Drill and temporarily fix one point first, put a spirit level on it, then mark out the other holes. The dish and LNB are best assembled at ground level, according to the manufacturer's instructions, and taken up the ladder, in one piece, for offering up to the mounting bracket. Be careful not to knock the feed-horn. This component is manufactured to the highest standards and any dent could render it useless. On occasions, where cement rendering masks the location of the brick courses, it is necessary to drill a small pilot holt to check the dust colour and drill resistance to locate brick rather than mortar; unfortunately a trial and error method is the only way to do this. Rendering is also used in some modern houses and extensions as a way of facing cheap breeze block construction. A small test bore will often indicate the structure of the underlying material so the appropriate fixing method can be selected.

Table 7.5 *Rawlok data*

Bolt size	M4.5	M6	M8	M10	M12	M16
Anchor length mm	26 38 58	42 42(SS) 66 92	48 48(SS) 75 100 58	70 70(SS) 98 126 64	108 142 84	114 158
max. fixture thickness** mm	5 9 27	9 9 35 60	9 9 36 60 9	22 22 50 80 13	55 90 25	57 100
Min hole depth* mm	22 30 30	35	40	50	55	60
Anchor hole diameter mm	6	8	10	12	16	20
Rec. tightening torque Nm	2.5	6.0	11.0	22.0	38.0	95.0

SS denotes stainless steel.
*Minimum recommended hole depth in concrete of 30 N/mm² compressive strength.
**Minimum hole depth for maximum fixture thickness; for thinner fixtures increase hole depth accordingly.

Loads at minimum hole depth

Size		M4.5	M6	M8	M10	M12	M16
Safe static load 30 N/mm² concrete							
Tension	kN	1.5	2.2	3.2	4.0	4.6	5.9
Shear	kN	1.8	2.5	3.4	5.4	7.8	14.5
Ultimate load 30 N/mm² concrete							
Tension	kN	5.9	8.9	12.1	16.0	17.9	22.7
Shear	kN	10.0	14.2	17.2	20.3	30.0	50.2
Minimum edge distance							
Tension	mm	60	75	80	100	110	120
Shear	mm	60	80	100	120	160	200
Min. spacing between bolts	mm	60	80	100	120	160	200
Safe static load for 20.5 N/mm² brickwork							
Tension	kN	0.6	0.9	1.2	1.6	1.9	
Shear	kN	1.4	1.5	1.6	1.7	1.9	
Ultimate load for 20.5 N/mm² brickwork							
Tension	kN	3.0	4.3	5.9	7.5	9.1	
Shear	kN	6.6	7.1	7.8	8.5	9.2	

Bolts above M12 are not suitable for brickwork

Loose bolt

Bolt size	M6		M8		M10	
Anchor length mm	45	70	55	80	60	75
Max. fixture thickness mm**	9	35	9	35	9	35
Min. hole depth* mm	35		45		55	
Anchor/hole diameter mm	8		10		12	
Rec. tightening torque Nm	6.0		11.0		22.0	

Loose bolt

Bolt size	M4.5				M6		M8	
Anchor length mm	34	58	76	98	60	86	74	102
Max. fixture thickness mm**	19	28	46	70	28	52	35	62
Min. hole depth* mm	16		30		35		40	
Anchor/hole diameter mm	6				8		10	
Rec. tightening torque Nm	2.5				6.0		11.0	

Round head

Bolt size	M4.5				M6		M8	
Anchor length mm	32	54	74	96	58	82	64	92
Max. fixture thickness mm**	16	25	45	67	25	50	25	52
Anchor/hole diameter mm	6				8		10	
rec. tightening torque Nm	2.5				6.0		11.0	

*Minimum recommended hole depth in concrete of 30 N/mm² compressive strength.
**Minimum hole depth for maximum fixture thickness; for thinner increase hole depth accordingly.
Source: The Rawplug Company

Cable entry holes

Cable entry holes should be a millimetre or two larger in diameter than the cable otherwise scuffing or damage to the outer sheath may result during installation. The hole should be drilled from the inside out and should have a downward tilt so that rain water will not drain into the house from outside. Large multi-sheath cables may be rolled up to feed through the appropriate hole. On completion of the installation, the hole should be sealed with a waterproof compound.

Large multi-sheath cables are best fed through walls. Thinner single coax cables can be conveniently fed through window frames, although this practice is not advisable with uPVC or metal window frames. For drilling the coax entry holes into wooden window frames a $^5/_{16}$ inch (8 mm) auger bit and hand brace does the neatest job. Angle the bit downwards and drill from the inside out to avoid splintering the wood on the inside. If a variable speed electric drill is acquired the hand brace is not needed. A stiff piece of wire is often useful to draw the cable through a cavity wall. Coax often tends to feed into the cavity if an obstruction, such as cavity wall insulation, is present. This material has an irritating habit of dropping down over the hole and blocking cable entry.

A useful tool for feeding multi-sheathed cables through cavity walls is a 500 mm to 1 m length of rigid plastic tubing, such as conduit. The cable is fed into the tubing until it jams solidly then the whole assembly is pushed through a slightly oversized pre-drilled hole in the cavity wall. The tube can then be pulled from the other side drawing the cable with it. I have found this method to be both quick and easy and acquiring an appropriately sized tube is well worth the effort in time saved on each job.

Tacking up the cable

Always agree the shortest route of the cable with the customer and try to make it as unobtrusive as possible by following natural lines of the building such as eaves or window frames; also avoid tacking cables close to entrances since small children often like to tug at them. Bending radii should be at least ten times the diameter of the cable in the absence of manufacturer's figures and the cable must not be allowed to scuff against sharp edges. Clips should be used of a size and type such that the cable is not deformed in any way; also perfectly regular tacking distances should be avoided. The recommended tacking intervals are less than 750 mm for vertical runs and less than 230 mm for horizontal runs. Fixings to poles should be less than 230 mm. For heavier multi-sheath cables the above intervals should be shorter as thought appropriate. Any extensions to cables should be done with the appropriate line connectors so that the impedance is within cable tolerance. They should also be over-wrapped with self-amalgamating tape and well supported if outside. Leave a sufficient length of cable at the head end to perform all the necessary wiring and to form a drip loop. Finally form the cable into a drip loop at the point of entry so that any water droplets form at a lower level.

Where standard wall mounting is not possible

In a small number of cases a suitable wall for mounting may not be available for any one of the reasons outlined in Chapter 6. As indicated there, the solution is often a pole mounted antenna attached to the wall with a T & K bracket. A typical example is shown in Figure 7.5 where the antenna needs to be lifted above the eaves to avoid signal blockage. Two-piece T & K brackets are available in 12, 18 and 24 inch lengths to suit a range of common eave widths. The T & K bracket is bolted to the wall using M8 and M10 Rawlbolts. A four or five feet length of pole is attached to the antenna using heavy duty U-bolts. The whole assembly is then taken up the ladder and attached to the T & K bracket using heavy duty U-bolts. Unfortunately, as mentioned in Chapter 6, some manufacturers' package system antennas are unsuitable for pole mounting whilst some others have optional pole mounting kits with elevation adjustment brackets (azimuth is simply set by rotating the antenna on the pole).

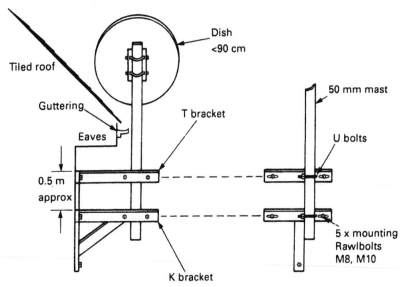

Figure 7.5   *A typical pole mount using T & K brackets*

Due to wind loading effects, even on relatively small antennas of 65 cm, a pole with a thick cross-sectional wall of at least 6 mm ($^1/_4$ inch) and an outside diameter of 50 mm (2 inch) should be chosen. This may be made of galvanized steel or alloy specifically manufactured for the purpose. Poles may also be embedded in concrete for location in gardens. A typical 'planting depth' for the pole is between 0.6 and 1 m into a 0.5 m diameter concrete-filled hole. The pole itself may be filled with concrete for extra resistance to whip and bending. All poles must be set perfectly vertical by performing multiple checks with a spirit level. Nothing looks worse than a pole that is not vertical from all viewing directions.

Another method of pole mounting in awkward situations is to place the pole in the attic attached to joists and exit the pole through a tile in the roof. Special weatherproof tiles have been developed to encase the exit. The antenna is then mounted to the pole outside in the usual way. Figure 7.6 shows this little used arrangement. A wide range of alternative bracket and pole assemblies designed to overcome awkward situations are available from many manufacturers. Figure 7.7 shows a typical range of mounting arrangements. Chimney mounting should never be used for antennas greater than 60 cm and should be avoided if at all possible. However, with some types of building, such as post-war prefabs, this may be the only low-labour and low-cost solution. The outer walls of such buildings are often made of either corrugated iron or asbestos and a wall fixing is certainly not recommended!

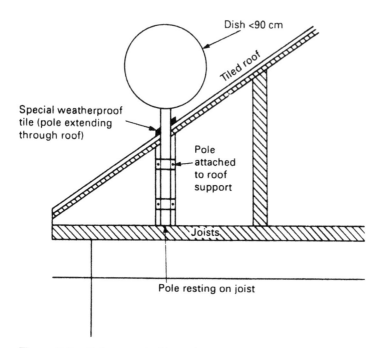

Figure 7.6   *Pole mounted in attic space*

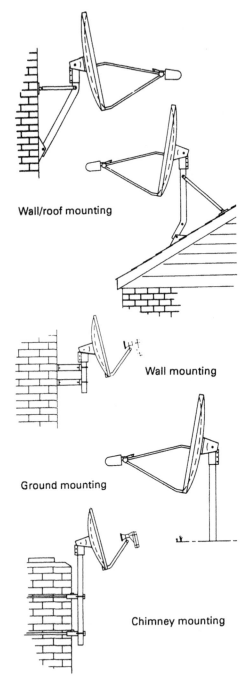

Wall/roof mounting

Wall mounting

Ground mounting

Chimney mounting

Figure 7.7   *Types of antenna mountings (Channel Master USA)*

Low profile installations

There are certain circumstances where an offset focus antenna may be mounted on its back so that a deceptively 'low profile' installation results. This method, shown in Figure 7.8, would be particularly suitable for use on balconies where the landlord, or town planning officer might otherwise exercise his jaw bones. This assumes of course that the landlord does not own a helicopter and megaphone! If exposed to the elements then a small hole should be drilled at the lowest point to drain away accumulated rain water. Make sure the hole is not larger then one-sixteenth of a wavelength of incoming signal in diameter.

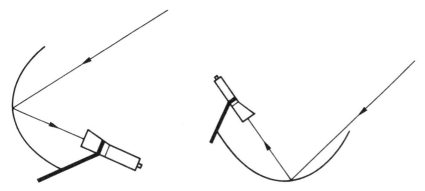

Figure 7.8   *Mounting an offset antenna on its back*

Mounting large antennas

S-band, C-band, and Ku-band dishes for out of footprint reception are much larger and can vary between 1 and perhaps 5 m or more. Above about 1.2 m wall mounting becomes inadequate due to the weight and sheer cumbersome nature of the equipment. The solution is to either ground mount the antenna or mount it on a flat roof. Although instructions and guides to mechanical installation are usually supplied with these larger antenna, a brief guide to the more popular methods is given below.

1   A mounting stand is bolted to perhaps three or four heavyweight concrete slabs on a flat roof. The mount's stability relies solely on the effects of gravity. (Use a layer of plastic or polystyrene foam under the slabs to protect roofing felt). If in doubt about the load bearing capacity of the roof consult a civil engineer.
2   Very large dishes over 3.5 m are usually mounted on a tripod tower stand bolted to a solid concrete base using specially made anchors.

3 Pitched roof mounting brackets. These tend to be OK at the smaller diameter end of the scale but may require the services of a civil engineer for structural surveys.
4 Perhaps the most common mounting for antennas less than 3.5 m is the deeply buried pole or mast, encased in concrete. The hole to be filled with concrete should be 3 or 4 times the diameter of the pole and at least 1 m deep. Ten per cent of this should be filled with gravel to aid drainage. For above ground pole heights exceeding 1.5 m the hole should be 20 per cent of additional height deeper than this. Poles should never protrude above about 5 m since the whip in wind might be excessive. Tall poles may be filled with concrete for extra strength. To embed the pole, use 3 parts gravel, 2 parts sand and 1.5 parts cement mixed fairly stiffly such that it just holds together when squeezed in the hand. A stiff mix allows the pole to remain vertically set, without temporary aids, until the concrete goes off by chemical reaction. Pound down the concrete with a piece of wood to ensure it is compacted well. Use a spirit level twice, at 90° intervals round the circumference to check for vertical alignment. Recheck this more than once.

For remote ground mounted installations the cable run may be fairly long, prone to damage, and perhaps dangerous if left to lie on the surface. The solution is to bury them or use an overhead span. Provided the cabling is polyethylene sheathed it may be directly buried. The depth of burial is not important provided the customer is aware of the route. Ducting or conduit may be advantageous for added protection and ease of cable replacement. A low cost, waterproof, and versatile, form of underground ducting is standard domestic bathroom waste pipe. It has the advantage that it is freely available and rarely out-of-stock. Always construct a 180° loop, at the point of emergence at the antenna end, so that rain water will not enter. In addition, seal the entrance with putty so that insects will not crawl along the pipe and into the customer's living area! As an alternative to underground cabling an overhead span may be used from a nearby tree or structure to the customer's residence. The cable needs to be supported by a galvanized stranded steel support wire attached to rigid eyelet wall fixings. The cable may be attached to the support wire with plastic cable ties. Ensure the overall sag at the centre lies between 1.5 and 2.5 per cent of the span length.

Earthing of metalwork

The premature failure of RF input stages in LNBs and tuner units is often attributed to a static electricity build up in the antenna. This can be discharged by earthing the antenna metalwork to a nearby earthing point such as an attic water pipe. In the absence of a suitable earthing point, connection to a grounding rod, buried at the foot of the antenna may be the solution. A conductor of minimum diameter 1.5 mm should be used for the purpose. Earthing of metalwork does not offer protection against lightning however, this can destroy the LNB whether earthed or not.

Assembly of antenna and head units

Most antennas arrive in flat packs and need to be assembled along with their mounting brackets. This is a fairly straightforward process since comprehensive instructions are included. Screwdrivers, molegrips and a set of ring spanners are the only tools normally needed to complete the assembly. The head units of some single satellite systems are often prefabricated in one piece but others consist of a feedhorn/magnetic polarizer assembly and LNB which need to be bolted together with an 'O' ring seal between them to prevent moisture ingress. The exit wire or wires for the polarizer embedded in the feedhorn usually exit downwards and indicate the orientation of the feedhorn relative to the ground. Feedhorn/polarizers with a rectangular waveguide must be matched dimension to dimension with the associated waveguide section in the LNB as shown in Figure 7.9.

Some head units consist of three components, a separate feedhorn, polarizer and LNB. The component parts all need to be bolted together using 'O' ring seals, again with attention to the alignment of the waveguide bores. Both these types often have a transition from a circular waveguide to a rectangular waveguide within the feedhorn or polarizer section.

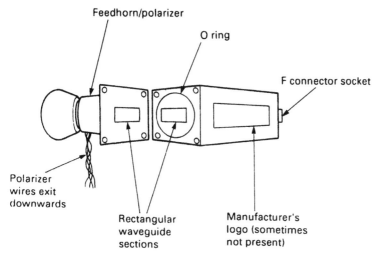

Figure 7.9 *Orientation of feedhorn and LNB assembly*

Assembling head units with ferrite polarizers

Magnetic ferrite polarizers are current-operated devices which usually operate on either ±35 mA current or 0 and 70 mA to twist polarity of the incoming waves to the orientation of the LNB probe. In the case of linear polarization, Figure 7.10 and 7.11 show how an LNB is orientated relative to polarity of incoming signal, depending on which of the above polarizer types is employed. The horizontal plane can be thought of as parallel to the ground. In practice I have found the ±35 mA variety is the more common.

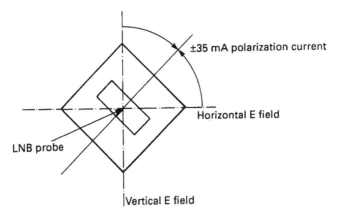

Figure 7.10   *Diamond orientation with ±35 mA magnetic polarizers*

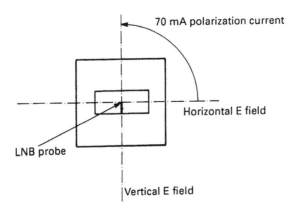

Figure 7.11   *Rectangular orientation with 0 and 70 mA polarizers*

8 Installation: antenna alignment and wiring

Instrumentation

Although it is possible to align an antenna by trial and error for DIY purposes, a considerable saving in time and energy can be gained by the use of specialist instruments. This chapter is a guide to the availability and use of such instruments along with the necessary wiring details for various systems.

Compass and inclinometer

A compass is an essential instrument used in both site survey work and antenna alignment. The cheap Christmas cracker variety is definitely out for professional work but may be used for one off DIY. A good sighting compass includes mirror, sighting lines and 360° bearing graduations and can be purchased from camping/sports shops. An inclinometer is essential for the measurement of elevation but some manufacturers stamp elevation scales on their mounting brackets for simple AZ/EL mounts. However, these are not too accurate unless the wall to which the dish is mounted is truly vertical. With offset focus dishes care must be exercised in the use of inclinometers since the offset angle needs to be taken into account. A temporary inclinometer, for DIY, can be made from a cheap protractor, piece of wood, nail and a plumb line.

Simple in-line peaking meters

Some form of signal monitoring is essential to optimize picture quality. Signal strength meters work by monitoring the first intermediate frequency (IF) signal level obtained after down-conversion in the low noise block (LNB). A simple, in-line, peak level meter of the type shown in Figure 8.1 is perhaps the cheapest available. These are simply connected in series with the LNB supply/coaxial lead using F connectors obtaining their power requirements from the LNB supply voltage. On some models there is the provision of an audio tone, pitch related to the signal strength.

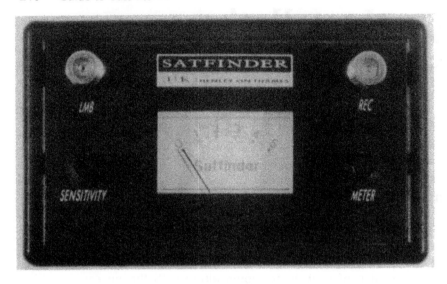

Figure 8.1   *An in-line signal strength meter (Satfinder (UK) Ltd)*

Advantages

1 Low cost, rugged and reliable.
2 Does not need internal batteries.

Disadvantages

1 The meter needs to be continually reset as full-scale deflection is reached and it does not give a relative reading of signal strength.
2 The display varies rather erratically with signal content and can confuse the installer.
3 Two temporary connections are necessary (wasteful of time).
4 Increased risk of damaging the satellite receiver by shorting the inner and outer conductors of the coaxial cable. LNB feed needs to be switched off while connections are made.
5 Alignment cannot proceed till the entire installation is completed.

Battery powered wideband signal strength meters

These instruments are an overall improvement on in-line types and are considerably more expensive. The LNB feed is supplied by internal batteries (optionally rechargable) and the display is smoother and more sophisticated. Switched attenuators are provided to set the sensitivity. The final attenuator position and the meter reading combine to give a relative indication of the received signal strength. A choice of analogue or digital displays are available depending on your personal preference. Figure 8.2 shows a digital display model, manufactured by Satfinder

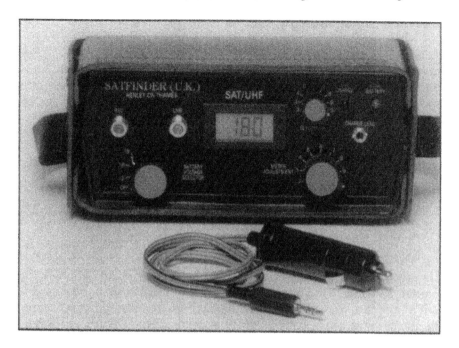

Figure 8.2 *A wideband signal strength meter (Satfinder (UK) Ltd)*

(UK), which has selectable LNB feed voltages of 12, 16 and 18 V to suit various polarizer types. The signal strength can also be monitored by audio tone.

Advantages

1 Internal batteries power up the LNB for immediate antenna alignment.
2 No risk of damaging the satellite receiver.
3 Wide-band, so easy to use for aligning a polar mount antenna. No need to selectively change frequency for different satellites in the geo-arc.
4 Relative signal strength figures can be obtained for comparison.
5 Auto-shutdown of battery power if the coaxial cable is shorted or if the unit is left switched on for a length of time.

Disadvantages

1 Batteries can go flat on the job!
2 Cannot select a particular channel (sometimes useful for satellite identification where another satellite is close by).

Selective signal strength meters

This type of instrument differs from the earlier types in that each down-converted channel from a particular satellite can be individually tuned and the signal strength measured to an absolute value. These portable instruments are expensive and elaborate and may incorporate the following features:

1 Selectable signal level measurement in dBμV.
2 Measurement of C/N ratio for each channel.
3 Direct channel or frequency input.
4 Composite video output.
5 Audio output and built-in loadspeaker allowing control of tone sub-carriers between 5 and 8 MHz.
6 Remote feed of LNB via internal rechargeable batteries.
7 Microcomputer system control.

An advantage of this type of meter is that of selectively identifying intended satellites. If you know the precise frequency or tuning spot of one of the required satellite's channels it may be used for identification purposes or at least, with experience, reduce the chances of peaking signals to an unintended adjacent satellite – an increasing risk now that the skies are becoming crowded. Another alleged advantage of this type is the peaking of polarization offset with low-cost linear polarization systems where the LNB is rotated in its holder to achieve minimum cross polarization and maximum signal strength. However, I have found this is fairly difficult to resolve in practice and a keen eye is needed. Although these instruments are ideally suited to single satellite installations they inherit a major disadvantage in that continual twiddling of tuning controls are needed for aligning polar mount antenna.

Advantages

1 Reduces the risk of aligning antenna to wrong satellite.
2 Multi-features useful for SMATV work.

Disadvantages

1 Not convenient for polar mount alignment.
2 Expensive.

Spectrum analysers

These instruments are considered the ultimate in portable signal monitoring and display the entire frequency spectrum of the satellite first IF band. They often incorporate other facilities such as a built-in TV reception circuitry to view the received carrier, and audio tones for 'point-by-ear' antenna alignment. The particular model shown in Figure 8.3 covers

Figure 8.3 *'ProPrint E' spectrum analyser (Applied Analogue Systems Ltd)*

the VHF and UHF bands as well as the FM satellite IF. Variable centre frequency and span controls allow the spectrum to be separated out if required. With this type of display the satellite channels can be seen as individual peaks.

Referring to Figure 8.4, the horizontal axis represents signal power density and the vertical axis frequency. This trace shows the complete satellite IF spectrum (0.9–2.0 GHz) and shows five active transponders.

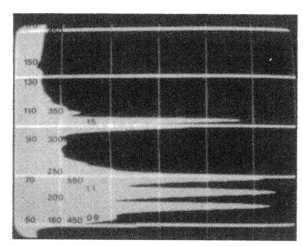

Figure 8.4 *Display of the satellite IF spectrum (Applied Analogue Systems Ltd)*

The noise floor of the LNB can be clearly seen at the second graticule (0.9–1.2 GHz), dropping to the first graticule (1.2–1.6 GHz).

Figure 8.5 shows the demodulated baseband sweep from 2.5 to 8.5 MHz. It shows a PAL colour signal at 4.43 MHz, the primary audio at 6.5 MHz, and four other audio sub-carriers at 7.02, 7.20, 7.38 and 7.56 MHz.

These battery powered units are particularly useful for optimizing cross-polar rejection since individual channel signal strength is the best clue to optimizing the skew setting of linear polarized signals. Spectrum analysers are particularly useful to SMATV and cable installers in tracking down sources of interference.

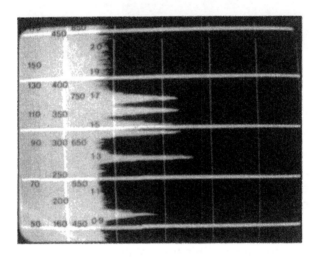

Figure 8.5   *Display of a demodulated carrier (Applied Analogue Systems Ltd)*

Portable test stations

A portable test station consisting of a portable television, 500 W mains-isolation transformer, a pre-tuned satellite receiver and an optional large-scale analogue meter connected to the AGC line is sometimes used for multi-satellite installations. Lugging all this lot around may indeed look impressive to the customer but not a lot of jobs are going to be completed in a day! In any case, modern small motorized systems are likely to be wall-mounted and it is doubtful if a portable television screen picture could be resolved by the human eye from the top of a ladder in bright sunlight. A practical solution using this method may be to use tower scaffolding and cart the whole lot up to the top! My advice is forget it – this method is really a relic of the past when signal strength meters were either expensive or difficult to obtain and high installation charges were the norm.

Feed centring

The first step in antenna alignment is to ensure the feed is accurately set to the focal point, otherwise the main lobe will not coincide with the look axis of the dish, not unlike a dipped headlight. This effect is particularly troublesome when aligning a polar mount, and can cause loss of efficiency. With small offset focus dishes this is not normally a problem since the boom is rigidly fixed. Very large dishes designed for S and C bands tend to be of the prime focus variety and it may be difficult to centre the feed in practice. A simple method to ensure this is done accurately is to tape two strings diagonally across the face of the dish. You can then adjust the feed, by sighting, to coincide with the intersection.

Outdoor cable connections (fixed satellite systems)

There appears to be no standardization with regard to cable requirements for various manufacturers' dishes. Some use single coax where the 1st IF signals are fed down the cable from the dish and a d.c. voltage is fed back up the cable from the receiver to supply the LNB. This d.c. voltage can be level varied to switch the solid state V/H switch type of polarizer. Other manufacturers use coax with an additional polarizer lead to alter the polarization sense of a magnetic polarizer. The LNB supply of 15 V is fed up the coax and an additional cable is used to feed current to the polarizer. This current is approximately 0 mA for vertical polarization and 40 to 80 mA for horizontal polarization. The current return is via the coax braid. The outdoor connection to the polarizer can be effected with a single section of connector strip or a 'Skotch lock' which on completion is overwrapped with self-amalgamating tape to prevent corrosion due to moisture ingress. Another often-used configuration uses coax with two separate polarizer leads. These magnetic polarizers use twin direction current flow, −40 mA and +40 mA approx, to select the polarization sense. Again two sections of common connector strip or a pair of Skotch locks can be used to provide the electrical connections. Remember to overwrap with self-amalgamating tape.

The coaxial cable/LNB connection is simply made with an F-connector. The twist-on variety are the simplest to fit. Leave sufficient cable to form a drop loop so that any water will drip from the lowest point, away from the connection itself. A weather-proof rubber boot or self-amalgamating tape should be used to finally seal the connection from the environment. Of the two, self-amalgamating tape is preferred since some makes of rubber boot can perish and crack with time.

Alignment of fixed AZ/EL antennas

Where time is not at a premium, it is possible to align a fixed mount antenna by trial and error with 'better or worse' feedback from a second person. Although perhaps adequate for a one-off DIY installation it is not to be recommended for trade practitioners. Alignment of a simple fixed antenna, known as a AZ/EL mount, is a fairly straightforward task and takes only a few minutes to perform with the right instruments. The necessary steps are given below:

1 Using Program 8.1, or by calculation using the equations provided in Chapter 5, obtain the necessary azimuth and elevation angle for the chosen satellite (or cluster) for your receive site. The azimuth angle should be corrected for magnetic variation (see Chapter 6) so giving a compass bearing. This will not vary much across your service area if small.

2 Tighten all the adjuster bolts to take up any slack but not so tight that the antenna cannot be moved with moderate effort.

3 Connect a stub of coaxial cable between the LNB and a battery powered signal strength meter. Alternatively use an in-line meter, but power down the receiver while making connections because an accidental short of the inner coaxial cable to the outer screen may damage it.

4 Power up the LNB; extra polarizer connections need not be connected in practice.

5 Set the elevation angle using an inclinometer or use the stamped graduations found on some antenna mounting brackets.

6 Set the meter attenuator switch, if fitted, to a low value until signals are detected.

7 Swing the antenna to the azimuth compass bearing. Signals may now be faintly detected. If not, slightly adjust the azimuth in either direction until signals are detected. If you still do not detect anything your elevation setting is incorrect.

8 Once a signal has been detected, fine adjust the azimuth and elevation adjusters while viewing the signal strength reading. Up the attenuator or sensitivity setting as appropriate until a maximum signal is obtained.

9 For linear polarized signals only, rotate the LNB in its holder for maximum signal strength consistent with minimum cross-polar interference. Although this angle may be calculated it requires a knowledge of the relative tilt of the electric vector at the spot beam centre relative to the sub-satellite point and the difference in longitude between the receive site and the satellite.

10 Grease all adjuster bolts to reduce corrosion.

11 Check reception (is it the intended satellite?) and waterproof all outdoor connections with self-amalgamating tape or purpose designed sealing compound. Avoid rubber boots, they do not last.

Sealing compounds

Never use sealing compounds which contain acetic acid as a curing agent to waterproof outdoor connections. It has been found this causes corrosion over a period of time, which eats into the internal electronic circuit boards of the LNB. If you are not sure about the chemical composition of the sealant then do not use it. It is better to use butylene self-amalgamating tape or a weatherproof boot bearing in mind that of the two, the former is recommended as rubber boots tend to perish and crack with time. However, now the problem has been identified, more compounds are being developed and manufactured for the purpose.

Outdoor cable connections (multi-satellite systems)

The cabling requirements of multi-satellite motorized systems are significantly more complex. The electrical connections are are follows:

Actuator motor power: Two heavy gauge wires for $+36$ V motor power.
Actuator position sensor: Three light gauge wires for $+5$ V, ground, and count pulses.
Polarizer: One, two, or three light gauge wires depending on type fitted. For motor driven polarizers, $+5$ V, ground and pulse. For magnetic polarizers one, or more commonly two, wires are needed.
LNB feed and signal out: Coaxial cable.

There seems to be no universal colour code for multi-satellite wiring so Figure 8.6 shows a suggested outdoor wiring scheme using cable type K1005 from Volex Radex. It is important to stick rigidly to the same colour coding for each installation otherwise confusion and mistakes may occur. Pay particular attention to the actuator motor power connections since mistakes here may cause considerable damage due to the relatively high voltage and current involved. The information contained in the particular manufacturer's installation manual should be read carefully before any wiring is attempted.

Any connections left exposed to the environment should be over-wrapped with self-amalgamating tape and rain drip loops formed to drain water away from them. The corresponding indoor connections to the positioner unit will be dealt with in Chapter 9.

The alignment of a modified polar mount

The alignment of a polar mount antenna is considerably more fiddly than simple AZ/EL mounts since accurate tracking of the whole visible

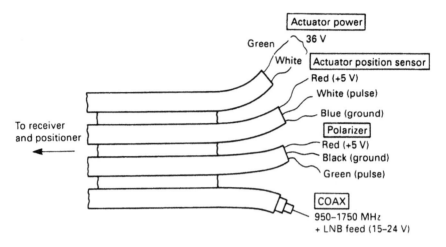

Figure 8.6 *Suggested wiring details for cable type K1005*

geo-arc is needed by one simple movement around the polar axis. Both polar mount, and the more popular derivative called the modified polar mount, geometry were discussed in detail in Chapter 2. Once set up, it can be either hand operated or motor driven by remote control. The point where the antenna's maximum or apex elevation is achieved should correspond to a true north/south line. At this central point the antenna is said to be in its apex position, and can be driven by an equal amount either eastward or westward thus accurately tracking the geo-arc. Table 8.1 shows the modified polar mount angles, over a wide range of latitudes, and are the elevation angles set when the antenna is in its apex position.

There are many different ways of tackling the alignment of a polar mount but one requirement that they all have in common is accurate setting of the dish north/south orientation in the apex position. Unless this is found very accurately the whole geo-arc tracking process becomes a failure. The apex setting can be thought of as the setting the dish would need for receiving signals from a hypothetical satellite due south of the receiving site, this corresponds to the highest point (apex) of the geo-arc. In general, the following rules are applied and should be remembered throughout the adjustment procedure.

1   Adjust the polar elevation angle to peak signals from satellites located at or near the geo-arc's apex. That is to say, the most southerly satellites.
2   Rotate the whole mount assembly around its mast or pillar when peaking signals from satellites far from the geo-arc's apex. That is to say trimming the north/south orientation.

One widely used alignment method which seems to work satisfactorily is described in the following section.

Table 8.1 *Modified polar mount settings*

(all angles are in degrees and apply to northern or southern latitudes)

Latitude	Polar axis	Polar elevation	Apex declination	Apex elevation	Declination offset
0.00	0.00	90.00	0.00	90.00	0.00
0.50	0.51	89.49	0.59	89.41	0.08
1.00	1.03	88.97	1.18	88.82	0.15
1.50	1.54	88.46	1.77	88.23	0.23
2.00	2.05	87.95	2.36	87.64	0.31
2.50	2.56	87.44	2.95	87.05	0.38
3.00	3.08	86.92	3.53	86.47	0.46
3.50	3.59	86.41	4.12	85.88	0.54
4.00	4.10	85.90	4.71	85.29	0.61
4.50	4.61	85.39	5.30	84.70	0.69
5.00	5.13	84.87	5.89	84.11	0.76
5.50	5.64	84.36	6.48	83.52	0.84
6.00	6.15	83.85	7.07	82.93	0.92
6.50	6.66	83.34	7.65	82.35	0.99
7.00	7.17	82.83	8.24	81.76	1.07
7.50	7.69	82.31	8.83	81.17	1.14
8.00	8.20	81.80	9.42	80.58	1.22
8.50	8.71	81.29	10.01	79.99	1.30
9.00	9.22	80.78	10.59	79.41	1.37
9.50	9.73	80.27	11.18	78.82	1.45
10.00	10.25	79.75	11.77	78.23	1.52
10.50	10.76	79.24	12.35	77.65	1.60
11.00	11.27	78.73	12.94	77.06	1.67
11.50	11.78	78.22	13.53	76.47	1.75
12.00	12.29	77.71	14.11	75.89	1.82
12.50	12.80	77.20	14.70	75.30	1.90
13.00	13.32	76.68	15.29	74.71	1.97
13.50	13.83	76.17	15.87	74.13	2.04
14.00	14.34	75.66	16.46	73.54	2.12
14.50	14.85	75.15	17.04	72.96	2.19
15.00	15.36	74.64	17.63	72.37	2.27
15.50	15.87	74.13	18.21	71.79	2.34
16.00	16.38	73.62	18.79	71.21	2.41
16.50	16.89	73.11	19.38	70.62	2.49
17.00	17.40	72.60	19.96	70.04	2.56
17.50	17.91	72.09	20.54	69.46	2.63
18.00	18.42	71.58	21.13	68.87	2.70
18.50	18.93	71.07	21.71	68.29	2.78
19.00	19.44	70.56	22.29	67.71	2.85
19.50	19.95	70.05	22.87	67.13	2.92
20.00	20.46	69.54	23.45	66.55	2.99
20.50	20.97	69.03	24.03	65.97	3.06
21.00	21.48	68.52	24.61	65.39	3.13
21.50	21.99	68.01	25.19	64.81	3.21
22.00	22.50	67.50	25.77	64.23	3.28
22.50	23.00	67.00	26.35	63.65	3.35
23.00	23.51	66.49	26.93	63.07	3.42
23.50	24.02	65.98	27.51	62.49	3.49

Table 8.1   *continued*

Latitude	Polar axis	Polar elevation	Apex declination	Apex elevation	Declination offset
24.00	24.53	65.47	28.08	61.92	3.55
24.50	25.04	64.96	28.66	61.34	3.62
25.00	25.54	64.46	29.24	60.76	3.69
25.50	26.05	63.95	29.81	60.19	3.76
26.00	26.56	63.44	30.39	59.61	3.83
26.50	27.07	62.93	30.96	59.04	3.90
27.00	27.57	62.43	31.54	58.46	3.96
27.50	28.08	61.92	32.11	57.89	4.03
28.00	28.59	61.41	32.69	57.31	4.10
28.50	29.09	60.91	33.26	56.74	4.17
29.00	29.60	60.40	33.83	56.17	4.23
29.50	30.11	59.89	34.40	55.60	4.30
30.00	30.61	59.39	34.97	55.03	4.36
30.50	31.12	58.88	35.55	54.45	4.43
31.00	31.62	58.38	36.12	53.88	4.49
31.50	32.13	57.87	36.69	53.31	4.56
32.00	32.63	57.37	37.25	52.75	4.62
32.50	33.14	56.86	37.82	52.18	4.68
33.00	33.64	56.36	38.39	51.61	4.75
33.50	34.15	55.85	38.96	51.04	4.81
34.00	34.65	55.35	39.52	50.48	4.87
34.50	35.16	54.84	40.09	49.91	4.94
35.00	35.66	54.34	40.66	49.34	5.00
35.50	36.16	53.84	41.22	48.78	5.06
36.00	36.67	53.33	41.79	48.21	5.12
36.50	37.17	52.83	42.35	47.65	5.18
37.00	37.67	52.33	42.91	47.09	5.24
37.50	38.17	51.83	43.47	46.53	5.30
38.00	38.68	51.32	44.04	45.96	5.36
38.50	39.18	50.82	44.60	45.40	5.42
39.00	39.68	50.32	45.16	44.84	5.48
39.50	40.18	49.82	45.72	44.28	5.53
40.00	40.69	49.31	46.28	43.72	5.59
40.50	41.19	48.81	46.83	43.17	5.65
41.00	41.69	48.31	47.39	42.61	5.70
41.50	42.19	47.81	47.95	42.05	5.76
42.00	42.69	47.31	48.51	41.49	5.82
42.50	43.19	46.81	49.06	40.94	5.87
43.00	43.69	46.31	49.62	40.38	5.93
43.50	44.19	45.81	50.17	39.83	5.98
44.00	44.69	45.31	50.73	39.27	6.03
44.50	45.19	44.81	51.28	38.72	6.09
45.00	45.69	44.31	51.83	38.17	6.14
45.50	46.19	43.81	52.38	37.62	6.19
46.00	46.69	43.31	52.93	37.07	6.24
46.50	47.19	42.81	53.48	36.52	6.30
47.00	47.69	42.31	54.03	35.97	6.35
47.50	48.18	41.82	54.58	35.42	6.40

Table 8.1 *continued*

Latitude	Polar axis	Polar elevation	Apex declination	Apex elevation	Declination offset
48.00	48.68	41.32	55.13	34.87	6.45
48.50	49.18	40.82	55.68	34.32	6.50
49.00	49.68	40.32	56.22	33.78	6.55
49.50	50.18	39.82	56.77	33.23	6.59
50.00	50.67	39.33	57.31	32.69	6.64
50.50	51.17	38.83	57.86	32.14	6.69
51.00	51.67	38.33	58.40	31.60	6.74
51.50	52.16	37.84	58.95	31.05	6.78
52.00	52.66	37.34	59.49	30.51	6.83
52.50	53.16	36.84	60.03	29.97	6.87
53.00	53.65	36.35	60.57	29.43	6.92
53.50	54.15	35.85	61.11	28.89	6.96
54.00	54.65	35.35	61.65	28.35	7.00
54.50	55.14	34.86	62.19	27.81	7.05
55.00	55.64	34.36	62.73	27.27	7.09
55.50	56.13	33.87	63.26	26.74	7.13
56.00	56.63	33.37	63.80	26.20	7.17
56.50	57.12	32.88	64.34	25.66	7.21
57.00	57.62	32.38	64.87	25.13	7.25
57.50	58.11	31.89	65.41	24.59	7.29
58.00	58.61	31.39	65.94	24.06	7.33
58.50	59.10	30.90	66.47	23.53	7.37
59.00	59.59	30.41	67.00	23.00	7.41
59.50	60.09	29.91	67.54	22.46	7.45
60.00	60.58	29.42	68.07	21.93	7.48
60.50	61.08	28.92	68.60	21.40	7.52
61.00	61.57	28.43	69.13	20.87	7.56
61.50	62.06	27.94	69.65	20.35	7.59
62.00	62.55	27.45	70.18	19.82	7.63
62.50	63.05	26.95	70.71	19.29	7.66
63.00	63.54	26.46	71.24	18.76	7.69
63.50	64.03	25.97	71.76	18.24	7.73
64.00	64.53	25.47	72.29	17.71	7.76
64.50	65.02	24.98	72.81	17.19	7.79
65.00	65.51	24.49	73.33	16.67	7.82
65.50	66.00	24.00	73.86	16.14	7.85
66.00	66.49	23.51	74.38	15.62	7.88
66.50	66.99	23.01	74.90	15.10	7.91
67.00	67.48	22.52	75.42	14.58	7.94
67.50	67.97	22.03	75.94	14.06	7.97
68.00	68.46	21.54	76.46	13.54	8.00
68.50	68.95	21.05	76.98	13.02	8.02
69.00	69.44	20.56	77.49	12.51	8.05
69.50	69.93	20.07	78.01	11.99	8.08
70.00	70.42	19.58	78.53	11.47	8.10
70.50	70.92	19.08	79.04	10.96	8.13
71.00	71.41	18.59	79.56	10.44	8.15
71.50	71.90	18.10	80.07	9.93	8.17

Table 8.1   *continued*

Latitude	Polar axis	Polar elevation	Apex declination	Apex elevation	Declination offset
72.00	72.39	17.61	80.58	9.42	8.20
72.50	72.88	17.12	81.10	8.90	8.22
73.00	73.37	16.63	81.61	8.39	8.24
73.50	73.86	16.14	82.12	7.88	8.26
74.00	74.35	15.65	82.63	7.37	8.28
74.50	74.84	15.16	83.14	6.86	8.30
75.00	75.33	14.67	83.65	6.35	8.32
75.50	75.82	14.18	84.16	5.84	8.34
76.00	76.31	13.69	84.66	5.34	8.36
76.50	76.80	13.20	85.17	4.83	8.37
77.00	77.29	12.71	85.68	4.32	8.39
77.50	77.78	12.22	86.18	3.82	8.41
78.00	78.27	11.73	86.69	3.31	8.42
78.50	78.75	11.25	87.19	2.81	8.44
79.00	79.24	10.76	87.69	2.31	8.45
79.50	79.73	10.27	88.20	1.80	8.46
80.00	80.22	9.78	88.70	1.30	8.48
80.50	80.71	9.29	89.20	0.80	8.49
81.0 0	81.20	8.80	89.70	0.30	8.50

Modified polar mount alignment

1   Using Table 8.1, Program 8.1, or using the equations given in Chapter 2, obtain the modified polar mount angles for the receive site latitude. The angle terminology used here can be seen by examining Figure 8.7. A simple angular relationship can be seen

Polar elevation = Apex elevation + Declination offset

and

Polar axis = Apex declination + Declination offset.

2   Again using Table 8.1 or otherwise, find the elevation angle of a convenient receivable satellite fairly low down in the geo-arc and far from its apex, say 25° or more to the east or west. Call this satellite SAT1.
3   Ensure the mounting pole or mast is vertical (most important) and set the antenna to its apex position facing as near due south as possible. If using a compass, remember to include the magnetic variation value.
4   Precisely set the polar elevation angle or the corresponding polar axis angle whichever is the most convenient to measure with an inclinometer. The end result is the same.

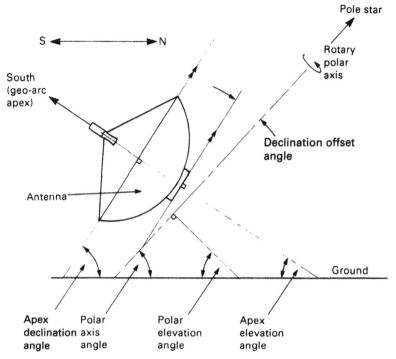

Figure 8.7   *Angle terminology used to describe polar mounts*

5   Precisely set the apex elevation angle or the corresponding apex declination angle whichever is the most convenient to measure with an inclinometer. The declination offset angle will automatically be correctly set using this method. If an antenna is an offset focus type subtract the offset angle from the apex elevation (or add the offset angle to the apex declination angle as convenient) to find the resultant 'offset' dish elevation. Typical offset angles are in the region of 22–27°. Any antenna manufacturer worth its salt will include the necessary offset angle in the antenna technical specifications sheet. Alternatively, they may provide a special corrective measurement point for elevation.

6   Connect up a signal-strength meter to the LNB. Using a non-tunable, wideband model is best since any transponder activity in the satellite IF will be detected. This alleviates the need for continual retuning to different transponders. An in-line type may be purchased so cheaply these days that it is feasible even for one-off DIY use.

7   Monitoring the resultant elevation angle (the actual elevation of the antenna), rotate the dish around the polar axis until the measured elevation matches the computed elevation angle for SAT1. Hold this position with the actuator arm. Slowly rotate the whole assembly around its mast or pillar until maximum signal strength is obtained from SAT1. Temporarily tighten the

clamp. This procedure effectively trims up the north–south orientation of the antenna by exploiting the station keeping accuracy of
SAT1. It may also be convenient to optimize the feed focal length
and centring at this stage.

8   By using the actuator, move the antenna back around the polar axis
until signals are detected from a satellite close to the geo-arc apex.
Call this SAT2. Very finely trim the polar elevation (or polar axis) for
maximum signal strength from SAT2. This should not be necessary
if your initial set-up angles were accurate and the mounting pole
plumb. Lightly pulling or pushing the lower rim of the dish and
noting whether the signal strength increases or decreases will help
in deciding if this is necessary.

9   Drive the antenna back to SAT1 and, if necessary, further trim for
maximum signal strength by slightly rotating the whole assembly
round its mast, thus trimming the north–south orientation as in 7.

10 Repeat steps 7 to 9 as often as required for consistently peaked
signals from both satellites. It may be necessary to temporarily
tighten adjuster bolts at each stage. If difficulty is experienced then
go back to 1 and check your initial set-up angles.

11 Check the received picture quality and tracking over a number of
satellites and when satisfied fully tighten and grease adjuster bolts.
Monitor signal strengths while doing this in case final tightening
pulls the alignment out slightly. Repeat steps 7 to 9 if this is the
case.

Universal angle finder program

Program 8.1 is a short 'no-frills' multi-purpose angle finder tool written
in BASIC. The program calculates azimuth and elevation angles for
any location and satellite combination in the world along with precisely
calculated modified polar mount settings. The program incorporates
nearly all the geometric equations presented in the book into an easy
to use package. By restricting the program to a common subset of
the BASIC language it should work on any computer with a BASIC
interpreter. This particular 'program' was written using Microsoft
QuickBASIC shipped with MS-DOS. For more elaborate software see
Appendix 2.

```
10 REM      ****************************************
20 REM      *       GLOBAL SATELLITE LOCATOR        *
30 REM      *               AND                     *
40 REM      *    MODIFIED POLAR MOUNT CALCULATOR     *
50 REM      *                                       *
60 REM      *        (UNIVERSAL BASIC SUBSET)        *
70 REM      *                                       *
80 REM      *            DJ STEPHENSON               *
100 REM     ****************************************
110 REM
120 PI = 3.141593
130 R = 6378000
140 H = 3.5765E+07
150 NG = 1E-12
160 DR = PI / 180
170 RD = 180 / PI
180 M = 6.608
190 CLS
200 PRINT "SATELLITE LOCATOR PROGRAM": PRINT
210 PRINT "-VE VALUES: LONGITUDES WEST,LATITUDES SOUTH,MAG.VARIATION WEST"
220 PRINT
230 INPUT "  ENTER LATITUDE OF SITE .......... ", LT
240 INPUT "  ENTER LONGITUDE OF SITE ......... ", LR
250 INPUT "  ENTER LONGITUDE OF SATELLITE .... ", LS
260 INPUT "  ENTER MAGNETIC VARIATION ........ ", MV
270 IF (LS < 0) THEN SE = 360 + LS
280 IF (LS >= 0) THEN SE = LS
290 IF (LR < 0) THEN RE = 360 + LR
300 IF (LR >= 0) THEN RE = LR
310 B = (RE - SE) * DR
320 M = 6.61
330 A = LT * DR
340 IF A = 0 THEN A = NG
350 IF (A = NG AND B = 0) THEN EL = 90: PD = 35786: GOTO 380
360 EL = ATN((M * COS(A) * COS(B) - 1) / (M * SQR(1 - COS(A) ^ 2 * COS(B) ^ 2))) * RD
370 IF (EL < 0) THEN PRINT "SATELLITE IS OVER HORIZON": END
380 PD = 35786 * SQR(1 + (.42 * (1 - COS(A) * COS(B))))
390 AZ = ATN(TAN(B) / SIN(A)) * RD
400 IF (LT >= 0) THEN AZ = AZ + 180
410 CP = AZ - MV
420 IF (CP >= 360) THEN CP = CP - 360
430 IF (CP < 0) THEN CP = CP + 360
440 LT = ABS(LT)
450 B = LT * DR
460 IF (B = 0) THEN B = NG
470 T = RD * ATN(SQR((R + H) ^ 2 - (R ^ 2 * COS(B) ^ 2)) / (R * SIN(B)))
480 D = ABS(90 - T)
490 C = T - (RD * ATN(((R + H) - (R * COS(B))) / (R * SIN(B))))
500 AD = D + C + LT
510 AE = 90 - AD
520 PA = LT + C
530 PE = 90 - PA
540 PRINT : PRINT "SATELLITE LOOK ANGLES (DEGREES)": PRINT
550 PRINT "  ELEVATION           = "; EL
560 PRINT "  TRUE AZIMUTH        = "; AZ
570 PRINT "  COMPASS BEARING     = "; CP
580 PRINT "  DISTANCE (Km)       = "; PD: PRINT
590 PRINT "MODIFIED POLAR MOUNT ANGLES (DEGREES)": PRINT
600 PRINT "  APEX ELEVATION      = "; AE
610 PRINT "  APEX DECLINATION    = "; AD
620 PRINT "  POLAR AXIS          = "; PA
630 PRINT "  POLAR ELEVATION     = "; PE
640 PRINT "  DECLINATION OFFSET  = "; D: END
```

Program 8.8 *Satellite finder and modified polar mount calculator*

9 Installation: indoor work

Introduction

Indoor work can be as time consuming as all the other work put together. This is particularly so if the satellite receiver output is to be fed to a distribution amplifier for multi-room reception along with normal UHF TV signals. This chapter is intended to be a guide to most aspects of indoor work and covers most situations likely to be encountered in the domestic environment, such as the interconnections and tuning arrangements to be found in both older and modern ancillary equipment. It is vital to leave all equipment switched off until final wiring is completed and checked.

Polarizer connections

For many single satellite packages using simple V/H switched types of polarizer, no separate wiring is necessary because the d.c. switching voltages are fed up the coaxial cable to the outdoor head unit from the receiver. Where magnetic polarizers are employed, one or two extra wires need to be connected to the satellite receiver (some designs use the 0 volt outer coax braid as the polarizer's earth return). A few receivers, increasingly rare, are equipped to control mechanical polarizers, and have +5 V, pulse and ground connections for that purpose. Refer to the manufacturer's manual for detailed connection data bearing in mind the polarizer type. With some models a suitable plug is provided for soldering onto the polarizer leads.

Actuator connections (multi-satellite systems)

Actuator requirements consist of two motor power cables (36 V) and three-way transducer leads (+5 V, pulse, and ground). These are connected either to the receiver or positioner unit (some receivers have built-in antenna positioner circuitry). It is vital that no mistakes are made with this wiring; in particular check that the motor power and position sensor connections are not crossed or permanent damage may result. Figure 9.1 shows the general wiring arrangement for multi-satellite systems, but it is essential to carefully read the manufacturer's installation instructions and recheck your work before switching on and testing.

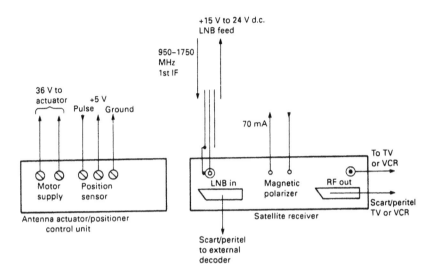

Figure 9.1 *Guide to multi-satellite receiver/positioner connections*

Signal interconnections

Before making any connections, ensure equipment is switched off. D.C. voltages to supply the head unit may be present even when the satellite receiver is in standby mode. Any shorts in the coaxial wiring can result in blown fuses or the destruction of special fusible safety resistors. The latter must be replaced with an identical type and not with with standard resistors, to comply with current safety regulations. In my experience these are unlikely to be stocked by most dealers so extreme care must be exercised. It is wise to check for possible shorts across the coaxial cable before powering up.

Prepare the coaxial cable and fit the connector (usually an F connector) to the satellite antenna input socket. This is usually labelled 'LNB IN' or marked with a graphic dish aerial symbol. There are many possible wiring configurations used with satellite receivers, TV, and video recorders and the following is a representative sample.

RF modulator configuration (simplest method)

This method is the most basic and versatile wiring arrangement and can be used with old or modern TV sets. The RF output signal from the satellite receiver (Sat Rx) is similar to that of a transmitted terrestrial channel and a spare channel on the TV must be tuned accordingly to process the signal. In the UK, the output RF modulators are normally preset to channel 38 or 39 in the UHF band. To assist the tuning of ancillary equipment a test signal, internally generated in the satellite receiver, is usually provided. Figure 9.2 shows the basic wiring arrangement. Most satellite receivers have a built-in 'loop-through' facility which provides for the connection of an additional terrestrial UHF aerial. Where this is not present it will be necessary to use a combiner as shown in Figure 9.3.

Connecting up

1   Plug the terrestrial aerial into the appropriately labelled input socket (where fitted).
2   Connect a coaxial flylead cable (usually supplied) between the satellite receiver RF signal output socket (labelled TV/CR or RF out) and the television receiver aerial socket.
3   A spare channel selector will need to be tuned to the satellite RF output (see later).

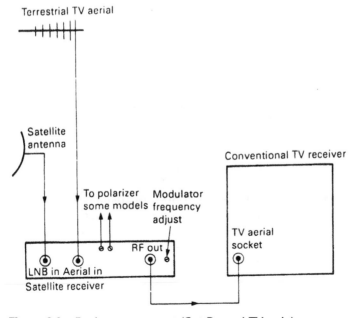

Figure 9.2 *Basic arrangement (Sat Rx and TV only)*

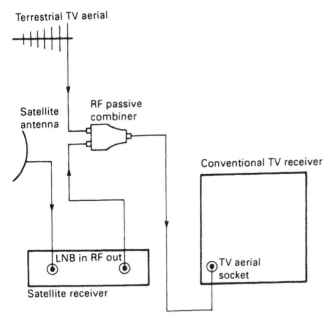

Figure 9.3 *Using a combiner where no loop-through facility exists*

Including a video recorder

Figure 9.4 shows the basic RF wiring arrangement with a video cassette recorder (VCR) in circuit. It is possible to record a selected satellite channel while watching a terrestrial channel, or vice versa, although recording a satellite channel whilst watching another satellite channel is not possible at the present time with any wiring interconnection. The RF output of a VCR is normally tuned to channel 36 in the UK.

Connecting up

1 Connect the terrestrial aerial into the appropriately labelled input socket.
2 Connect the RF output socket of the satellite receiver to the RF input socket of the VCR using a flylead.
3 Connect the RF output socket of the VCR to the aerial socket of the TV.
4 Switch on TV and VCR.
5 Tune a spare channel of the TV to the satellite receiver RC output.
6 Select the VCR playback channel on the TV and tune a spare VCR channel to the satellite receiver RF output. If patterning is experienced the chances are that there is a clash between the satellite

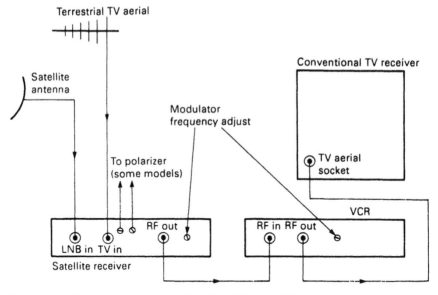

Terrestrial TV aerial

Satellite
antenna

Conventional TV receiver

Modulator
frequency adjust

To polarizer
(some models)

TV aerial
socket

VCR

RF out

RF in RF out

LNB in TV in

Satellite receiver

Figure 9.4 *Basic arrangement (Sat Rx, TV and VCR)*

receiver and VCR RF modulator frequencies. In such a case see the RF modulator adjustment section at the end of this chapter.

SCART/PERITEL alternatives

Alternative wiring arrangements using 21 pin SCART/PERITEL television connectors can provide slightly improved picture and sound quality with modern equipment by cutting out the satellite receiver RF modulator section to either the VCR or TV. Additionally, stereo sound can be utilized with suitably equipped satellite receivers and TV sets. Figure 9.5 shows the connection to a TV set which also alleviates the need for TV tuning. Satellite signals are automatically present when the auxiliary or AV channel is selected on the TV. Terrestrial TV viewing is effected by normal TV channel selection.

Figure 9.6 shows one method of interconnecting a satellite receiver, VCR and TV using SCART/PERITEL connectors. This arrangement allows the recording of terrestrial TV whilst watching satellite TV but disallows the recording of satellite TV whilst watching terrestrial TV. Direct audio and video signals from the satellite receiver are available when the auxiliary or AV input is selected on the TV.

The foregoing limitation can be overcome by the loop-through arrangement shown in Figure 9.7. A spare VCR channel is tuned to the satellite receiver RF output for the recording of satellite programmes.

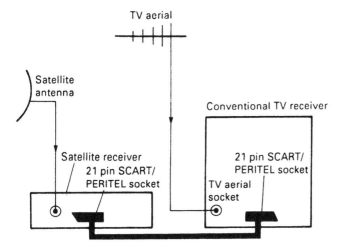

Figure 9.5 *SCART/PERITEL connections to a TV set*

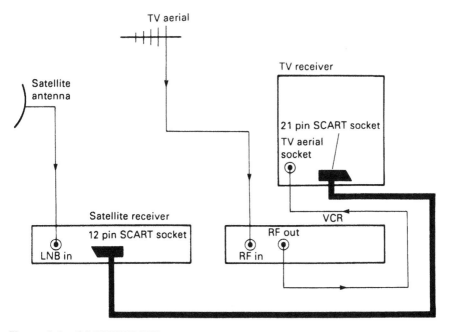

Figure 9.6 *SCART/PERITEL connections (terrestrial TV record only)*

Figure 9.8 shows another method of interconnecting a satellite receiver, VCR and TV using SCART/PERITEL connectors. This configuration allows the recording of satellite TV whilst watching terrestrial TV but disallows the recording of terrestrial TV whilst watching satellite TV. The VCR must be switched to auxiliary input or AV for recording or viewing the selected satellite channel. When viewing or monitoring satellite programmes, the VCR playback channel on the TV is also selected.

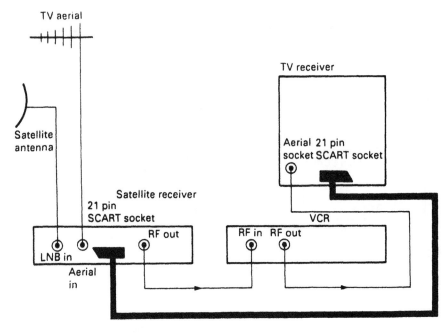

Figure 9.7   *SCART/PERITEL connections (terrestrial or satellite recording)*

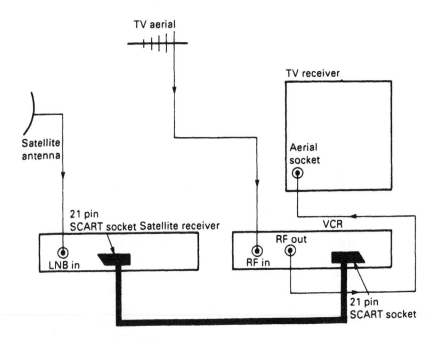

Figure 9.8   *SCART/PERITEL connections (satellite TV record only)*

The configuration shown in Figure 9.9 overcomes the previous limitation by using the RF loop-through facility to allow recording of terrestrial programmes. The VCR is switched to auxiliary input or AV only for recording satellite programmes. A spare TV channel needs tuning for the satellite receiver RF output for viewing while a terrestrial channel is being recorded. This method needs both the TV and VCR to be tuned to the satellite receiver RF output for maximum flexibility.

Other direct audio and video connections

Many TVs and VCRs, particularly older models, do not have SCART/ PERITEL sockets. Instead they use a combination of other sockets to provide input and output of direct audio and video signals. Video signal sockets can be BNC, Phono, PL259, of 6 pin DIN. Audio signal sockets can be Phono, 5, 6 or 7 pin DIN, and in a small number of cases a 3.5 mm jack socket. These may be specially wired to interface with a SCART socket where needed. The connection data for SCART sockets was given in Chapter 4 along with the subminiature D type socket. Figure 9.10 shows the pin numbering and connection data for the various DIN sockets likely to be encountered. The outputs from satellite receivers are all mains isolated so it is permissible to directly connect the appropriate audio outputs to the audio input of a hi-fi sys-

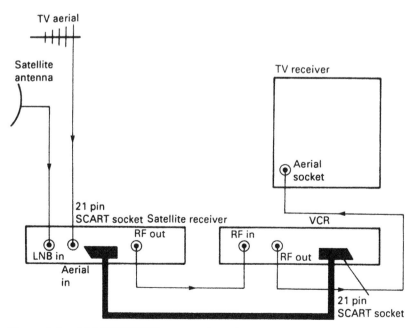

Figure 9.9 *SCART/PERITEL connections (satellite or terrestrial recording)*

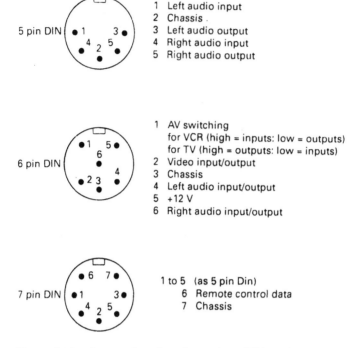

5 pin DIN

1 Left audio input
2 Chassis .
3 Left audio output
4 Right audio input
5 Right audio output

6 pin DIN

1 AV switching
   for VCR (high = inputs: low = outputs)
   for TV (high = outputs: low = inputs)
2 Video input/output
3 Chassis
4 Left audio input/output
5 +12 V
6 Right audio input/output

7 pin DIN

1 to 5 (as 5 pin Din)
     6 Remote control data
     7 Chassis

Figure 9.10   *Connection data for various DIN sockets*

tem. Where no direct audio or video input sockets are fitted to a television receiver, never make any direct connections to the internal circuitry. This can be very dangerous and may damage both television and satellite receiver as many television chassis, particularly older models, are at half-mains potential relative to ground or earth. Manufacturers only fit video and audio sockets on mains-isolated TV sets which either have a large internal isolation transformer or a mains-isolated switched-mode power supply.

Distribution amplifiers

UHF distribution amplifiers already exist in the lofts of many customers' houses for multi-outlet terrestrial channel viewing so it is not unreasonable to expect that the customer will want the selected channel on his/her satellite receiver to be sent to all TVs and VCRs in the residence. An extra coax cable (normal UHF quality) needs to be run up from the RF output of the satellite receiver to the input of the distribution amplifier. In addition all the TVs and VCRs will also need to be tuned to the satellite receiver RF modulator output. In view of all the extra work involved, it is wise to enquire about such a possibility at the site survey stage so an extra charge may be quoted. Much swearing and cursing often results when the dreaded words 'it's a distribution amp job' are interchanged by an unprepared installation team arriving at five o'clock in the afternoon. The prospect of crawling about in a grubby attic with 25 grams of loft insulation down your trousers is not one of the trade's most pleasant tasks.

The satellite receiver RF output and the terrestrial TV aerial are both connected to a standard UHF 2-way combiner/splitter which is then plugged into the aerial input of the distribution amplifier. Figure 9.11 shows the basic arrangement and from this realization of the extra work involved will become apparent. A complication may arise, particularly in bad terrestrial TV reception areas, where a mast head amplifier is connected to the UHF aerial. Its 12 V power requirements are supplied via the coaxial cable from the distribution amplifier itself. If a passive combiner is used the supply voltage to it may be substantially reduced and thus the gain will dramatically fall. A simple way out of this problem is to power up the mast head amplifier from a separate 12 V supply unit (obtainable from TV aerial suppliers) and use the alternative configuration shown in Figure 9.12.

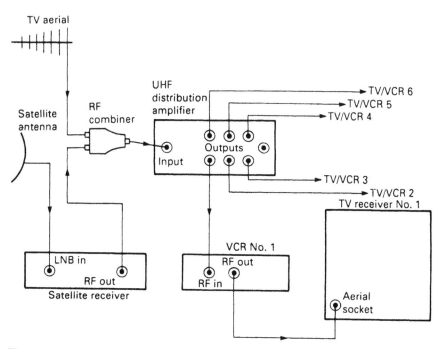

Figure 9.11 *UHF distribution amplifier with satellite channel*

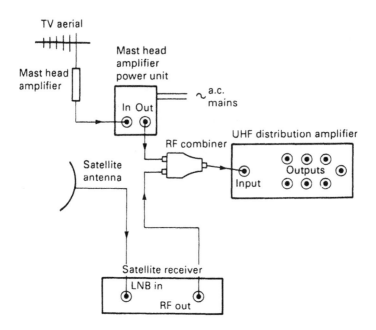

Figure 9.12   *Possible configuration with a mast head amplifier*

Remote control extensions

A common customer requirement is to be able to control a lounge-based satellite receiver from a remote location such as a bedroom. The RF out signal from the satellite receiver can be split to the two locations as previously described, but what about being able to control the equipment from both locations? The solution is to use remote control extension equipment. One such example is the Xtralink remote control extender. A small box is placed near the TV set in the bedroom and the infra-red (IR) remote control signals, from the normal satellite receiver handset, are converted to electrical pulses which are passed along the coax RF feed cable. A special remote control repeater unit placed in the lounge reconverts electrical signals back to IR and thus completes the control chain. A tiny infra-red light emitting diode (LED) must be placed near the satellite TV receiver's IR window. In high ambient lighting conditions the placing of the repeating IR LED may be critical. Figure 9.13 shows the overall arrangement and equipment is relatively inexpensive.

Tuning systems adopted with TVs and VCRs

The actual sequence of button pushing for tuning every make and model of TV and VCR that has ever been sold would necessitate the

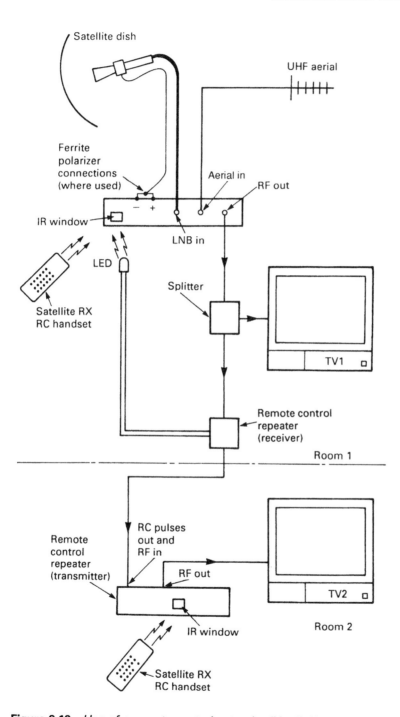

Figure 9.13 *Use of a remote control extender (Xtralink)*

space of a few volumes. However, all the various tuning sequences can be broadly grouped into three basic types. A VHF/UHF tuner unit in a typical TV or VCR requires a specific and very stable voltage to be stored whose level corresponds uniquely to the channel tuned. In other words all modern TV and VCR tuner units employ voltage controlled tuning. The object of all currently encountered tuning systems is to store and recall on demand various tuning voltages. TV and VCR tuners incorporate variable capacitance diodes (varicaps) as part of their tuned circuits and as their reverse bias is changed the capacitance and hence the tuned frequency alters.

Preset potentiometer tuning

This is the simplest method of tuning and can be one of the easiest to use. Preset potentiometers are adjusted until the required channel is tuned. The low and high limits of the frequency range correspond to the endstops of the potentiometers. In effect this method memorizes the tuning voltage by the position of the wiper on the resistive track. The disadvantages of this system is that moisture, dust and poor wiper contact can lead to instability and poor reproducibility of tuning voltages. Continual trimming is needed as wear progresses and reliability can be poor. This type of tuning arrangement is found on relatively old equipment but occasionally it is still adopted on budget-priced equipment. The number of potentiometers fitted varies but is usually in the range six to sixteen. Switch the AFC button (where fitted) to the OFF position while tuning and switch the AFC switch to ON when completed.

Electronic scan tuning

Electronic scan tuning is effected by storing and recalling the digital representation of a specific analogue tuning voltage level. For each numbered programme channel in turn, the tuning range is 'swept' and stops at the first channel found. Either the opportunity to store the digital representation of the tuning voltage level can be taken or the appropriate tune button pressed to restart the scan from where it left off. A variety of different methods have evolved and range from early dedicated hardware systems to modern microcomputer types which have inbuilt tuning algorithms. Some versions have an additional fine-tune option. Electronic tuning is far more reliable than the potentiometer method since no moving parts are needed and the number of available channel stores can be increased. Earlier models allowed scanning in one direction only, with a wraparound facility but modern versions allow scanning in both directions which is an advantage if you overshoot the required channel. The disadvantages of this type of tuning is the tediously long time it takes to set up initially. The current sweep position within the tuning band is often not displayed, particularly on VCRs. TV tuning is sometimes helped by the 'on screen' appearance of a horizontal bar whose instantaneous length is a measure of the sweep position within the tuning band.

Direct channel entry or frequency synthesis

The frequency synthesis system allows for direct key entry of the required channel. For example if UHF channel 38 is to be tuned, all that is needed is to open the memory, punch in the keys 38 and press the memory button within a preset time limit. This tuning method is fast to set up but a knowledge of locally received channel numbers is needed. A few TVs and VCRs of this type also have additional scan up and scan down buttons similar to the above system for fine tuning. For equipment without a numerical key pad (or remote control), channel entry is usually performed by a channel 10s button and a channel 1s button to input the tuning information, and a dual seven segment display mirrors the information keyed in. More modern equipment has a comprehensive 'on screen' display to assist tuning and other preset conditions. The procedure is repeated for each programme channel as required.

RF modulator adjustment

On occasions, particularly with RF interconnections, interference may occur due to the settings of the RF modulators on the satellite receiver and a connected VCR. This effect may be further compounded by local terrestrial TV transmitters operating in the same narrow frequency band. This can be overcome by fine tuning the modulator frequency presets which are usually located on the back panel of the equipment. The adjustments are made to minimize interference patterning between any of the contributing signals. It will be necessary to retune the TV and/ or VCR after each adjustment to check the result. By trial and error, a combination of settings will be quickly arrived at. However, in some service areas this may be particularly difficult and it may be necessary to resort to direct AV connection between two items of equipment to exclude a modulator.

If the customer has a VCR and two separate satellite receivers all using RF modulators, it can be very difficult to obtain interference-free use of all three, particularly if terrestial TV signals are also received in the same limited frequency slot. Many cheap RF modulators are only adjustable between UHF channel 30 and 40 (UK) and may be poor in quality. In difficult cases it may be necessary to resort to use of the same channel for each satellite receiver so that either receiver A or receiver B can be used at the same time. That is to say no recording from one receiver may be permitted while the other is active, the unused receiver set to standby mode.

In the UK a new terrestial TV station, Channel 5, is threatening to use channel 36 in the UHF band as a national transmission frequency. This is likely to complicate the situation even further and interfere with many VCR and satellite TV receiver modulators set at channel 36 and channel 38 respectively. A considerable amount of work will be needed, resetting modulator frequencies to counteract this potential interference problem. Some manufacturers have already identified this problem and provide modulators with a wider range of frequency adjustment. As more satellite tuners are incorporated into television receivers such complications will be eliminated.

Tuning satellite receivers

Most satellite receivers come factory pretuned, however in some cases it may be necessary to trim the tuning slightly. Most satellite receivers employ three buttons for fine tuning either on the receiver or via the remote control handset. These are typically called scan up, scan down and store. The optimal tuning point for each channel is found with scan up and scan down buttons by locating the point midway between on screen appearance of sparklies and 'darklies' (white or black noise spots). With poor reception and equal on screen distribution of sparklies and darklies is the optimal tuning point. Pressing the store, sometimes called memory button memorizes the new setting. Another possibility giving rise to sparklies with some models equipped with ferrite or motorized polarizers, is that the programmable polarization offset or skew settings are not set correctly for each channel. Again adjust for minimum sparklies and store the setting in memory.

AFC offset adjustment

With fixed satellite dish systems dedicated to say, Astra, the whole downconverted block of first IF channels can vary in frequency within a small tolerance from LNB to LNB. Normally this does not cause much trouble, but sparklies or darklies may appear due to a slight frequency shift mismatching of the first IF tuner in the satellite receiver and the incoming block of first IF channel frequencies. This may only be noticeable when receiving signals from weaker transponders and is often encountered after changing a faulty LNB. The adjustment to correct this effect is known as AFC offset. On some models it can be adjusted from the remote control handset after first entering the set up code, on others adjustment is an internal preset resistor located within the satellite receiver case. If trouble like this is encountered it is worth checking this adjustment rather than condemning a perfectly good LNB. Early Amstrad and Ferguson Astra systems employing the Marconi Star LNB often suffered this problem. The Amstrad's AFC offset can be adjusted internally and the Ferguson's by punching in code 258 and tuning in the normal way from the handset.

10 Repair of satellite equipment

Introduction

Sooner or later most domestic electronic equipment breaks down and the initial supplier or a technician is expected to perform a quick and reliable repair. Larger retail outlets are normally not interested in repairs once the equipment sold is out of guarantee. This is put into practice by pricing themselves out of the market with high call-out charges that customers are unwilling to pay. It seems that they have neither the resources nor the manpower to effect a fast and efficient repair to all the equipment they have sold. Fortunately, this leaves a large pool of repair work available to independent dealers or self-employed technicians who have much lower operating costs and can respond more quickly. However, in many cases, this may be difficult since the relevant service information is often unavailable for some time after the release of new equipment.

This chapter is intended to be a guide to the servicing of satellite receiving equipment and the treatment will be biased towards the vast majority of equipment in service in the domestic environment – namely, the fixed antenna type which targets one or a group of satellites in a single orbital slot. Practically, the only difference with multi-satellite motorized types, is the restriction to magnetic or mechanical polarizers in conjunction with additional antenna positioning circuitry. Video and audio circuitry also tends to be more complicated to comply with the variety of signal formats used in current satellite transmissions.

The repair of satellite equipment is fairly straightforward and in a lot of ways is considerably easier than TV or VCR repairs. The familiar set top receiver system can be broken down into five main areas or sections.

1 *The mechanical area* – The mounting itself, the antenna, the head unit support and the feedhorn assembly.
2 *The polarizer circuit* – The polarizer, wiring and polarization control circuitry.
3 *The actuator circuit* – The outdoor actuator unit, wiring and the associated control circuitry in the positioner or receiver. This only applies to motorized polar mount antennas.

4 *The signal path area* – The low noise block, tuner/demodulator, video and audio processing circuitry. This would include any external decoders that are fitted.
5 *The power supply area* – This is the main area where a fault is likely to be encountered.

Most service calls can be prevented, particularly in the initial stages of customer ownership, over the telephone. They invariably involve 'finger trouble' with the customer controls, leads accidentally pulled out while cleaning behind the indoor unit or 'little Johnny' twiddling the various tuning controls. However, if the customer is local, it is often quicker to call and check the system out yourself since it is virtually impossible to obtain any detailed information from some customers. Never listen to claims that 'no-one has touched it' or 'it just went off' and always check the customer control settings in a logical manner before embarking on more involved diagnosis.

Once a system is installed the customer is usually pleased or even surprised with the results initially, but after a week or two has passed, they compare the picture quality with that of terrestrial TV and become more critical. The odd sparklie or minor patterning gives rise to a moan or two and much time can be fruitlessly wasted checking the alignment of the antenna. Floods of service calls come in after a heavy rain storm due to sparklies appearing but, of course, there is nothing that can be done about it. It is important to warn customers about this possibility at the installation or sales stage so that such calls may be avoided or considerably reduced.

Fault symptoms

On a service call the first thing to do is to interrogate the customer as to the exact problem he or she is experiencing. The following is a guide to the sort of questions to be asked.

1 Is the fault always present or intermittent?
2 What are the fault symptoms and when did they first appear?
3 Did the fault get progressively worse with time?
4 Has any other technician called recently to repair or attempt to repair the equipment? (Man-made faults are the worst to find!)
5 Has there been a recent thunderstorm?
6 Can all intended satellites be received well (polar mounts only)?
7 Is picture quality poorer on some channels than others?

Once the fault symptoms are established the area or subsystem where the fault lies can be quickly isolated by a few quick and easy measurements.

Test equipment

For on-site servicing, the following test equipment is all that is necessary to locate the vast majority of faults. The last two items can be considered optional due to their expense.

Signal strength meter

A lot of information about a fault can be obtained from a signal strength meter provided it has some form of calibration. This is useful for detecting head unit faults or antenna alignment problems.

Multimeter

A multimeter is one of the electronic technicians basic test instruments. It enables voltage, current and resistance to be accurately measured or checked. The battle still rages on between the exponents of analogue versus digital displays but, for this work, the increased ruggedness of a hand-held digital multimeter is by far the major consideration. The more expensive varieties come with a special impact-resistant holster which, it is alleged, will protect the enclosed instrument from damage when dropped from the top of a telegraph pole onto concrete. The viewing angle of a digital display is also less critical and consequently the instrument is potentially safer to use when working up ladders. The type with an audible continuity tester can be particularly useful when checking for open circuit cables.

Spectrum analyser

If expense is no object, a spectrum analyser is a good diagnostic tool. Depending on the polarization selected on the receiver, the entire range of co-polarized channels in a block may be displayed at the same time and any source of interference or unwanted attenuation can be detected. Faulty LNBs and polarizer faults can be quickly identified. The instrument is also ideal for programming optimal skew with frequency settings for each channel (where the facility exists). The basic operation of this instrument was described in Chapter 8.

Oscilloscope

An oscilloscope is useful for the diagnosis of video or audio signal path faults and the measurement of position sensor pulses from an actuator unit. A mains operated unit should be accompanied by the use of a portable isolation transformer. However, a number of good specification oscilloscopes are available with internal battery power and these are ideal for on-site servicing. As we will see waveform checks at certain key areas in the baseband path can quickly isolate a faulty stage. The standard oscilloscope used for general electronic servicing is a 20 MHz dual beam type.

Customer controls and indoor connections

The most common problem encountered will be tuning problems associated with the satellite receiver itself or ancillary tuning of the TV or VCR. If suspected perform the following checks:

1 Check that both the TV and VCR input stages are tuned to the RF output of the satellite receiver; check that the TV is tuned to the VCR RF modulator output.
2 Once (1) is checked then attention can be paid to the settings of the satellite tuner itself.
3 With models incorporating magnetic or mechanical polarizers with programmable skew setting, check that each channel is set to its optimum polarization.
4 With V/H switched types of polarizer check that the correct polarization sense is programmed for each required channel.
5 Check the connections at the rear of the equipment. Are all the cables present and connected correctly?

Visual check of the outdoor unit

A quick visual inspection of the outdoor unit is often a first step if you have good reason to suspect it. This would be when, say, sparklies were corrupting the picture quality or when FM noise only appears on the screen. In the case of multi-satellite motorized systems the actuator might be inoperative if only one satellite can be received. The main points to check are as follows:

1 Check that the antenna is pointing in the right general direction. In the absence of a neighbour's antenna as a guide, use a sighting compass.
2 Check for signs of vandalism. It is not uncommon for local 'drunks' or 'yobs' to alter antenna pointing or pull at cables 'for a laugh'. On occasions satellite antenna may be used for target shooting practice with air guns, catapults or even more exotic weapons. Another possibility is that a drawing pin or the like has been pushed into the cable thus creating a short.

3   Check cable condition and connections outdoors for corrosion, water ingress, poor contact or shorts. Check also whether the cable has been chafing on a corner stone, roof tile or guttering.
4   Check the tightness of the antenna mounting bolts. Does the antenna move in any direction with minimal effort?
5   Has the position of the head unit shifted in either focal length or rotation?
6   Has the ground been disturbed in the area of buried cables? If so a cable might be severed or damaged.
7   Check that the feedhorn cover or cap is not punctured, distorted or missing. If damaged or missing check that signs of insects or moisture are eradicated before refitting a new cap.
8   Check that water is not getting in between the connection of the feedhorn flange and the LNB. This often occurs if the bolts are not tight and the 'O' ring is not sufficiently compressed. If this is the case dry out with a hairdryer and reseal. This is a common cause of 'sparklies' with some systems.

Checking the subsystems

If all the above checks do not bear fruit then you have a genuine fault which must be tackled in a logical and systematic way. The first job is to isolate the subsystem which is causing the trouble. The following technique is as good as any and can usually be performed without removing the cover of the satellite receiver.

1   Check the output of the LNB with a signal strength meter; if a good reading is obtained then the LNB is probably satisfactory but it may still be low gain unless you have an absolute standard by which to assess it. This facility is normally only present on the more expensive signal strength meters. Another signal strength monitoring method is to measure the AGC line with a multimeter. On some models a convenient measuring point is brought out to the rear panel and this can also be used for antenna alignment. If in doubt about the gain of an LNB it is a fairly quick matter to substitute another without disturbing the antenna alignment.
2   If no output is obtained then check that the LNB supply voltage, typically in the range 12 to 24 V, is present both at the receiver LNB input socket and at the LNB end. When installing equipment this can easily be lost by carelessly shorting the coax outer braid to the central conductor while the equipment is switched on. If this voltage is missing then the likely fault is that an internal fuse has blown or a safety resistor has burned out. These resistors can easily be identified, since they are normally stood well off from the circuit board. Do not use standard resistors for replacement, use only the approved type recommended by the manufacturers or

the equipment will not conform to the recognized safety standards. It is also important that the correct type of fuse is fitted; these can be either quick blow types identified by an F prefix to the value, or time-lag (anti-surge) prefixed by the letter T. Remember that the technician concerned will be held legally responsible for any incidents occurring from careless servicing.

3 Check the polarization control circuitry. If signals are either missing or weak, or signals of only one polarization sense can be resolved, then it is possible that a fault in the polarization section is responsible. For V/H switched types, such as the Marconi unit fitted to many Astra packages, check that the voltage level shift of the LNB feed voltage, which switches the polarization, is 13 V for vertically polarized channels and 17 V for horizontally polarized channels. If these voltage level shifts occur and only one polarization sense can be received then suspect the head unit. With magnetic polarizers, which are current driven devices, insert a multimeter switched to the 200 mA range and check the current through the magnetic polarizer. Some designs select zero current for vertical polarization and between 40 mA and 80 mA for horizontal polarization. Others use a plus or minus current method to alter the polarization sense and often require −40 mA or +40 mA to select either vertical or horizontal polarization. Presence of the correct currents is normally sufficient to establish whether the subsystem is faulty. If a fault is detected then the most likely problem will be in the wiring. Finally, with the increasingly rare mechanical type of polarizer it is necessary to check that the 5 V supply is present, and check that control pulses are being sent to the servo motor. If the skew control still has little or no effect then suspect either that the probe has been bent or distorted or that the control circuitry is faulty. The chip normally used to provide the pulse output is derived from an inexpensive 555 timer chip.

4 Check the actuator circuit by observing that the antenna moves while the actuator is operated. If not check that the 36 V motor supply is reaching the actuator. Two people will be necessary for this task, one to measure the voltage at the actuator motor and the other to activate the positioner unit. Check that the position sensor is wired up correctly at both the indoor and outdoor connection points. Multi-satellite system cables are colour coded to aid this task. Incorrect wiring can burn out key components in the position sensor and this is often apparent by visual inspection. Persistent blowing of the actuator motor fuse indicates that either the motor power cable is short-circuited or more likely that the motor windings have shorted turns, or that the assembly is seized up. Also check that water is not getting into the actuator assembly and causing problems.

5 Check for continuity of cables. This can be performed by utilizing another length of cable. For example, to check the continuity between the ends of a long length of coax cable all that is needed is to bridge the inner conductor to the coax braid at one end with a

short lead and two crocodile clips. Measuring the resistance between the central conductor and the braid at the other end will determine if either conductor is open circuit. A good cable should read nearly zero ohms. A similar method can be used for other long runs of cable by 'borrowing' another length of cable used for another purpose. In this way, two lengths of cable can be checked at the same time. A shorted coaxial cable can be detected by checking the resistance between the central conductor and the outer braid (without the crocodile clip lead attached). If zero ohms is measured then a short circuit cable or connector is the fault.

Power supply faults

If the faulty section has still not been located by the above checks, or supply voltages/currents are found to be missing, the fault is most likely to be in the power supply of the receiver. In general, this area is the most unreliable section of most electronic equipment and with satellite receivers, this is no exception. This section of the receiver generates the supply voltages for its internal circuitry and also external feeds to the LNB, polarizer and actuator (where fitted). The first items to check are fuses and fusible safety resistors; these can go faulty at the drop of a hat and their failure does not necessarily indicate that a fault is present. Refitting a new one and switching on will determine if a fault is present. If it does not, the probability is that a transient pulse from the mains supply has caused it to blow. It is also not uncommon for fuses to fracture due to mechanical mishandling or faulty manufacture. In these cases a visual inspection of the fuse is often insufficient, since no detectable fracture is visible. The solution is to check each fuse, out of circuit, with a multimeter set to the ohms range. It should read about zero ohms. Figure 10.1 shows a representative power supply circuit. This particular one is fitted in the 'Tatung 1000 series' Astra satellite receivers.

Repairing a faulty power supply

If experience is gained in the area of diagnosis and repair of faulty power supplies then you will be equipped to deal with the majority of faults likely to occur in satellite equipment. The following is a guide to the repair of the power supply section.

Regulator chips

As can be seen from Figure 10.1 the power supply section of a typical satellite receiver consists of fixed voltage regulator chips, rated at 1 A,

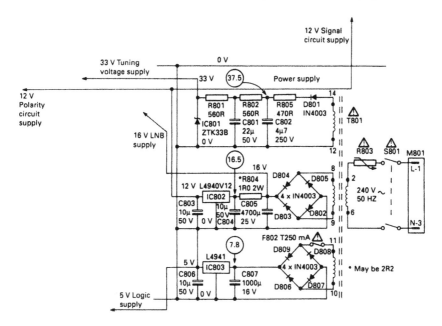

Figure 10.1 *Power supply used in the Tatung 1000 series Astra satellite receiver (Source: Tatung (UK) Ltd)*

which are mounted on heatsinks or the outer casing. Receivers designed to control actuators will have an additional 36 V higher current supply. By measuring the input and output voltages a faulty chip is easily identified. If the input voltage is present and no output voltage is present the likelihood is that the chip is faulty. However, some of these chips have short circuit protection so it could be possible for a short to be across the output. A resistance check from the output to chassis can be used to test for this. A 5 V regulator is often used to supply internal digital chips such as tuning processors and to supply the position sensor in motorized polar mounts. A 12 V regulator is often used to supply the video and audio processing circuitry and higher voltage regulators (15 V to 24 V) are used to supply LNB requirements. In the case of the circuit given in Figure 10.1 the LNB feed is unregulated and is taken direct from the reservoir capacitor C805. Where V/H switched types of polarizer are used, such as Astra receivers designed to work with the popular Marconi unit, a level-switched LNB supply is needed. In these cases a programmable regulator chip (1.2 V to 37 V programmable output) is employed which can be switched to produce either 13 V or 17 V LNB feed voltages. A possible fault, which is not uncommon with all regulator chips, is that they become temperature sensitive. That is to say they work satisfactorily for an hour or two and then shut down. If this is suspected a hairdryer can be used to accelerate the occurrence of the fault. When the fault appears, spraying the suspect chip with freezer often restores normal operation which indicates that the chip is faulty. By way of a guide, Figure 10.2 shows the pinout connections for a typical 1 A fixed regulator chip of the 78 series. Most other equivalent regulators have the same pinout connections.

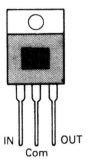

IN   OUT
Com

Figure 10.2 *Pinout connections for 78 series or equivalent regulators*

Power supply diodes

Another fault that can occur, often with dramatic results in terms of fuse blowing, are short circuit bridge rectifier diodes such as D802 to D809 in Figure 10.1. This is primarily caused by transient pulses arriving from the mains supply possibly due to lightning. These can normally be checked with an analogue multimeter switched to the ohms range. The good diode should read approximately 1000 ohms in one direction and open circuit (i.e. infinite resistance) in the other direction. Some digital multimeters have a special diode check range, in which case a reading of 400 to 800 in one direction indicates a good diode.

Another type of diode which often crops up in power supplies is the zener diode which, when reverse biased, has a specific breakdown voltage. These diodes are usually labelled with their specific voltage and are used to stabilize a d.c. voltage very accurately. If the voltage across a zener diode exceeds its specific voltage value it can be deemed to be open circuit and in need of replacement. They can also become permanently short circuit although this is less likely to occur in practice.

Transformers

Mains transformers often give trouble; the most common fault is that the primary winding becomes open circuit. If this is suspected there will be no a.c. output voltage on the secondary windings. To confirm that it is the transformer at fault, ensure the mains is disconnected and measure the d.c. resistance across the primary winding with a multimeter switched to the ohms range. It should read just a few ohms if satisfactory. The secondary windings are rarely found to be faulty.

Power supply resistors and capacitors

Any component with a little triangle with an exclamation mark inside it (sometimes a diamond shape with a letter 'S' inside it) indicates that the component has been specially selected to conform with current safety regulations. Such marked components must be replaced by one of an identical type supplied by the manufacturers. Experience has shown that these become weak links in power supply circuits and can fail at the drop of a hat. When confronted with a power supply fault, experience has shown that a safety resistor being open circuit is the most likely culprit. Personally, I like safety regulations because it provides 'built in unreliability' which of course is the service technician's bread and butter (or should I say wholemeal and polyunsaturated spread). Any resistor connected in series with a supply line is prone to failure and should be checked. Some receivers, such as the Tatung 1000 series, have a safety thermistor (temperature sensitive resistor) with a positive temperature coefficient (PTC). This component, R803 in Figure 10.1, is located on the primary side of the mains transformer. The purpose of this resistor is to provide a low resistance at normal working temperatures, but when a fault is present the increased current flow causes the PTC to heat up, resulting in a rapid increase in resistance which limits the current to a safe value. These components often replace the conventional mains input fuse but they can go open circuit permanently. It is important that they be replaced with an identical type.

Electrolytic capacitors have a tendency to become open circuit particularly if old or exposed to heat and high voltages. A quick practical method of checking them is to 'rock them'. If they are dried out the chances are that one of the connection pins rots and can easily be detached exposing evidence of 'capacitor excreta' underneath. Distortion or shrinkage of the outer plastic sheath is also a tell-tale sign of a faulty electrolytic capacitor. If either of these methods fail then they can be removed and checked on a capacitance meter if suspected. An open circuit reservoir or smoothing capacitor is often detected by a significant reduction in the expected d.c. voltage across it.

Switched-mode power supplies

Power supplies incorporating regulators, although simple to diagnose and repair tend to run very warm. To reduce this problem and improve efficiency an alternative method of deriving the required voltages is sometimes used, called a switched-mode power supply. Most modern TV sets derive their power supplies by a similar method and most TV technicians will be familiar with them. Quite high voltages (mains potential) are present on the primary side of the circuit and consequently servicing should be left to suitably qualified personnel.

Dabbling could lead to major damage or possible electrocution. A common switched-mode power supply circuit is based around the Siemens TDA 4600 series control IC and is often found in satellite receivers of European origin. The essential features of this type of power supply, fitted in the Grundig STR20 series of satellite receivers, is shown greatly simplified in Figure 10.3.

Basic circuit operation

When the chopper transistor T634 is switched on, current flows in the primary of TR601 thus energy is temporarily stored in the transformer. The secondary windings can only draw current related to this stored energy. When the chopper transistor is switched off the magnetic field collapses and current flows through the rectifiers in the secondary circuits (to the right of the transformer) thus charging the associated reservoir capacitors. The time the chopper is switched on determines the maximum secondary current. Feedback pulses are derived from pin 9 of the transformer and these are fed to the control IC at pin 2. The result is that internal logic within the chip causes the chopper transistor to be switched on when the magnetic field has fully collapsed. Regulation is provided by controlling the time at which the chopper transistor is to switch off. A feedback voltage, proportional to the current drawn, is derived from the winding connected between pins 7 and 9 of the transformer, rectified by D647 and fed to pin 3 of the IC. Internal control circuitry within the chip controls the time that the chopper is to switch off. The basic switching frequency is controlled by R646 and C646 which are supplied from a mains rectified 300 V line. This provides stabilization against variation of mains input voltages. Internal circuitry within the control IC discharges C646 during the chopper transistor's off time. The output voltages are set by the adjustment of R647 which controls the feedback to pin 3 of the IC, and should be set up whilst monitoring the +15 V line.

Overcurrent protection is incorporated in the control IC by counting the variable mark to space pulses. An increase in the designed working current, due to a possible fault condition, results in the frequency of oscillation falling. When this falls to beyond a certain preset value then internal fixed switching within the chip takes over and thus limits the current.

Overvoltage protection is provided by an increase in the frequency of operation and would happen in cases where, say, a secondary supply line is open circuit and less current is being drawn.

An initial start up supply is provided by D616 and this should be at least 7 V. Once oscillation has started the chip supply is fed from a transformer-derived voltage of about 12 V via D633/C633. If the mains input is too low, protection is provided by the start up voltage being below the 7 V threshold thus causing the power supply not to start up.

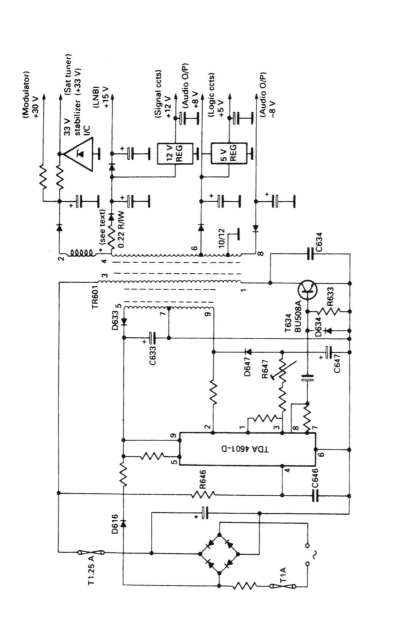

Figure 10.3 *TDA 4600 switched mode power supply (simplified)*

Similarly, the voltage present at pin 5 of the IC should be greater than 2.2 V or the chip shuts down.

Typical faults

If R646 goes high in value or C646 is open circuit the chopper transistor will blow (become short circuit). If failure of the chopper transistor is experienced always change R646, even if it appears to read correctly on an ohmeter. Failure to do this will almost certainly lead to a recall at some later date with another blown chopper. It is very rare these days for choppers to blow unless some underlying fault is also present. In general, failure of high value resistors in power supplies is often the cause of repeated failure of the chopper transistor.

If the adjustment of R647 has no effect on the output voltage then the chances are that C647 is open circuit or dried out. If the power supply fails to start on occasions, check C633 by substitution. If the power supply will not start up at all, check that the rectifiers in the secondary circuit are not short circuit and that at least 7 V is present at pin 9 of the IC at switch on. If this value is well over 12 V then the chip is faulty. Other diagnostic checks that can be made are the presence of a 4 V square wave at pin 1 of the IC and square wave drive of at least 1 V amplitude at pin 7. Likewise this waveform should be present at the base of the chopper transistor. If T634 is even found to be faulty always check the components D634, R633 and C632 and replace R646 before fitting a replacement transistor.

All the supplies fed from the secondary winding of the transformer usually incorporate low value safety resistors, fitted before the rectifiers, these can go open circuit. The one fitted in the LNB supply line usually blows if a prolonged short is present on the coaxial cable during installation. This particular component is marked with an asterisk on Figure 10.3

Dry joints can be a problem in these power supplies, particularly on the pin out connections of wound components. This is because they tend to vibrate at the working frequency which often results in solder joints cracking in a circular pattern round the pins.

Faults in the head unit and tuner/demodulator

If the LNB is found to be faulty, repair should not be attempted and it should be sent back to the manufacturer or supplier for an exchange unit. The compact layout of components using surface mounted technology necessitates specialist repair. Likewise, the polarizer/feedhorn assembly is not serviceable since, with magnetic polarizers in particular, the windings are embedded into the feedhorn itself. Each component part of the head unit can thus be considered as a module for replacement only.

The tuner/demodulator can, as its name implies, contain the 1st IF tuner unit (950 MHz to 1750 MHz) and the FM demodulator. Again, the layout is critical and a lot of surface mounted technology is employed necessitating complete replacement as a unit rather than repair. In any case, the time spent diagnosing and replacing an internal faulty component will probably cost more than the replacement of the entire unit. The module's many pins will need to be unsoldered and the unit removed from the receiver for replacement. This should not be attempted unless a special desoldering tool is to hand, or damage to the printed circuit will result. A 'one shot' suction type desoldering tool with a plastic nozzle can be acquired quite cheaply to remove the molten solder from each connection. However, the plastic nozzles are short lived and experience has shown that a better type to use is one with a built in soldering iron which allows single handed removal of solder from the many connections. The nozzle, made of metal, is also the bit of the soldering iron and subsequently has a long working life.

Magnetic polarizer control

The purpose of a magnetic polarizer control circuit is to maintain a constant current through the polarizer windings over a wide range of outdoor temperature conditions. With low cost receivers designed to receive single satellite transmissions this is often all that is provided. However higher priced receivers and multi-satellite equipments provide for additional fine trimming of the polarizer current whose value can be indirectly stored to take into account skew with frequency effects. The reason for this is that the degree of wave-twisting is not constant and varies with frequency. With simple receivers the physical polarizer orientation must be set at a compromise position for best cross polar rejection over the required frequency band.

One method for achieving a constant current through the polarizer is shown in Figure 10.4 and is used in the Tatung 1000 series of Astra receivers.

The polarizer is required to select either vertically or horizontally polarized signals. The direction of current flowing through the magnetic polarizer, and thus polarization selection, is controlled by the output from pin 13 of the microprocessor IC102: a logic 'high' for vertical and 'low' for horizontal. This output is used to switch TR104 which in turn controls TR106 and TR108. These two transistors are used to switch the current direction and select the polarity of the received signals.

For horizontally polarized signals TR105 and TR106 are switched on and TR107 and TR108 are switched off. Current flows out through

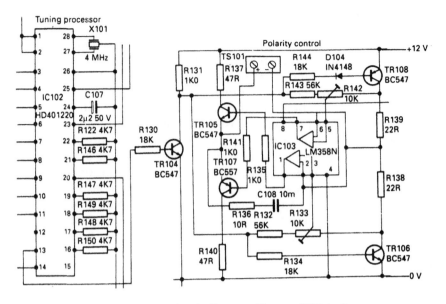

Figure 10.4   *Polarity control circuit (Source: Tatung (UK) Ltd)*

TR105's emitter, through the polarizer winding and back to 0 V via TR106. The required current value, preset by R133, is compared with that measured through R138 and any correction to the current is controlled via IC 103 output on pin 1, by TR105.

For vertically polarized signals TR107 and TR108 are switched on and TR105 and TR106 are switched off, thus current flows in the opposite direction out through the polarizer via TR108 and TR107. The corresponding current level is set by R142. The required current value, present by R142, is compared with that measured through R139 and any correction to the current is controlled via IC103 output on pin 7, by TR107.

The most likely faults to be encountered in this type of circuit are failure of the switching transistor/s or faulty presets R142 and R143.

V/H switched polarizer control

A typical circuit used to control 14/18 volt feeds of simple V/H switched polarizers is shown in Figure 10.5. The H/V select signal, from the microprocessor is used to select either 14 V for vertical or 18 V for horizontal polarization. An LM317 programmable regulator is used to provide both voltages which are preset by a pair of presets VR401 and VR402. Although the LM317 chip has internal short circuit protection it is common, in many receivers, to find an additional low value safety resistor in the transformer secondary circuit. This gives additional pro-

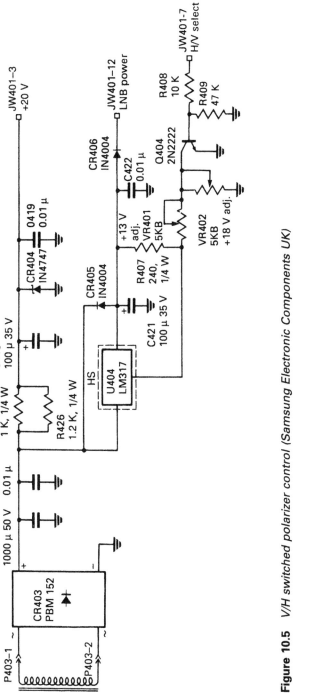

Figure 10.5 *V/H switched polarizer control (Samsung Electronic Components UK)*

tection and safety should the chip fail due to a prolonged short on the LNB supply feed. In practice, this fault is probably the most common encountered.

Baseband, video and audio processing circuits

For fault-finding in this area an oscilloscope is useful. By the strategic monitoring of waveforms at signal inputs and outputs of the various stages the fault can usually be quickly isolated to a particular stage and the appropriate action taken. Alternatively, a low cost signal injector can be used to inject a signal loaded with high frequency harmonics at the input/outputs of various stages. The resulting video patterning on the screen or audio breakthrough (or lack of it) can often be used to deduce the faulty stage. Components most likely to fail are electrolytic capacitors and semiconductors.

Digital and remote control circuitry

Digital circuits, such as microprocessors and remote control circuitry are very reliable in practice and are rarely found to be faulty. In cases where a fault is suspected in this area, the first thing to do is to check that the various supply voltages to the chips are present. Secondly, check the clock frequencies with an oscilloscope or frequency counter since crystals, used to keep the frequency of oscillation stable and accurate, can either be off frequency or drift with temperature, replacement being the only cure. It is very rare for microprocessors to go faulty, so suspect these chips last.

Remote control circuitry can sometimes give trouble, but 99 times out of a 100 it is the handset which is faulty, not the receiving circuitry. An infra-red remote control tester is a valuable addition to the toolbox, so that each function button of a suspected handset can be quickly checked for both operation and range. The most common cause of handset failure is tea, coffee, or other liquid being spilled, in which case it is necessary to dismantle and clean up with methylated spirit and a brush. However, with the membrane type of keypad replacement is normally necessary.

Twiddling

When confronted with a faulty receiver it is tempting to start the whole-sale twiddling of presets and other adjustable components. This should be avoided like the plague, unless you have good reason to suspect maladjustment. If for any reason you do, always mark its original adjustment position so that it may be reset precisely to its original position afterwards. Out of control twiddling can render a whole receiver useless or require extensive manufacturer's set up procedures to be followed which may be both tedious and need expensive equipment. A golden rule: never twiddle unless you know precisely what the adjustment is for.

11 Digital satellite TV systems

Introduction

In recent years there has been a steady growth in the use of digital techniques in telecommunications. Digital communication has been with us for some time now: smoke signals and drums have been used to convey signals over distance for thousands of years. Even the telegraph (Morse) used in the last century is essentially a digital system. Any system is pronounced digital if it can be conveyed using a finite number of discrete states as opposed to analogue systems which are represented by a continuously varying quantity.

Need for compression

One of the drawbacks of implementing digital television in the past was the bit rate, hence bandwidth, required to provide studio quality pictures. By way of an example, samples of analogue signals must be taken at over twice the highest frequency present. The former CCIR recommended that samples be taken at 13.5 MHz for the luminance component and 6.75 MHz each for both the R-Y and B-Y chrominance signals. This leads to 13.5 × (2 × 6.75) = 27 Msamples/s. Assuming 10 bits per sample (studio equipment standard), this results in a bit rate of at least 270 Mbit/s. The transmission problems alone at this kind of bit rate would be bad enough, but the storage problems would be even worse. For example, a typical feature film runs for about 90 minutes (5400 seconds). A 4 Gb hard disk as fitted in a personal computer has a capacity of just 32 Gbits so it would require 46 such hard disks to store one feature-length film! It is clear that video compression is not just advantageous – it is a commercial necessity.

A 33 MHz transponder can handle a bit rate of about 55 Mbit/s, so bit rates of this magnitude will just not fit into a standard satellite transponder's available bandwidth without some form of compression. Without compression, standard analogue FM would be a clear winner on bandwidth usage.

Even more startling, without compression, a studio quality HDTV signal would require about 1 Gbit/s for interlaced pictures and 2 Gbit/s for non-interlaced! Far too high for a practical satellite delivery system, bearing in mind the expected demand for such services and a choice of programmes. Consequently, a certain amount of compression coding will be necessary in order to reduce the transmission bit rate to a practical level. All the bit-rate reduction systems have inherent errors called artefacts or defects, which are divided between those received and those due to the coding process itself. Experts in the field are in agreement that a 'virtually transparent' W-HDTV picture will need a transmission bit rate at about 110 to 120 Mbit/s to be free of perceptible coding artefacts. It is also argued that lower bit rates of 30 Mbit/s or lower may be tolerated, but this would lower the quality from 'virtually transparent' to simply 'good'.

A DCT (discrete cosine transform) algorithm, coupled with simple quantization schemes is chosen to achieve the data compression. Further bit-rate reductions can also be achieved using a technique known as motion compensation.

The modulation system (see below) adopted for MPEG-2 transmissions is a compromise based on the following requirements and considerations:

1 Tolerant of high levels of noise to conserve transponder power requirements and allow acceptable sizes dishes for reception.
2 Tolerant of high levels of interference to allow efficient use of the available spectrum.
3 Satellite transponders use TWTs (travelling wave tubes) driven close to saturation for efficient use of energy. Unfortunately, this is not a linear region of operation.
4 Simplicity of modulation system.
5 Complexity of error-correction system.
6 Conservation of bandwidth (maximize the number of bits coded per signalling state).

Digital modulation systems are generally more robust than their analogue counterparts in the presence of noise and interference. The modulation adopted is usually coupled with some form of concatenated channel coding to reduce bit errors. The precise combination and details are not particularly illuminating when visualizing the workings of a particular system, so this can be thought of as a 'black box' area.

Bit-rate reduction techniques

Commercial terrestrial television pictures are transmitted as *analogue* signals and are likely to remain so for a few years yet due to the cost of upgrading and the difficult decisions that will have to be made on international standards. But satellite services are going over to *digital* transmission because it is relatively noise free and, provided the information is suitably *compressed,* can offer high quality television and audio services with a wider choice of channels at a reasonable cost.

Commercial satellite TV services, as well as terrestrial, must remain economically viable, so they must make the most efficient use of available bandwidth and storage capacity. The choice lies between a few high quality transmissions, each occupying a wide bandwidth, or a greater number of transmissions of an acceptable quality. To this end, techniques of video compression are all based on the concept of redundancy.

Redundancy

A famous line in a certain country's constitution reads, *'All men are born equal'*, although some wit later added, *' – but some are more equal than others'*. The same cynical view can be taken with television pictures. Although all parts of a picture may be important from the aesthetic viewpoint, as far as the eye of the average television viewer is concerned, some parts are more important than others. Channel bandwidth is at a premium, so the parts that convey low grade information can be considered redundant and, in the interests of efficiency, can be removed from the image signal before it is transmitted. Digital computers, which became cheap and widely available in the mid-1970s, helped to spur the advance in digital compression, but the underlying principles of information theory were set out almost three decades before.

Information theory

In 1948, Claude E. Shannon, an electrical engineer in the USA, placed the whole subject of information on a scientific basis when he presented a paper entitled, 'The Mathematical Theory of Communication'. He showed that information can be manipulated by mathematics like mass, energy or any other physical quantity. In addition, the paper revealed that not all components of a message are of equal importance – some letters can be missed out and yet the message can still remain intelligible. In the, somewhat crude, example below, all the vowels have been omitted and yet it will still be understood by most people:

MST PPL HV LTTL DFFCLTY N RDNG THS SNTNC

In terms of information theory, we could say that consonants are more important than vowels – in relative terms, vowels may be considered redundant. In general, the status of a given piece of information is proportional to its *rarity*. Whether digital or analogue transmission is used, redundancy exists in two forms:

Statistical redundancy the level of a signal at any moment can, to a certain extent, be predictable from its value in the past because picture sample-values are related to each other, not only in the same line but also in previous frames.

Subjective visual redundancy depending on the picture content, the human eye can tolerate a certain amount of distortion without being aware that parts of the picture have suffered from the ravages of redundancy.

The price to be paid for the efficiency of redundancy-reduced transmission is an increased tendency for channel distortion. This is as equally true for bit errors in digital transmissions as it is for impairment of picture quality in analogue transmissions. Any practical system must therefore leave in, or reintroduce, some controlled redundancy in the signal.

Picture acceptability

Compression coding can be *lossy* or *lossless*. Lossless compression, vital in computer data storage systems, guarantees that in the absence of channel errors, the coding at the transmission end and the decoding at the receiving end are identical bit for bit. In video applications, strict accuracy of information transfer is considered not essential so lossy compression is considered essential for maintaining commercial viability. Lossy coding which is visually indistinguishable from the studio quality original is said to be *virtually transparent* and has the advantage of reducing the bit rate down to between a quarter and an eighth of the corresponding lossless rate.

At one time, it was considered that digital transmissions should always produce noiseless and distortion-free pictures, even those which made heavy demands on scene content. This can only be achieved if the channel capacity is high enough to cope with worst-case pictures, even if they only occur rarely – say, less than 1% of the time. These high ideals have now been discarded in the interests of commercial pressure and it is now considered reasonable to accept some degree of distortion on rare scenes and use the bit capacity saved to achieve more resolution on the majority of scenes. The optimum trade-off requires detailed statistical analysis of broadcast criticality as well as some subjective testing of the viewer's tolerance to selective distortion. More sophisticated methods are under way to assess the average viewer's subjective acceptance of redundancy-reduced distortion. At present, the responsibility rests with viewing panels, so their decisions do lack precision.

It seems clear from all this that digitally coded pictures are markedly different from conventional PAL or MAC analogue systems. As far as analogue transmissions are concerned, the picture quality can be predicted from the noise level and channel bandwidth, but where digital transmissions incorporate redundancy, the channel must be regarded as *information*, rather than *bandwidth,* limited. Thus, after redundancy is removed from some scenes, the coding system must consider whether the average bit rate is less than that of the channel's upper limit. If this is the case, it can be assumed that distortion-free transmission of those particular scenes is possible. If this is not the case, then some additional distortion must be introduced to lower the information content. The designer of the system ensures that the viewer can accept the subjective annoyance caused by redundancy distortion.

Analogue-to-digital conversion

Sound and picture information is essentially analogue in character, that is to say, the waveform amplitudes vary in a smooth and continuously variable manner, so the first step is to convert this smooth outline into a series of *sampling pulses*, each of a height (voltage) corresponding to the instantaneous amplitude of the waveform. The voltage level of each pulse is then a simple *number* which, in the raw state, will almost certainly be in decimal format. Although decimal numbers, which use the ten characters 0, 1, 2, ..., 9 are eminently suitable for humans, they are less suitable for electronic circuits. *Binary* is the simplest numbering system, because it uses only two characters, '1' and '0', each known as a binary *bit*. Whereas decimal numbers are based on powers of *ten*, binary numbers are in powers of *two*, so, instead of counting in steps of 1, 10, 100, 1000, etc., they proceed in steps of 1, 2, 4, 8, 16, 32, etc. Electronic circuits that discriminate between ten different voltage or current levels require careful design and high tolerance components. In contrast, binary circuits only have to distinguish two levels so comparatively crude switch-action is all that is required to avoid ambiguity. The following table shows the relation between decimal and binary for those who are unfamiliar with this method of counting:

Decimal	Binary				
	(16	8	4	2	1)
1					1
2				0	1
3				1	1
4			1	0	0
5			1	0	1
6			1	1	0
7			1	1	1
8		1	0	0	0
9		1	0	0	1
10		1	0	1	0
15			1	1	1	1
16	1	0	0	0	0
17	1	0	0	0	1

Note that it takes three bits to represent binary numbers up to decimal 7, and four bits to represent binary numbers up to decimal 15. In general, to express any decimal number in binary, it must be put in the form 2^n. For example, decimal 256 is 2^8 so eight bits are required. (Electronic circuitry exists for converting decimal values to their binary equivalent.)

The Nyquist rate

The reference above to *sampling pulses* invites the obvious question – how many are needed? It would seem that the more sampling pulses we use, the more closely they will represent the original waveform, but this will be at the expense of increased *bandwidth.*

The American scientist, Harry Nyquist, investigated the conflicting requirements of sampling frequency and bandwidth and came up with the following far-reaching statement:

'Any analogue waveform can be uniquely represented by discrete samples spaced no more than one over twice the bandwidth apart.'

The statement can also be expressed in the following alternative form:

The sampling frequency should be, at least, twice the highest frequency present.

Example
In telephone practice, the bandwidth (B) is usually fixed at around 3000 Hz. So, samples must be, at most, 1/6000 seconds apart. Or, conversely, the sampling *frequency* (the Nyquist rate) must be, at least, 6000 Hz. Figure 11.1 shows, in outline form, how sampling is carried out.

The analogue signal is sampled periodically by a switch which rotates at the Nyquist rate – or slightly faster to be on the safe side! In practice, of course, the mechanical switch shown would be replaced by an electronic version.

Binary encoding

The voltage output levels from the sampling switch appear as simple decimal numbers, so they require conversion to binary before they are ready for transmission. Figure 11.2 shows, again in outline form, the steps in the conversion process.

The numbers 7, 6, 5, 6, 5 (representing the relative amplitudes) have been arbitrarily selected and, for simplicity, it is assumed that only eight discrete levels of the waveform are recognized, decimal 0 to decimal 7. To distinguish each of these eight levels requires only 3 binary bits although, in practice, limiting to 3 bits would give rather crude results. In theory, there is no limit to the number of levels which can be used but,

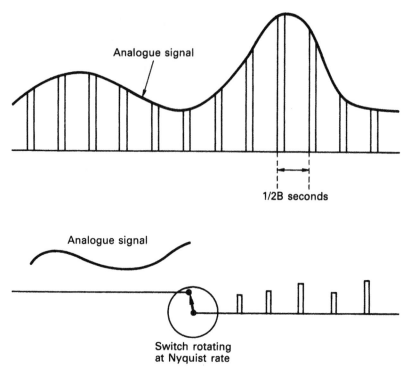

Figure 11.1 *Pulse sampling of an analogue signal*

as mentioned above, the more bits used to distinguish each level, the greater the bandwidth – a parameter which is always at a premium! The operation for analogue to digital conversion is shown in Figure 11.3.

Source image coding

Figure 11.4 shows the central idea behind current ideas of image source coding. The first block receives the raw input picture samples and subjects them to statistical redundancy removal. For example, a clear overhead sky contains multiple redundancies because the individual elements are everywhere similar, but the scene below may show cows nibbling away at the grass and perhaps the odd farm labourer or two munching sandwiches.

The output from the first block is at a *variable* data rate. The second block stores a few frames and, by *averaging* the capacity, converts them to a *fixed* data rate. This allows the bit allocation to be varied so that some of the bits which make up the relatively featureless overhead sky can be 'saved' and, instead, 'spent' on the more interesting bits which go to make up the picture of the cows and farm labourers. Note that the block is furnished with a *negative-feedback* loop to control the relative distribution of the distortion. If the buffer contents begin to rise,

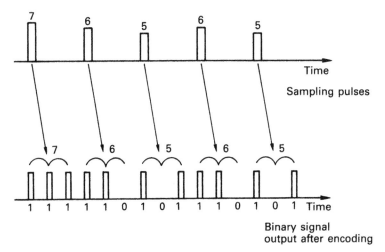

Figure 11.2 *Sample pulses undergoing conversion to 3 bit binary*

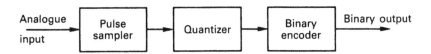

Figure 11.3 *Analogue to digital process*

it indicates that the scene contains too much information for the channel to handle, so the feedback mechanism introduces more of the subjectively tailored distortion which, in turn, will tend to reduce the volume of data entering the buffer. It follows, therefore, that critical scenes which are active over the whole screen will contain more distortion than scenes which are less critical.

Pre-processing

The video signal is often pre-processed to remove any awkward to code information which is relatively unimportant in the reconstructed picture.

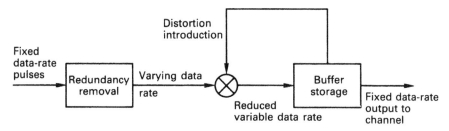

Figure 11.4 *Digital source image coding*

This is usually performed by a combination of spatial and temporal non-linear filtering, the details of which are not particularly informative in this context.

Redundancy-reduction coding

Redundancy-reduction coding may be implemented throughout the range of television systems including 16 : 9 standard format and W-HDTV at 140 megabits per second. Most of these systems use a mathematical device which is blessed with the title of *discrete cosine transfer coding* (DCT for short). The subject is made even more complex and confusing by the many variations in use: different quantizers, processing block sizes, variable-length codes and even the specific DCT transforms adopted, while the details of the system implementation are buried inside a number of chips. The computer simulated examples which follow avoid the mind-bending complexity of full system specifications without clouding the essential principles.

Intrafield DCT coding

In this form of coding, no use is made of the information in previous frames. The block of samples most commonly used is in the form of an 8 × 8 or 16 × 16 matrix. Table 11.1 shows the original 8 × 8 pixel block followed by the various processes acting on it. The original block sample values are arbitrarily chosen so have no significance.

This block is then processed by the above-mentioned discrete cosine transform (DCT). The mathematics of this transform is a little tedious to follow, but fortunately engineers have produced a lightening fast chip which performs the DCT transform and another for converting it back (the inverse transform).

The effect of passing through the DCT chip is to transform the raw samples of the original block to a new set of coefficients which represent the various spatial frequencies present. The top left-hand corner number of the DCT processed block, shown in Table 11.1, represents the *DC level* of the original block. Horizontal resolution (spatial frequency) increases to the right with increasing vertical resolution to the bottom. So the rather less important higher resolution components increase progressively towards the bottom right-hand corner. Note particularly that the coefficients are now far from uniformly distributed because the transform has concentrated the energy into the *top left-hand coefficients* – which represent the amplitudes of the various frequency components in the original sample block. The top left-hand coefficient, 295, represents the DC component of the original block and is about twice the mean value of the other members of the block. The bottom right-hand quadrant has very few coefficients of any significance. The primary ob-

Table 11.1 *DCT processing of an 8 × 8 pixel block (forward and reverse)*

Original pixel block

131	134	139	143	144	144	144	144
134	141	143	146	149	146	146	146
143	145	150	153	148	146	146	146
148	151	152	151	150	148	148	148
150	150	151	152	152	145	145	145
152	151	151	151	150	147	146	145
151	152	153	153	151	148	147	147
151	152	153	153	151	149	149	148

DCT processed block

294.94	0.32	−3.08	−1.37	0.44	−0.02	−0.42	0.19
−5.81	−4.07	−1.19	−0.41	−0.15	−0.15	−0.08	−0.11
−2.62	−2.40	−0.34	0.09	−0.29	0.01	0.06	−0.17
−2.06	−0.19	0.10	0.65	0.31	−0.17	0.00	0.21
−0.25	−0.20	0.28	0.11	0.00	−0.01	0.02	0.16
−0.26	0.13	0.32	−0.20	−0.21	0.31	0.33	−0.22
−0.27	0.35	0.12	−0.00	0.12	0.66	0.28	−0.19
−0.31	0.36	−0.50	−0.32	0.52	0.37	0.07	−0.01

Thresholding

294.94	0.00	−3.08	−1.37	0.00	0.00	0.00	0.00
−5.81	−4.07	−1.19	0.00	0.00	0.00	0.00	0.00
−2.62	−2.40	0.00	0.00	0.00	0.00	0.00	0.00
−2.06	0.00	0.00	0.00	0.00	0.00	0.00	0.00
0.00	0.00	0.00	0.00	0.00	0.00	0.00	0.00
0.00	0.00	0.00	0.00	0.00	0.00	0.00	0.00
0.00	0.00	0.00	0.00	0.00	0.00	0.00	0.00
0.00	0.00	0.00	0.00	0.00	0.00	0.00	0.00

Quantized coefficient block

295	0	−3	−1	0	0	0	0
−6	−4	−1	0	0	0	0	0
−3	−2	0	0	0	0	0	0
−2	0	0	0	0	0	0	0
0	0	0	0	0	0	0	0
0	0	0	0	0	0	0	0
0	0	0	0	0	0	0	0
0	0	0	0	0	0	0	0

Reconstructed block

131	134	139	142	144	144	144	144
136	139	143	146	147	146	146	145
143	145	148	150	150	148	147	147
148	150	152	152	151	149	148	147
150	151	152	152	150	148	146	146
150	151	152	152	150	148	146	145
151	152	153	152	150	148	147	147
151	152	153	153	151	149	149	148

ject of the coding scheme is to convey to the receiver a representation of this transformed block so that it can perform the inverse transform to reconstruct the picture or, at least, a close approximation of it. The DCT has highlighted redundant data by redistributing the coefficients (i.e. signal power) into lower frequency components.

Thresholding

It is possible to increase redundancy even further by discarding coefficients which fall below some arbitrarily defined threshold because they would contribute little to the final picture quality. This pre-quantization process is known as thresholding. See the block labelled 'Thresholding' in Table 11.1. In this example, all coefficients with an absolute value (neglecting sign) less than 1 are discarded from the DCT processed block. There is now a higher proportion of zeros and a clear cluster of the more significant coefficients at the top left-hand corner of the block. Thresholding is not always used but can be implemented in a dynamic fashion so that the distortion is only introduced when demands on system loading are high.

Quantization

After the various transforms and motion compensation (see below) have concentrated most of the signal power into the top left-hand corner of the block, it is possible to reduce the information content even further because it is not necessary to preserve the exact precision of the DCT coefficients in order to obtain good quality reproduction at the receiver. The range of encoded values can be tightened up still further by reducing the precision with which the coefficients are described by a process called *quantization* which reduces the number of coded binary bits used to represent the coefficients. The average TV viewer is fairly tolerant of quantization noise in the high frequency coefficients at the bottom right corner of the block, so a quantization law can be applied to each coefficient. For example, suppose only rounded integer values are allowed as shown in the quantized coefficient block in Table 11.1. This will have a significant effect on information reduction, since only integers rather than floating point values need be coded. Other quantization schemes may only allow values to be mapped to certain discrete integer values, say spaced three apart so that only values -6, -3, 0, 3, 6, 9, etc. are allowed. The reduced number of possible discrete values would clearly result in a lower number of bits required to encode them. The DC term in the top left-hand corner is often excluded from the latter type of quantization, otherwise a chunky or blocky look may be given to the picture. Again the quantization is usually dynamically variable to allow a trade-off between the maximum transmission bit rate allowed and the resulting picture quality.

Scanning blocks

At some point the two-dimensional blocks must be serialized into a linear single-dimensional stream before they are in a fit state to be channel coded for transmission. Because the significant coefficients are compacted at the top left-hand corner, it is advantageous to employ a zigzag serializing scan as shown in Figure 11.5. As an aid to visualization, I have shown all blocks of coefficients in two-dimensional form throughout. In practice the zigzag scan is often performed directly after the DCT transform stage, thus allowing thresholding and quantization to be performed when the data is in a more suitable serial form. However, this makes no difference to the final outcome.

In this conceptually simplified example, zigzag scanning is left to the end, so the scanned order is 295, 0, −6, −3, −4, −3, −1, −1, −2, −2, EOB. Note that no further transmissions are necessary after the last coefficient −2, because the following numbers are all zeros and therefore contain no additional information. Because of this, the special code for end of block (EOB) can be placed after the last significant coefficient, −2. Saving 54 zeros out of a block total of 64 represents a huge saving. For reasons of clarity, the examples given in Table 11.1 are an extreme case where maximum redundancy due to heavy system loading is required. Although not shown here, there is often odd significant coefficients, surrounded by zeros, trapped within the body of the block. Any long strings of zeros remaining in the block, before the EOB, could be handled by using a special code for each string.

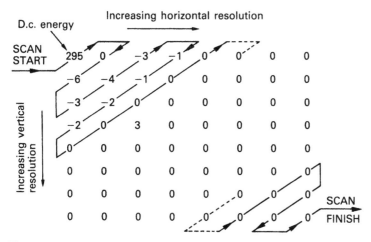

Figure 11.5 *Serial scanning method*

Coding patterns

The number of bits required to represent each sample can be further reduced by sending coded, rather than direct, binary strings. The traditional Morse code is essentially binary in character because it only uses two symbols, the 'dot' and the 'dash'. The principle behind the code is to use the shortest patterns for the most frequently recurring letters – this is why old Mr Morse chose the single 'dot' for the letter 'e'. The *Huffman* code is just one of the many possible binary equivalents of the Morse code and may be used to increase redundancy. Yet another dodge, called *run length coding* (RLC), is to replace long runs of the same value by a single multiplying suffix. For example, a sequence like 5, 5, 5, 5, 5, 5, 5 can be shortened by coding it as 7, 5 although it is pointless using RLC for very short runs.

Interframe coding

Although most scenes contain moving objects, the majority of the background information remains unchanged, so why send every complete picture frame? All that is required is to detect any change between one frame and the next. The system in use is called motion-compensated interframe coding. It is DCT-based and is similar to intrafield coding except it is formed from the error resulting from an attempt to predict sample values in a block *from the contents of the previous frame.* The method is to offset any motion (change) and to use a shifted block from the previous frame as the *prediction.* The task of detecting the changes, if any, which have occurred between the current block and the previous frame is called *block-matching.* If any change is discovered, its coded representation (called the *displacement vector*) is sent to the decoder, together with the DCT-coded prediction error, thus enabling it to recover the transmitted picture. Figure 11.6 shows the complete system in outline form.

The encoded image data, A, is subtracted from the input image data to produce the *difference* image, B, which is then processed by the DCT stage. It is then scanned in a zigzag manner to convert the two-dimensional block coefficients to a single serial bit-stream and, after passing through the quantization stage, is sent to the transmission encoder. The fixed and variable stores, together with the motion-compensation unit, minimizes any differences between consecutive frames by taking the reconstructed image from the output of the inverse DCT processor and searching for the best match. Only the best possible match is selected for entry to the subtraction stage, A. The displacement vectors, generated by the motion-compensation unit, are sent to the transmission unit.

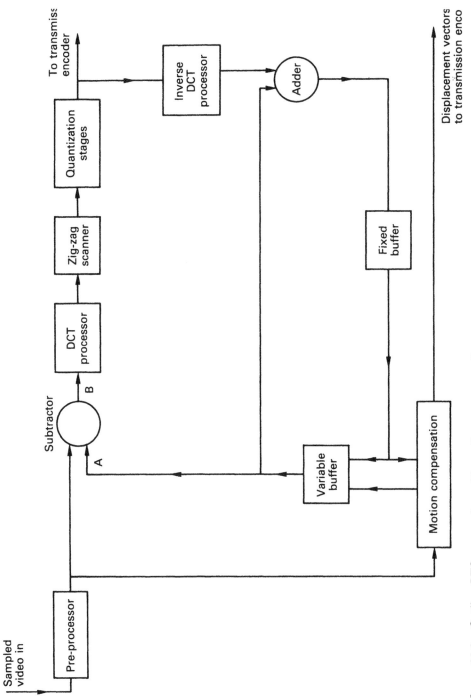

Figure 11.6 *Outline DTC processing with motion compensation*

Adaptive coding

The decision whether to use intrafield or interframe coding depends, to some extent, on the frame contents. In most cases, the interframe coding method is the best choice but where the scene is erratic or some sudden background has appeared, the intrafield option is preferable because it provides a lower data volume output.

Some systems include circuitry that decide, on a block-by-block basis, which of the two coding schemes is preferable in any given circumstance. A single control bit is sent between coder and decoder to indicate which mode has been chosen for that block.

The intrafield mode is deliberately chosen, on a periodic basis, for a number of blocks in each field to sweep out any persistent error effects which might otherwise continue from frame to frame. This 'flushing' action is made to occur at regular intervals of one or more seconds so that every block position is coded by the intrafield method.

A further trick, called *post-processing*, is to tone down the visual impact of very high distortion levels that may be introduced from time to time during the coding process.

Decoding

Signals that have been subjected to redundancy reduction will be virtually unintelligible to humans at the receiving end until, at some stage, it is restored to health by a *decoder*. The composite signal received contains important information on the redundancy steps that were taken before transmission. For example, if it has been subjected to any of the DCT transforms, the decoder will obligingly produce the inverse transform. This information will, in most cases, be a kind of digital shorthand: a few 1's and 0's here and there to assist the decoder in its primary task of restoring the picture to its former glory – well almost to its former glory because some of the redundancy coding was based on the *subjective acceptability* of the viewer (in other words, the average viewer can easily be hoodwinked!). The final reconstructed block at the receiver is also shown in Table 11.1. By comparing the reconstructed block to the original block in Table 11.1 you may see there is very little information lost in the process.

Digital modulation methods

Nowadays, because of the jargon of computer hacks, the term digital is often synonymous with the two-state binary signals used in present day computers. However, it is important to remember that, in the general case, a binary system is just one particular subset of digital systems.

Other digital systems include three state, *ternary,* widely used in pulse code modulation (PCM), four state, *quaternary,* and 8, 16 and even 256 state used in digital data transmission. We will see later that digital TV is transmitted using a multi-state system of modulation.

Binary message signals

Binary systems use two states to represent the data they convey. These are referred to as logic state 1 and logic state 0. The logic 1 state is represented by V_1 and the logic state 0 by V_0. The choice of actual values for V_1 and V_0 is arbitrary but, by convention, if $V_1 > V_0$, the system is said to use positive logic. Conversely, if $V_1 < V_0$, the system is said to use negative logic. Figure 11.7 illustrates this point more clearly.

Signalling rate and data rate

One of the fundamental parameters of a serial digital transmission system is the *signalling rate* measured in *states per second* or, after J. M. E. Baudot, *bauds* . That is to say, 1 baud = 1 state per second or 1 symbol per second. Only for two-state binary systems is the signalling rate equal to the data rate, so it is understandable why data rate, measured

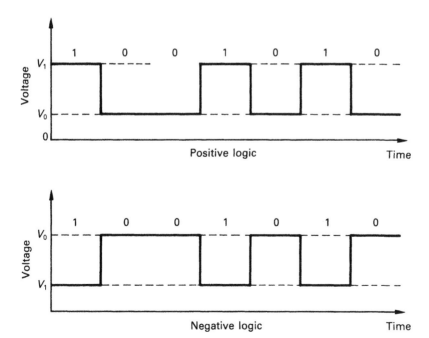

Figure 11.7   *Example of negative and positive logic*

in bits per second, and bauds are so often mistakenly interchanged. If we take a four-state transmission system, shown in Table 11.2, and conveniently label the four states A, B, C and D it is easy to see the essential difference. The data transmitted consists of a serial stream of binary digits. The stream is handled in pairs of two digits at a time, of which there are four possible combinations of digit pairs. Each digit pair, called a *dibit* or *2-bit symbol*, represents a different state, so a 2-bit symbol may be transmitted using a single state. It follows that the signalling rate in bauds is half that of the data rate in bits per second. You may ask, why use such a complicated multi-state system? The reason is that the bandwidth needed for data transmission depends more on the signalling rate than the data rate. Systems designed to maximize the data rate within a given bandwidth will have a lower signalling rate than a simple binary system where the two rates are the same.

Table 11.2   *Four-state digital transmission*

Dibit (2-bit symbol)		Signalling state
0	0	A
0	1	B
1	0	C
1	1	D

Digital frequency modulation or frequency shift keying (FSK)

For analogue satellite TV systems, frequency modulation (FM) is the method invariably used to modulate a carrier with signals such as PAL, SECAM or NTSC, and is described elsewhere in this book. When a binary digital signal is input to a frequency modulator, its output results in a sine wave at one of two frequencies as shown in Figure 11.8. The result is that the frequency shifts in accordance with the input data and the demodulation system is much the same as that for analogue systems. In this context, and for historical reasons, this form of digital frequency modulation is referred to as FSK or *frequency shift keying* (a remnant from the days of Morse telegraph systems).

Digital phase modulation or phase shift keying (PSK)

To operate at the high data rates required for digital satellite TV, either the number of cycles per signalled state has to be decreased or the

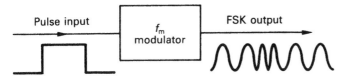

Figure 11.8   *Frequency shift keying (FSK)*

signalling frequencies have to be increased. Clearly, it is unwise to increase the latter since there is a limited bandwidth. There is also a limit to the former since simple demodulators, such as phase locked loops, have a limited transient response and may take several cycles to lock on to a new frequency. There is simply not enough time available since each digit will need to be translated to as little as one cycle, so we can conclude that FSK would be impractical for satellite TV as it would need far too much bandwidth. A digital transmission system is needed which increases the data rate without creating the need for more precious bandwidth. The modulating signal itself can be compressed by various techniques before transmission, but the problem still remains that a more efficient method of modulation than FSK is required. An improvement can be made by the use of a single-frequency variant called *digital phase modulation* also known as *phase shift keying* (PSK). Here the phase of the carrier is changed rather than the frequency.

Phase modulation (PM) is closely related to frequency modulation and is well suited to multi-state digital transmission. As with FM, spectral analysis is notoriously difficult and the two spectra would appear very similar. The basic process of phase shift keying is shown in Figure 11.9(a). The phase of the carrier is altered with respect to the digital message signal. In this particular example, a binary '0' is transmitted as a 0° phase shift of the carrier and a binary '1' is represented as a phase shift of 180°. A phase shift of 180° may be performed on the carrier by multiplying it by −1 (inverting), so if negative logic conversion is used on the message signal to change the 1 and 0 binary states to −1 and +1, respectively, two-phase PSK could be produced using a simple multiplier as shown in Figure 11.10(a). The phase change introduced by the message signal is called the *phase deviation* and can be varied by changing the modulator sensitivity. In general, frequency modulation can be performed by integrating the message signal and applying it to a phase modulator. Conversely phase modulation may be produced by differentiating the message signal and applying it to a frequency modulator, hence the similarity.

The demodulation process in the receiver, shown in Figure 11.10(b) could be performed by using a product detector which effectively multiplies the received PSK signal by the locally generated reference carrier, thus recovering the original message signal.

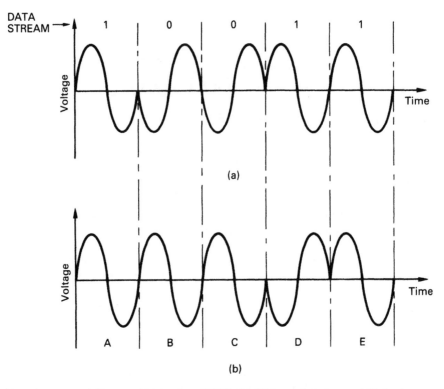

Figure 11.9   *(a) Phase shift keying (PSK). (b) Differential phase shift keying (DPSK)*

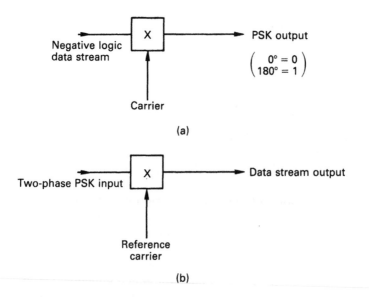

Figure 11.10   *(a) Modulation of PSK. (b) Demodulation of PSK*

Differential phase shift keying (DPSK)

During demodulation, it would be difficult to accurately generate the reference carrier shown in the previous conceptual example of PSK since the phase relationship at any frequency, due to down-conversion, may slowly vary over the link. Frequency shifts are simply continuously varying phase shifts, so clearly data might be misread because of the varying phase shifts introduced by the link. The solution is to use differential phase shift keying (DPSK) where phase changes are introduced with reference to the phase of the previous signalled state. The principle is shown in Figure 11.9(b) for comparison with simple PSK. The carrier reference frequency during demodulation is reconstructed only from the previously received signalled state and thus largely eliminates the effects of unpredictable phase variations over the link. Referring to Figure 11.9(b) the system works as follows. The carrier reference phase for signal B is signal A. The carrier reference phase for signal C is carrier B, and so on. In reality, phase shifts of 0° would be avoided so the receiver would always receive phase shifts at the data rate. For example +90° and +270° for '0' and '1' might be used instead of 0° and 180° to avoid long periods of unmodulated carrier which may lead to too much energy being concentrated in certain areas of the spectrum leading to interference.

Quaternary phase shift keying (QPSK or 4-PSK)

Quaternary phase shift keying (QPSK) is a further development of PSK where, for a given carrier frequency, the data rate is effectively doubled without increasing the signalling rate. The penalty being that the S/N ratio on demodulation is lower. With QPSK each signalled state can encode a dibit (2-bit symbol). Typically, the four 90° phase shifts used are +45°, +135°, +225° and +315°. Remember, that 0° is rarely used in practice to prevent long periods of unmodulated carrier. The upgrade from a two-state signalling system to a four-state one means that the data rate in bits per second is greater than the signalling rate in bauds. Table 11.3 shows the QPSK system where four dibits are encoded by four phase shifts. Figure 11.11 shows how the basic modulation of a QPSK signal may be performed. Two carriers, of the same frequency, with a 90° phase shift between them are fed to a pair of multipliers. Each multiplier is fed with identical rate, digital input signals of +1 (binary 0) or −1 (binary 1) using negative logic as before. The outputs derived from the multipliers thus render the same encoding as in the simple case earlier. That is to say, a binary zero is represented by a 180° phase shift and a binary zero by a 0° phase shift. The main difference over simple PSK is how these outputs are combined by the adder. An adder combines the outputs and produces a final output waveform corresponding to the four possible combinations of message signals as

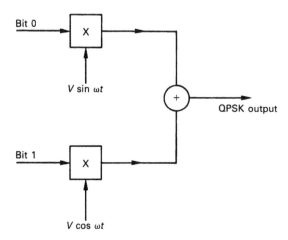

Figure 11.11 *Production of QPSK*

shown in Table 11.3. The phasor diagram (Figure 11.12), shows Table 11.4 in phasor form and clearly shows how the four phase shifts, or code vectors, of +45°, +135°, +225° and +315° represent the dibit from the additions of the two modulated outputs.

Figure 11.13 shows the reverse demodulation process. The incoming signals are connected in parallel to two product detectors and the reference carrier generator. The reference carrier is reconstructed from the received signal at the data rate, in the way described previously for DPSK. This carrier is fed directly to one phase detector and via a 90° phase shifting network to the other product detector. The dibit is recovered by examining the size of the output from each product detector at the data rate according to Table 11.4.

QPSK has the advantage that it can operate with transponder power close to saturation (maximum power), so is very energy efficient. In

Table 11.3 *Phase changes produced by signalling states in QPSK*

Dibit (2-bit symbol)		Modulator A message phase (bit 1)	Modulator B message phase (bit 0)	Output phase (degrees)
Bit 1	*Bit 0*			
0	0	+	+	+45
0	1	+	−	+135
1	1	−	−	+225
1	0	−	+	+315

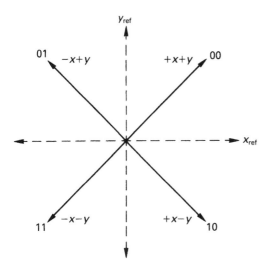

Figure 11.12 *Phasor diagram for QPSK*

Table 11.4 *Demodulator output table*

Product detector A (bit 1)	Product detector B (bit 0)	Dibit recovered	
+	+	0	0
+	−	0	1
−	−	1	1
−	+	1	0

addition, the method is well suited to dual polarization strategies since it has very low sensitivity to interference from other digital systems.

8-PSK and 16-PSK

An eight-phase modulation scheme called 8-PSK is a further refinement of QPSK. Here eight code vectors are used rather than the four with QPSK. The spacing of the vectors could be 45° apart, but to avoid un-modulated carrier conditions, the spacing normally chosen is that shown in Figure 11.14. The signalling rate is increased since a 3-bit pattern, known as a *3-bit symbol*, is encoded using this method. A corresponding reduction of S/N is also experienced. Optimum performance

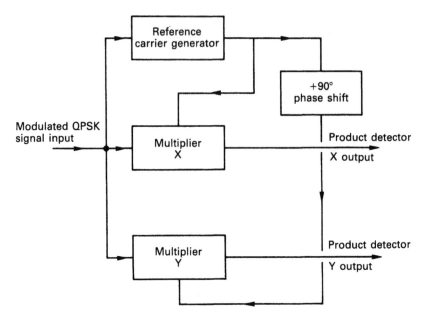

Figure 11.13   *Demodulation of QPSK*

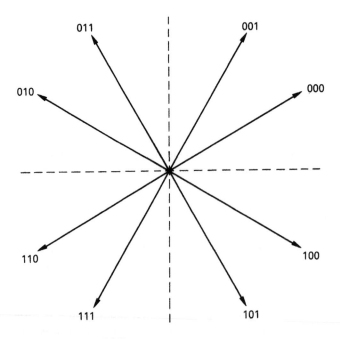

Figure 11.14   *8-PSK vectors*

of 8-PSK occurs with a transponder output back-off value of about 0.5 dB. This makes it slightly less power efficient than QPSK but it still retains low sensitivity to interference from other digital systems. A further development is the logical extension to 16 phases in 16-PSK.

Quadrature amplitude modulation

Quadrature amplitude modulation (QAM) is essentially an extension of 8-PSK with a further hybrid introduction of amplitude modulation to allow a higher order digital modulation scheme. A particular species suitable for digital television transmission is known as 16 quadrature amplitude modulation (16-QAM). The usual implementation works like this. Eight carrier phases, similar to 8-PSK, are used with the spacing shown in Figure 11.15. An extra four vectors are introduced spaced 90° apart as in QPSK. These extra quadrature vectors are further amplitude modulated with one of two discrete levels. This results in a total of 16 unique signalling states, thus allowing a 4-bit symbol (nybble) to be encoded per signalling state. There is, of course, a penalty to pay for using 16-QAM over its PSK counterparts, and that is a corresponding decrease in the S/N ratio on demodulation, so a higher C/N would be required for a satellite service. Another disadvantage is that 16-QAM, due to its inherent amplitude modulation, shows high sensitivity to non-linearities in satellite transponders so the power is usually backed off by about 7 dB, giving lower power efficiency than QPSK or 8-QPSK. 16-QAM is also said to be unsuitable for dual polarization use, in the same service area, due to its relatively high sensitivity to interference. There is

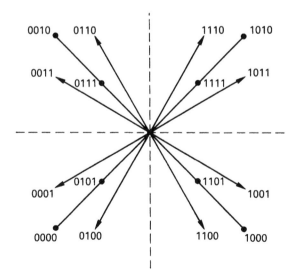

Figure 11.15 *16-QAM vectors*

an alternative configuration for 16-QAM which uses four quadrature vectors similar to QPSK, but each is amplitude modulated with one of four discrete levels. This configuration is claimed to have a lower sensitivity to interference due to a more even spectral distribution. The main advantage is that the bit rate is doubled over 8-PSK, for a given signalling rate and thus bandwidth.

Future developments

The practical problems of developing and implementing an increased number of signalling states over 16-QAM are legion. This would require either more phases or more discrete amplitude modulated levels, or both, and may prove difficult to resolve accurately in practice. However, it is not unreasonable to assume that 64-QAM or perhaps 256-QAM will be successfully developed. The latter implementation would allow 256 unique signalling states corresponding to an 8-bit symbol per signalling state. However, the S/N ratio would likely suffer using such a system.

DVB/MPEG-2

International standards were approved for digital video broadcasting in late 1994: these are ISO/IEO 13818-1 (MPEG-2 Systems), 150/IEC 13818-2 (MPEG-2 Video) and ISO/IEO 13818-3 (MPEG-2 Audio). MPEG is short for Moving Picture Experts Group.

MPEG-2 defines a system for the all digital transmission of broadcast TV quality pictures at coded bit rates of between 4 and 9 Mbit/s and is also efficient at the higher bit rates required for HDTV (high definition television).

Today's analogue TV is interlaced and contains both fields and frames. Redundancy blocks are chosen to be either field or frame based, as previously described, depending on the picture content or motion activity at the time. Decoder manufacturers are free, within the constraints of the system specification, on how to implement the MPEG-2 system. Some implementations, of course, may work better than others.

There are certain elements of the DVB philosophy which are common throughout the system. These are:

a Systems are composed of fixed-length packets allowing flexible combinations of MPEG-2 video, audio and data.
b All systems use the common MPEG-2 transport stream multiplex.
c All systems use the service information (SI) system providing details of programming.
d All systems use a common (188, 204) outer Reed Solomon forward error correction code.

e   Modulation and additional channel coding systems are chosen by the requirements of the service provider.

f   All systems adopt a common scrambling and conditional access system.

DVB-S

The DVB-S system is designed for use with transponder bandwidths (26 to 72 MHz) which encompass the vast majority of current TV satellites, such as the Astra and Hotbird clusters.

DVB-S is a single-carrier system best described by the ubiquitous onion analogy. The core of the 'onion' represents the program material (payload) and the outer layers are built up on top of it to protect it from errors during transmission. The official name for this onion is the 'MPEG System Layer'.

The payload may consist of many channels time division multiplexed on to a single carrier in multi-channel per carrier (MCPC) mode or a single channel using satellite transponder sharing in single carrier per channel (SCPC) mode. MCPC maximizes the utilization of the transponder and allows the operator to transmit at full power. The SCPC system allows diverse uplink sites to access a common or shared transponder. An offset QPSK modulation scheme (OQPSK) is used for SCPC to reduce intermodulation interference between carriers sharing a single transponder.

Video, audio and other data, as appropriate, are arranged into fixed-length packets, or containers, of 188 bytes, including a 4-byte header. The following steps are then applied:

a   Every eight-packet header has its syncronization byte inverted.
b   The packet contents are randomized in a pre-defined way.
c   An efficient system-wide outer code comprising (188, 204) Reed Solomon forward error correction is applied to the packetized data, adding a fixed 16-byte overhead of just $(204-188) \times 100/188 = 8.51\%$.
d   Convolutional interleaving is applied to the packet contents.
e   A second, or inner, code is added. This punctured convolutional code can have the amount of overhead adjusted according to the needs of the broadcaster.
f   The signal is modulated on to the satellite broadcast carrier using quadrature phase shift keying (QPSK). Many channels may be time division multiplexed (TDM) into a single MPEG-2 transport stream.

The system adapts automatically to the error characteristics of the channel. The broadcaster can juggle with two variables, the total size of the structure and the inner error correction code within the link budget. The receiver will automatically lock on to the transmitted parameters by rapid trial-and-error. For example, the broadcaster may choose to use a $\frac{3}{4}$ convolutional code rate for the inner code.

DVB-SI

MPEG-2 provides program specific information (PSI) which is intended to help the decoder lock on and decode the MPEG-2 packet structure. This data is transmitted along with video and sound to automatically configure the decoder and allow it to reconstitute the transmitted video information.

MPEG-2 also allows a second, but open, service information (SI) system to be used. Decoder configuration data is available in the PSI, but additional DVB-SI adds additional information which allows the receiver to automatically tune itself. This is necessary due to the very large number of channels that will be available in the future. Clearly the user will need help! A further use for DVB-SI is the development of electronic program guides (EPG).

DVB-SI is composed of four main tables and a further number of optional ones. These are:

a NIT The network information table which groups services from a particular broadcaster. It contains all the tuning information for the IRD (integrated receiver/decoder). This is also used to signal changes in tuning information.
b SDT The service description table tabulates names and assorted parameters associated with a particular service.
c EIT The event information table is used to relay information about the technical parameters of the MPEG-2 multiplex such as FEC, modulation used, etc.
d TDT Current time and date to update the IRD's clock.
e BAT Optional bouquet association table providing groups of services associated with a particular service provider (e.g. Sky multi-channels package). A particular service may belong to more than one bouquet.
f RST Optional running status table provides detailed information about the current program or service and perhaps even information on others.
g ST Optional stuffing tables which, as the name suggests, is used to replace, invalidate or modify other SI tables.

MPEG-2 audio

The MPEG Layer 2 specification (MUSICAM) is used for audio. MUSICAM is a welcome acronym for the following distasteful mouthful: 'Masking pattern adapted Universal Sub-band Integrated Coding And Multiplexing'. The system provides near CD quality at very low data rates and is flexible in that mono, stereo, surround sound or multi-lingual audio may be transmitted. This digital compression technique takes advantage of the domination of one sound element against those of nearby but lower level background sounds or noise which would not be heard even if reproduced with high fidelity. The redundant information is thus not coded.

MPEG-2 video

MPEG-2 video currently comprises four source formats:

1 Low level   This is typical VHS quality or better at about 2 to 4 Mbit/s.
2 Main level   Studio quality ITU-R BT601 around 9 Mbit/s.
3 High-1440 level   This is a high definition format with 1440 samples/ line.
4 High level   This is a high definition format with 1920 samples/line.

MPEG-2 video also comprises a flexible system of compression tools divided into profiles. Each profile defines a different set of compression tools increasing in sophistication and implementation cost. Five such profiles are outlined in (a) to (e) below but the uptake will depend on market demand. Each profile is completely backward compatible with previous profiles in the list (e.g. a main profile decoder will also decode simple profile encoded pictures).

a The simple profile: This uses the fewest compression tools and would be used in the lowest cost receivers.
b The main profile: Has all the tools of (a) plus bi-directional prediction. This gives better quality than the simple profile for the same bit rate. A main profile decoder will also decode simple profile encoded pictures, but at higher decoder cost. Line sequential colour difference signals may also be added to this profile at a later date.
c S/N scalable profile: This profile is intended to provide additional tools to enhance the signal to noise ratio (S/N) in certain difficult cases and is not currently supported in DVB due to the increased decoder complexity required.
d Spatially scalable profile: This profile has similar objectives to (c) and for similar reasons is not currently supported in DVB.
e High profile: This is the Rolls-Royce of profiles and has the capability to code line simultaneous colour difference signals and is intended for the most prestigious applications using very high bit rates.

The first generation of European IRDs will cater for main level, main profile broadcasts with 625 line studio quality in 4 : 3, 16 : 9 or 20 : 9 aspect ratios. Broadcasters can decide on using either variable or constant bit rates bearing in mind that higher bit rates, in general, show pictures with less coding artefacts. Today's PAL and SECAM picture quality can be achieved with bit rates of about 5 to 6 Mbit/s. The MPEG system is particularly suited to film at 24/25 frames per second since it is easier to code than videotape and looks good even at low bit rates. A typical arrangement for a complete transmission and reception system is shown in Figure 11.16.

DirecTV

DirecTV, a US operation, is claimed to be the first to supply digitally compressed television to the home in 1994. The system, using all 32 (or more) licensed DBS frequencies at 101°W, was designed to distribute 150 channels to consumers equipped with a small 45 cm dish and a suitable IRD. The uplink frequencies are in the 17.3 to 17.8 GHz band and the downlink frequencies in the 12.2 to 12.7 GHz band. Each transponder is capable of supplying four live action TV channels or eight TV film channels.

The specifications of each satellite are as follows:

Azimuth location   101° W longitude
Number of transponders carried   16
Each transponder, power   120 W
Each transponder, bandwidth   24 MHz
Polarization   circular
Footprint   the entire area of the United States.

The compressed video information, from whatever source, is *time division multiplexed (TDM)* into a single serial data stream that allows the number of bits per second allocated to each source to be varied according to content or channel format. Before compression, the signal data rate is 270 Mbit/s but, after MPEG compression, this is reduced to the range 3.75 to 7.5 Mbit/s, representing a compression ratio of between 36 : 1 and 70 : 1. Time division multiplexing allows the transmission rate to approach 30 Mbit/s, the equivalent of four to eight TV channels, depending on content.

The integrity of the DBS signal is maintained by a powerful error-correction system, based on a mixture of Reed Solomon and convolutional coding. The DBS signal is thus fully protected on its journey to the customer's decoding equipment and virtually guarantees 99.7% 'average year' availability. Channel coding, of which there are many species, are rather complex mathematically. Fortunately, like DCT, it need not concern us since the details are buried within a chip. The data stream is modulated on to the carrier by QPSK.

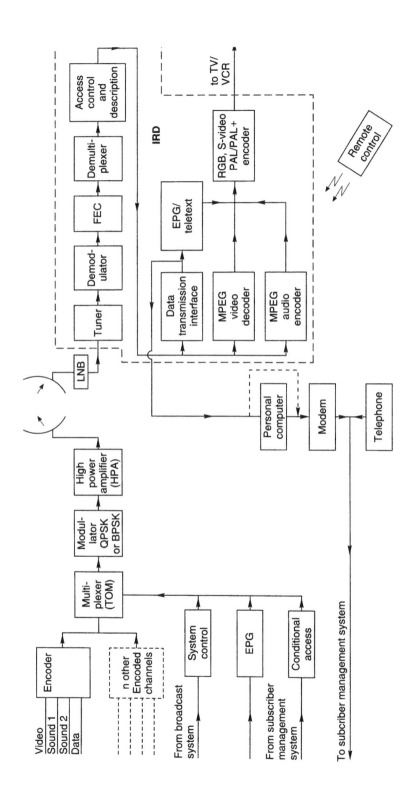

Figure 11.16 *Typical arrangement to transmit and receive DVB via satellite*

12 Satellite IF distribution

Introduction

The satellite IF is the intermediate frequency output from a frequency down-converter (low noise block (LNB)) and is distributed via coaxial cable, within the range 950–1750 MHz, to one or more satellite receivers. IF distribution describes the method by which this group of frequencies is delivered to two or more receivers.

From the installation point of view, perhaps the simplest dish sharing scheme is that shown in Figure 12.1. It provides two independently switched V/H outputs and is ideal for houses where just two receivers are required, or where adjacent semi-detached householders agree to use the same dish. A LNB with dual V/H switched outputs is central to the scheme. It is important to be aware that another type of dual LNB exists where the two polarizations are permanently fixed simply to elim-inate the need for an orthomodal transducer (OMT).

Definition of technical terms

It may help to glance over the following definitions before reading the text. Pay particular attention to synonymous expressions.

Active device: Any distribution component that provides amplification. That is to say provides a net gain rather than a loss. Some components such as active splitters only provide sufficient gain to compensate for their inherent splitting loss, so are deemed lossless.
Branch: (See *Tap line*.)
Coupling loss: (See *Tap Loss*.)
Decoupling: (See *Isolation*.)
Insertion loss: (See *through loss*.)
Isolation: Passive devices are normally directional, in that signals are allowed to pass easily in one direction only. The ratio of forward to reverse transmission expressed in decibels is known as the 'isolation' or 'decoupling'.

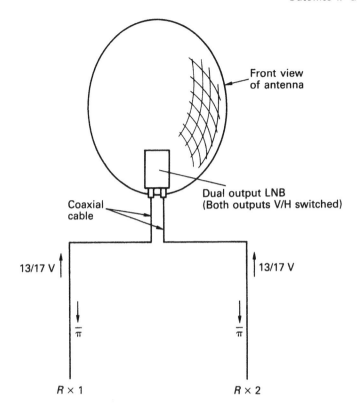

Figure 12.1 *A simple dish sharing system using a twin V/H switched LNB*

Passive device: any distribution component that does not provide extra amplification to compensate for its inherent loss (e.g. capacitors, inductors and resistors, but not semi-conductors).

Return loss: Any device placed in a transmission line will reflect a certain amount of signal back toward the source. The ratio of the signal reflected back to the incident signal, expressed in decibels, is known as the 'return loss'.

Star wired: Single cables extending outwards from a common point.

Tap line: The cable branch from the trunk to the subscriber's wall plate.

Tap loss: The attenuation of the signal in decibels incurred between a trunk level input and a tap line.

Through loss: The attenuation of the signal in decibels due to the presence of a passive (non-amplified) tap in the trunk.

Tree wired: Cables forming a central trunk with branches at various points along its length.

Trunk: The main distribution cable of a tree network which is tapped off at various points to subscribers' dwellings (akin to a water main in a plumbing system).

Side loss: (See *Tap loss*.)

IF distribution v channelized distribution

Channelized distribution is where a limited number of individual satellite channels are demodulated using a bank of satellite receivers at the head end. The channels are then remodulated and combined, as standard VHF or UHF TV signals for direct tuning by domestic television sets.

Existing terrestrial TV distribution systems slot into the frequency range 40–860 MHz. Since these elderly systems were not designed for the higher frequencies involved with satellite IF, the cables, taps and splitters, etc., will inherit high attenuation figures and will clearly be unfit for the job. This leaves two possible alternatives:

1 Add a channelized satellite TV system to the existing terrestrial system.
2 Strip out the entire system and start again with an IF distribution approach capable of handling the range 40–2060 MHz.

There are significant advantages in using the IF distribution approach where the number of subscribers is relatively small, say, below 32. This number is not a maximum, just that the cost/performance trade-off needs examining before decisions are made. To help in this decision, compare the main advantages and disadvantages by examining Table 12.1. For new small distribution systems there is no contest!

Satellite IF distribution methods

Neglecting novel hybrid combinations, there are three distinct network wiring configurations:

1 *A loop wired network* where each subscriber wall plate (outlet) is passed by one or two cables (depending on the number of polarizations required) in a loop-through fashion. Loop wired systems usually employ a launch amplifier and specially designed wall plates with stepped, but high value, coupling losses to prevent oscillator interference between connected receivers. The wall plates have a low through-loss of around 1–1.5 dB and a high isolation so that signals will not pass easily in the reverse direction. Loop wired systems have a central power supply so no d.c. control voltages from receivers are to be allowed back into the system via the wall plate. The LNB supply from the receiver should also be switched off or disabled. Although relatively easy to wire and consuming a minimum of cable, the method is not currently a recommended method. A rewire to a tapped tree or star-wired system should always be considered. The obvious disadvantage with loop wired systems is that wall plates are often disconnected by subscribers when redecorating and unwittingly cut off services to other subscribers.

Table 12.1 *Comparison of IF distribution and channelized distribution*

IF distribution	Channelized distribution
• Probably better quality since signals are processed by subscribers' own equipment as with direct from dish reception	• Prone to interference problems if not carefully designed and set up
• Viewers free to select any satellite channel from a large number. Important if a fixed satellite cluster is targeted, e.g. Astra 1 (A to E)	• Viewers restricted to a few channels, chosen by the installer, from the total available
• Radio, stereo, and multi-language sound sub-carrier channels available	• Usually sub-carrier channels are not available
• Each viewer provides their own independent satellite receiver with decoders of their choice (unless supplied by installation company on contract)	• No extra equipment is required by viewers, but more needs to be provided by installation company (added cost and complications)
• Low cost for small systems, but less so as the number of subscribers increases	• High cost for small systems, but becomes more economical as more subscribers are connected
• Re-cabling may be necessary for existing systems	• No re-cabling necessary
• Easy to install, set up and maintain with low cost signal monitoring equipment	• Difficult to install, set up, and maintain

2 *A tree network* where one or two cables (depending on the number of polarizations distributed), called 'trunks' are run on the outside of the building, convenient corridor, or stairwell. Passive taps are connected at intervals along the length to form branches to subscribers' dwellings. A satellite IF launch amplifier is used at the head end to compensate for losses incurred with passive taps and cable attenuation. In two-cable systems, designed to distribute two polarizations simultaneously, a V/H switch is also needed at each pair of tap points unless the receivers are equipped with two IF input sockets. Although not common in the UK, such receivers are widely available in continental Europe. The taps used have a low trunk loss of 1.5–5.5 dB, depending on the quality and manufacturer. Tap losses are available in stepped values to allow for decreasing trunk signal levels. Tree networks, used in conjunction with repeater amplifiers, would be suitable for large systems in excess of 32 outlets. They can be used effectively for smaller systems, but it is important to remember that equipment costs would be higher compared with the alternative of a star-wired system. However, cable usage is lower using tree systems, and the system tends

to look more aesthetically pleasing if tacked on the outside of a single building.

3 *A star network* where a separate cable is installed from the head end to each subscriber. Such systems rely heavily on the use of splitters and, where two polarizations are required, V/H switches mounted at the head end. The latter are available in convenient matrices called 'multi-switches'. From the head end distribution complex, the cables branch out in a 'star-like' fashion to each subscriber's wall plate. A clear advantage in star-wired systems is the central localization of distribution components which makes fault tracing considerably easier. At the head end, VHF/UHF terrestrial signals in the range 40–860 MHz may be combined with the satellite IF signals using diplexers and carried on the same cable to the subscriber where the signals are separated out at the wall plate.

Distributing a single polarization

The simple case of distributing a single polarization, say either left- or right-hand circular polarization (LHCP or RHCP) from a direct broadcast service (DBS) satellite, requires the use of splitters for star networks or a single trunk for tree networks. Figure 12.2 shows a simple star-wired network suitable for distributing a single polarization along with terrestrial signals. It is always better to use lossless active splitters with d.c. bypass through any of its outputs for star-wired systems. The system

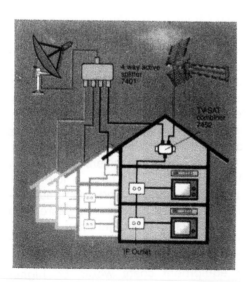

Figure 12.2 *Star network distributing a single polarization plus terrestrial signals (Televes)*

works like this. D.C. voltage from whichever satellite receiver happens to be switched on at the time powers both the active splitter and the LNB. The combiners (diplexers) must be fitted after the splitter, as shown in the diagram. One diplexer is needed for each subscriber cable. The active splitters may be cascaded to form more outlets if required. Tree networks, using launch amplifiers and taps can also be used for single polarization distribution as shown in Figure 12.3.

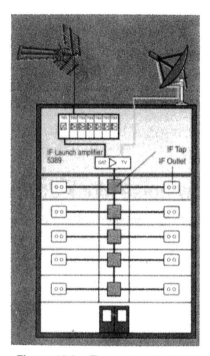

Figure 12.3 *Tree network distributing a single polarization plus terrestrial signals (Televes)*

Distributing two polarizations

Planning an IF distribution system for two polarizations is complicated by the fact that each connected subscriber (viewer) will need to select independently signals of either polarization where a dual polarization scheme is adopted on a target satellite.

The outdoor unit

In order to distribute the two polarization senses simultaneously to a number of separate satellite receivers the following equipment is needed:

1   An antenna slightly larger than that needed for normal direct-to-home (DTH) use, to compensate for the extra noise introduced from splitters, switching systems and cables. As a bonus the protection against rain fades is also increased.
2   A feed capable of receiving the two polarization senses.
3   An orthomodal transducer to separate out the two senses of polarization.
4   Two separate LNBs, one to receive each sense.

Note: items 2, 3 and 4 may be replaced with low-cost dual output LNBs designed for the purpose. Such units have a significant advantage in as much as the likelihood of rain water penetration is less than for a multi-component arrangement with many flanges.

Using a single cable star-wired network

Signals of the two polarizations are split at the head end using a number of splitters and V/H switch poles equal to the number of subscribers. V/H switch units are available as discrete switches or in matrices called 'multi-switches'. Multi-switches are available which have integrated all the necessary diplexers for terrestrial TV thus providing a common input. Some models have power injectors built in but some do not. Figure 12.4 shows a typical arrangement using Televes components to supply

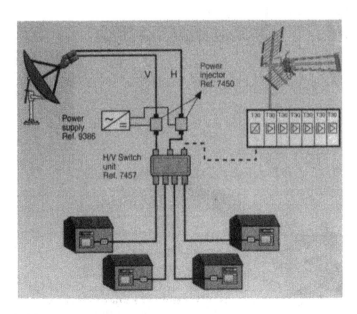

Figure 12.4   *Star network distributing two polarizations plus terrestrial signals to four dwellings (Televes)*

four outlets with satellite IF and terrestrial TV. Two cables are fed from a pair of LNBs, one carrying horizontally polarized signals and the other carrying vertical ones. The LNBs are powered from a separate power supply via power injector modules (also known as 'd.c. blocks'). A method is needed for each subscriber's satellite receiver to select (switch in) the choice of signal polarization. The polarization selection is done by V/H multi-switches. The standard 13/18 V polarization selection circuitry built into each compatible satellite receiver is used to control the multi-switch rather than the polarizer directly. Thus if vertical polarization is selected on the receiver, 13 V is produced across the coaxial conductors, and the multi-switch would respond by switching through vertically polarized signals similarly if horizontal polarization is selected (17 V) the multi-switch would respond by routing through horizontally polarized signals. The system can grow by simply adding extra splitters and multi-switches according to the general plan shown in Figure 12.5. Note that the use of a terrestrial distribution amplifier eliminates additional splitters needed to feed the multi-switches.

We need not concern ourselves with a polarizer specifically, since we will not be using one (i.e. the two polarizations are derived using two orthomodal LNBs or a twin LNB), but we need to be careful over compatibility between the receiver polarizer control method and that of the

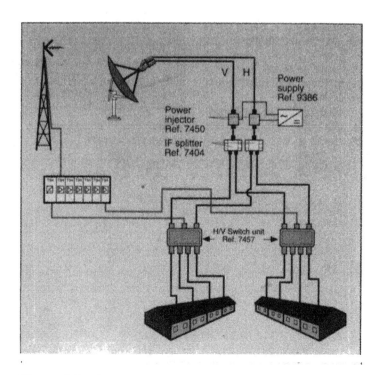

Figure 12.5 *Star network distributing two polarizations plus terrestrial signals to eight dwellings (Televes)*

multi-switches. As detailed in previous chapters there are three basic types of receiver polarizer control circuitry, 13/17 V level shift for V/H switched, constant current for electromagnetic, and pulse for mechanical polarizers. As with the polarizers they effectively replace, multi-switches are also available in the same three basic control types. However, the 13/18 V level shift type is by far the more common, since no extra wiring is needed per subscriber. Dual sheathed figure-of-eight cable with two or three extra conductors would be needed to control the others and would complicate the system unnecessarily. Since 95% of receivers currently on the market are capable of simple 13/17 V control it would be unwise to base a distribution system on the other more complex methods except in very small domestic systems of perhaps two to four outlets. However, odd incompatibility cases, with individual receivers, can usually be rectified with a suitable subscriber outlet interface.

Using a twin cable tree-wired network

Tree networks employ two cables to distribute the two polarizations along two separate trunks, as shown in Figure 12.6. If satellite receivers with two IF input sockets are to be used, the two branched tap lines, one per polarization sense, may be directly connected via a suitable wall plate to the satellite receivers. Receivers with a single IF input require an additional V/H switch unit to be fitted before the outlet as indicated in Figure 12.6. The IF taps are available for both interior use, say, where a central stairwell is used for the cabling or exterior use where the cables are tacked onto the outside of the building. An IF launch amplifier is used both to compensate for losses due to the use of passive IF taps and, of course, cable attenuation which is high at IF frequencies. Systems employing launch amplifiers are designed using stepped tap losses with the highest being the first after the amplifier. The launch amplifier may include an integral power supply to power up the LNBs. If not a separate head end power supply and power injectors would be needed.

Accessories and components

This section reviews some of the basic components of a satellite IF distribution system and forms a guide to the function and selection of components.

Satellite IF/terrestrial diplexers

When designing an IF distribution system, try to arrange that as much of the distribution components as possible are at a common location,

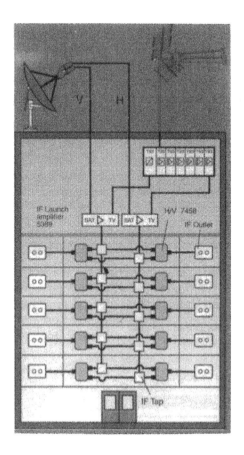

Figure 12.6 *Tree network distributing two polarizations plus terrestrial signals (Televes)*

preferably indoors, and include terrestrial signals such as VHF/UHF TV and FM radio (i.e. 40–2060 MHz). The higher limit is to allow for future expansion. If AM radio is also required the lower limit will need to be extended down to 0.15 MHz. A component known as a 'satellite IF/terrestrial diplexer' allows satellite IF signals and VHF/UHF terrestrial signals to be combined to a single output. These units are often waterproofed for outdoor fitting and employ high isolation bandpass filters between the input ports to prevent LNB noise interfering with terrestrial signals. Some have an option, with power supply and d.c. path for the connection of an external mast head VHF/UHF amplifier. If a satellite IF launch amplifier is used the diplexer is fitted after the output unless a specific terrestrial input is provided. In tree networks, only two diplexers of this type are needed, but in star networks each subscriber cable needs its own diplexer at the head end. Multi-switches and diplexers are frequently integrated into a single unit so are 'transparent'. At each subscriber outlet the reverse process is installed to separate the VHF/UHF terrestrial signals and the satellite IF. In two cable tree networks an alternative is to combine vertical polarization with FM radio on one trunk and horizontal polarization with VHF/UHF TV on the other.

Splitters

Splitters are characterized by the number of equal loss outlet ports provided from a single input; thus if four outputs are provided it would be called a 'four-way splitter'. There are two types: passive and active. Passive types have inherent losses of typically 4 dB for a two-way, 8 dB for a four-way and 11 dB for an eight-way splitter and are also available in doubles, that is to say two passive splitter modules in one case. Active splitters are effectively lossless since amplification is provided to compensate for the losses incurred. High-quality versions have a sloped gain to compensate for cable loss at the higher frequencies. Most active splitters have a d.c. bypass facility through any of its outputs for powering an LNB from any receiver that may be switched on at the time. With passive splitters it is common for just one port to pass d.c. from the input.

V/H switches

These are a single switch to select between two polarizations either from the same satellite of different satellites. Most are waterproof for mounting outside near the antenna. The signals and d.c. supply to the two LNBs are switched together so that the receiver powers up a single LNB at a time. This has the added advantage of eliminating possible interference between the two LNB local oscillators. Types are available to interface with most receiver polarizer control systems (i.e. pulse, electromagnetic, or 13/17 V). V/H switches are used a lot in two-cable tree networks and are often mounted alongside the two taps to each subscriber.

V/H multi-switch units

Multi-switches are matrices of V/H switches in the same case. They allow many receivers to select independently a fixed number of polarizations. They are characterized by the number of polarizations input to the number of outputs to receivers. Inputs of two polarizations, usually horizontal and vertical from the same satellite, are the most common, so if four outputs to receivers are provided then it would be called a '2 in, 8 out' multi-switch. These are controlled by standard 13/17 V or 14/18 V switching. Multi-switches with four input polarizations (i.e. 4 in, 4 out) are also available for dual satellite use but require receivers capable of providing 0, 3, 6 and 9 V control. Multi-switches are effectively lossless and have a typical isolation between polarizations of 30–40 dB. In conjunction with splitters they can be cascaded to provide quite large distribution systems suitable for blocks of flats and hotels. If terrestrial signals are required then multi-switches with inbuilt diplexers may be used in the design of single cable star networks.

Satellite and terrestrial multi-switches

Switches known as 'diplexed multi-switches' are similar to the above in operation but allow two satellite polarizations and terrestrial signals to be distributed. High isolation filters are used to diplex terrestrial signals onto both vertical and horizontal polarizations. This is a considerable advantage since, as mentioned earlier, a separate diplexer would otherwise be needed for each subscriber cable of a single-cable star network. Low-cost versions have a typical 15 dB loss for terrestrial signals, but higher quality units have an internal amplifier to compensate and are effectively lossless.

Power injectors

These units, in conjunction with a suitable supply, inject d.c. power into a descending coaxial cable towards the LNB whilst blocking d.c. from passing down into the rest of the distribution system or receivers. They have typical insertion losses of <0.5 dB and a pass band of 10–2060 MHz. Dual units are designed to power up dual output switching LNBs (or two fixed single LNBs) where one injector is arranged to provide 13 V and the other 17 V.

Satellite IF launch amplifiers

Satellite IF launch amplifiers are used to compensate for distribution losses in passive systems such as tree networks (one per cable). They often have a positive gain slope, of 2–5 dB, so higher frequencies have increased amplification to compensate for the frequency response of coaxial cables. Typical gains are 18–28 dB, although some go as high as 40–45 dB. If a special input socket is not provided, terrestrial signals should be combined after the output. In cases where this is not practical the IF amplifier must be bypassed using a pair of diplexers placed at the input and output. Terrestrial signals may be amplified separately if considered necessary. D.C. powering options are wide and internally switchable. For example: switchable 13/17 V supply to the LNB;

optional trunk power at the output for supplying additional line amplifiers; line power to terrestrial TV amplifiers; or powered externally by line power to either input or output sockets.

Line amplifiers

Line amplifiers (repeater amplifiers) are used to compensate for losses incurred between the LNB and receiver due to long cables or the use of passive splitters. They come in two qualities, those with sloped gain and those with a flat response; the latter are quite adequate for the vast majority of domestic installations. They should be positioned after some attenuation has occurred, say 15 dB. The output level of a line amplifier should not exceed that of the launch amplifier output or otherwise should not exceed the LNB output level by more than about 8 dB.

Taps

Taps are passive devices normally used in two-cable tree networks in conjunction with a launch amplifier and are used mainly for larger IF distribution systems. The bandwidth of taps is typically 40–2060 MHz and so can be used to distribute VHF/UHF TV at the same time. They come in a variety of tap loss (side loss) values (typically 30, 27, 20, 15, 12 and 10 dB) with much lower trunk loss (trough loss) values which can be anywhere between 1.5 and 5.5 dB depending on quality and manufacturer. For example, using a suitable launch amplifier it may be possible to arrange for 30 dB taps on the first few floors and then 27 dB taps for the next few, and so on. Professional installations use saddle and clamp connectors and have a very low trunk loss, but for domestic installations simple F connectors are the norm although they are slightly inferior in this respect. Multi-taps (up to four in the same case) are available which provide for four taps from the same trunk branch. These are ideal for four flats per floor installations. Special line end taps are available to terminate the trunks.

V/H switching taps

These taps are used in two-cable tree networks where satellite receivers are not equipped with two separate IF inputs but are capable of 13/17 V polarization selection via a single IF input socket. For example, multi-tap modules containing four switching taps from two separately connected trunks are ideal for four flats per floor installations (or use two for eight flats per floor). Like standard taps, they are also available in a variety of different side loss values to compensate for decreasing trunk levels and a choice of line end values to terminate the trunks.

Polarizer control interface units

In view of the differing polarizer types and the various controlling methods adopted in satellite receivers, specific interface units are available to match certain combinations. Interfaces are available which allow a receiver equipped with electromagnetic polarizer control to switch either a 13/17 V LNB or 13/17 V multi-switches. This is achieved by way of a d.c.-to-d.c. converter to increase the 15 V LNB supply voltage to the LNB (or multi-switch matrix). Another type allows a 13/17 V switching receiver to control an electro-magnetic (ferrite) polarizer. Such interfaces provide uni- or bi-directional current, the polarizer current being drawn from the LNB supply. Both types have a typical insertion loss of less than 1 dB.

Coaxial cables

Coaxial cables were covered in detail in Chapter 3. When upgrading a terrestrial signal to include satellite IF, it is important to check the overall condition, type and the attenuation per 100 m figure. Cables designed for terrestrial TV may have attenuation figures of around 40 dB per 100 m at 1750 MHz, whereas those for satellite IF (e.g. CT100) are typically 28.3 dB per 100 m at this frequency. Although it may be possible to get away with using TV grade cables in small systems with short run lengths, for larger systems a re-cable must be considered using double-screened low-loss cables. New systems should always be installed with double-screened cables.

Wall plates

Star networks employing multi-switches require that d.c. control voltages are passed to them via the wall plate (subscriber outlet). For two-wire tree systems, not using extra V/H switches, no d.c. is necessary since both polarizations are present simultaneously via two sockets. In such cases the wall plate should block d.c.

For single-cable networks there is a choice between: three-way wall plates which diplex the incoming signals into separate FM radio, TV, and satellite IF outlets; two-way wall plates that diplex to one FM/TV and one satellite IF output; and two-way wall plates that diplex to FM only and a wide band TV/SAT IF outlet. The last of these are ideal for TVs and VCRs with built in satellite tuners. For two-cable systems twin wide band outlets are available up to 1750 MHz for the flexible connection of TVs and VCRs with built in satellite tuners. Such equipment must have two separate IF input sockets fitted.

Band shifters

Where signals (or polarizations) from two or more different satellites are distributed (for example, Astra and any of the Eutelsat II series) the channel frequencies when down-converted may often clash leading to intolerable interference. The use of specially designed band shifters is one solution. One of the satellite IF feeds is fed through a band shifter to translate it to an alternative frequency range within the satellite IF band. These fixed units are only available for the more popular pairs of satellites. More versatile and adjustable systems based on this idea are under development.

Satellite receivers

It is important to match satellite TV receivers (or TVs or VCRs with integrated satellite receivers) to the distribution system. For two-wire systems, without additional V/H switches, they should have two IF input sockets. Additionally, they should have the option of disabling the LNB supply from the IF input sockets. Receivers capable of 13/17 V polarization selection are required to control most V/H switches via the coaxial cable.

Small integrated systems

Rather than design a system using discrete components there are many pre-wired solutions which contain all the necessary components of a star-wired switching system in a single 'lock-up box'. Such units, particularly suited to small blocks of flats and hotels, are capable of distributing terrestrial TV and satellite IF to between 4 and 24 dwellings. An advantage in using these systems is that the operator has total control over the system from a single location and the wiring is considerably simplified.

Satellite IF pre-processing

New developments in the satellite IF distribution field allow versatile pre-processing of the satellite IF band. Since the available frequency range of the IF band is fixed, it imposes a limit on the number of channels that can be provided on an IF distribution system from a number of satellites. Where signals (or polarizations) from two or more different satellites are required (say, Astra and any of the Eutelsat II series), the channel frequencies when down-converted may either clash or, if band shifted, exceed the range of the allowed IF frequency band. Between the various LNB feeds and the distribution system, equipment is installed whereby it is possible for the installer to choose or re-arrange in frequency the final composition of the distributed satellite IF band. This is performed by selecting frequency slices from any of the available feeds. For example, certain low-priority channels from one satellite may be ditched in favour of more popular ones from another. The use of the extended frequency range 10.70–10.95 GHz on Astra 1C to 1E, which down-converts to 700–950 MHz using a low side 10 GHz LNB local oscillator, is beyond the tuning range of many satellite receivers. This may also result in satellite IF frequencies clash-ing with terrestrial signals in some areas. With equipment such as this they may be slotted in elsewhere in the IF band, perhaps even between, say, 1700 and 2060 GHz. The overall effect is to allow control over both the frequency and content of the satellite IF band so the installer, rather than the satellite operator, may choose which channels appear on the system.

Designing simple systems

One of the beauties of working with logarithmic units, such as decibels, is that multiplication and division of gains and losses simplify to addi-tion and subtraction, respectively. Gains are additive in decibel terms and losses are subtractive for items such as cable runs, splitters, taps, etc. This makes designing a satellite IF distribution system a relatively easy task.

As a guide, the minimum and maximum input signal levels (margin of input) required for typical terrestrial and satellite receivers are given in Table 12.2. Say the signal level at the LNB output is 73 dBuV, it is ap-parent from Table 12.2 that we have about 20 dB to play with before losses start to get critical. Thus we arrive at a rule-of-thumb figure of about 20 dB for the maximum loss that may be introduced between the LNB output and any connected satellite receiver. If a launch amplifier is used in conjunction with taps then the maximum allowable loss between the launch amplifier and any connected receiver must not exceed the launch amplifier gain plus 20 dB. For example, if the gain of the launch amplifier is 25 dB the maximum losses must not exceed 25+20=45 dB to any receiver. If the signal is likely to drop below this figure, in large systems, repeater amplifiers (line amplifiers) may be used at suitable intervals along the trunks. Tap loss values are chosen such that the signal level does not exceed the maximum for any outlet. The latter is particularly important for the first few outlets nearest the launch amplifier. In reality, it is a good practice to aim at an LNB out-put level of about 78–80 dBuV to allow for losses from the LNB to the head end distribution equipment.

Cable attenuation should be calculated at 1750 MHz, the worst case condition. For CT100, and equivalent, this is 28.3 dB/100 m or 0.283 dB/m. When planning, allow about 5 m between floors which corre-

Table 12.2   Guide to the minimum and maximum signals levels required at wall plate outlets

Frequency (MHz)	Receiver margin of input (75 ohm)					
	Minimum			Maximum		
	dBuV	dBmV	mV	dBuV	dBmV	mV
40–860	60	0	1	80	25	10
950–1750	52	–8	0.4	73	13	4.5

sponds to a cable loss per floor using CT100 of $0.283 \times 5 = 1.42$ dB. A tap line from a trunk may be, say, 10 m per flat (measure it at the site survey stage); this would give a cable attenuation of 2.83 dB. Table 12.3 shows the attenuation for multiples of 5 m lengths of CT100 cable. For a tap line we also need to add in the losses for a diplexed wall plate which is typically 2.5 dB at 1750 MHz (sat.IF) and 1.5 dB at 860 MHz (terrestrial). Thus a 10 m tap line and wall plate would lose just over 5 dB in total. For simple star-wired systems using active splitters and lossless multi-switches the maximum cable length allowing a 2.5 dB loss for the wall plate is about 60 m. If passive distribution components were used this would be significantly less since their insertion losses would need to be subtracted from the maximum 20 dB allowance.

Example: calculating a single-cable star-wired system
Figure 12.7 shows a simple terrestrial and satellite IF distribution system for eight subscriber outlets using a pair of 2×4 satellite/terrestrial

Table 12.3   *Attenuation for various lengths of CT100 cable (or similar)*

Length (m)	Attenuation (loss)
1	0.28
2	0.56
3	0.85
4	1.13
5	1.42
10	2.83
20	5.66
30	8.49
40	11.32
50	14.15
60	16.98
70	19.81
80	22.64
90	25.47
100	28.30

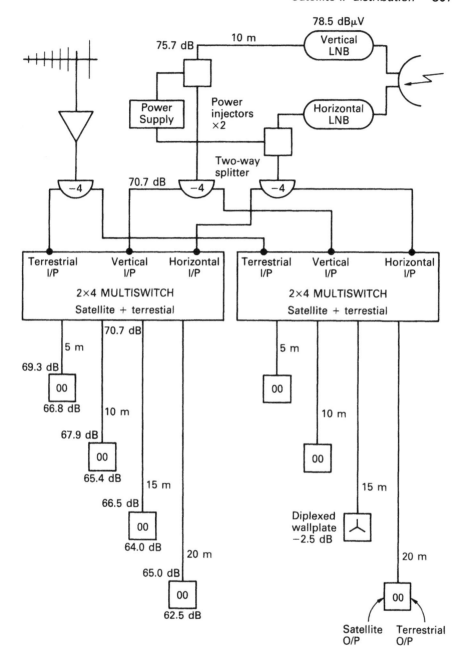

Figure 12.7   *An eight-outlet, star-wired system calculation*

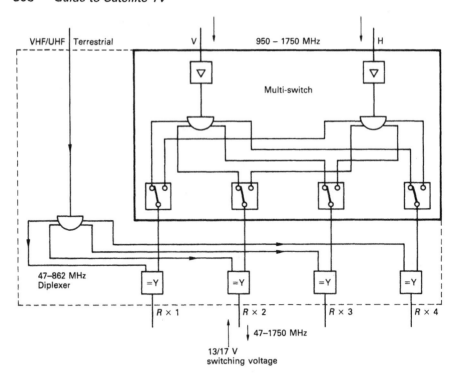

Figure 12.8 *Internal schematic of a multi-switch system*

multi-switches and two two-way passive splitters. Although only eight outlets are shown, systems can be enlarged in a similar fashion using more splitters and multi-switches as necessary. Values shown are typical and may vary from manufacturer to manufacturer.

The satellite IF signal levels at various points are shown, along with CT100 cable lengths, as a guide to the simple add/subtract calculation method. Terrestrial signal levels can be calculated in a similar way. The internal components of a satellite/terrestrial multi-switch are shown in Figure 12.8. The actual switching can be performed by analogue switch chips. Note that a separate satellite/terrestrial diplexer is needed for each output. If just the multi-switch section was used it would be necessary to have separate diplexers at each output – a considerable complication. In some versions the terrestrial splitters have internal amplification to compensate for the losses due to splitting and diplexing. Since the satellite polarizations are internally amplified within the multi-switch, there is no overall loss due to the switching matrix or the diplexers in the IF path.

Example: calculating a two-cable tree distribution system
Figure 12.9 shows an alternative two-wire terrestrial/satellite IF distribution system for a small hotel using a launch amplifier and stepped

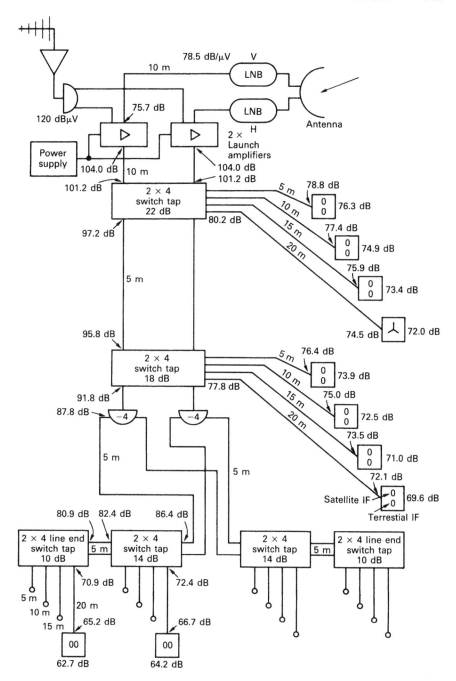

Figure 12.9 *Calculation of a 24-outlet, two-cable, tree-wired system*

switch taps. The signal levels obtained at various points are given along with CT100 cable lengths. The gains of the launch amplifiers are usually adjustable, so once you have calculated the various distribution losses at some arbitrary value, say 104 dB, you can adjust the whole lot upwards or downwards to get all levels within the receiver's margin of input. It is not usually necessary to calculate every branch as shown in the diagram. For outlets near the launch amplifier you only need to calculate the shortest branch to check that the output level at the wall plate is less than the receiver's maximum margin of input. Longer branches will always be a lower level. For outlets near the bottom of the trunk it is best to calculate the wall plate output for the longest branch to check that the signal level is above the receiver's minimum margin of input, shorter branches being at a higher level. Although a small distribution system is shown, much larger ones may be devised along the same lines. In such cases line amplifiers may be employed as repeaters to boost the trunk levels, provided about 15 dB losses have occurred since the launch amplifier. Taps, including switch taps, normally pass d.c. down the trunks to power repeater amplifiers. Note that passive splitters are used to diverge the trunks to suit the shape of the building and wiring convenience. In the diagram, the wall plate through-loss is taken as 2.5 dB, the passive splitter loss 4 dB, the tap through-loss 4 dB, and the stepped tap side losses 22, 18, 14 and 10 dB.

Design, installation and safety

When designing a system it is always advisable to stick with products from a single manufacturer of your choice. You will need their product/technical specification catalogue from which to obtain component data for your specific design. Product ranges are usually well thought out and components designed to complement each other in a predictable fashion. Since all the hard work has already been done, system level design then becomes a relatively easy task. If active splitters and loss-less multi-switches are used for smallish star networks you do not have to worry unduly about losses. Whether installing new systems or upgrading existing MATV systems it is important to check that all cables, splitters and taps are capable of operating at satellite IF frequencies (2060 MHz now seems to be the norm). If not, they should be replaced. If a d.c. supply is needed for LNBs or, in the opposite direction, for line amplifiers, it is necessary to check that they are capable of passing it safely.

The head-end equipment may be located outside if waterproof, but more commonly an attic space or utility room is preferred. Fortunately most modern buildings have suitable conduit or ducting already installed. For older buildings the cabling, and taps if required, may be tacked on the outside of the structure or perhaps within a central stairwell. A SMATV licence is not currently required in the UK to serve a pair of semi-detached houses, a hotel, or a building where all dwellings can be reached from a single entrance.

Distribution systems should be designed such that dangerous voltages cannot be passed from one outlet to another or back into the system itself. Safety requirements are prone to change so it is always wise to check the latest recommendations periodically. Membership of a recognized trade association helps here. At the time of writing the following were in effect:

1 The coaxial braiding of each outlet should be earth bonded (i.e. connected to a common central earthing point).
2 Special safety isolated wall plates should be used conforming to or the European standard CENELEC HD195 S6. Some are designed to pass d.c. control voltages in the opposite direction whilst at the same time preventing dangerous voltages which could occur under certain receiver fault conditions.

Selecting a suitable dish size

To select a suitable dish size (on paper) to feed an IF distribution system it is advisable to perform a link budget calculation as outlined in Chapter 5. In addition it may be wise to check that the LNB output is greater than about 78.5 dBuV using a method such as that outlined below. See Appendix 2 for time-saving computer software.

Step 1: perform the link budget as outlined in Chapter 5 for the dish size and LNB noise figure you propose to use to check that signal availability is at least 99.7% of an average year (99% worst month).

Step 2: calculate the spreading loss, L_S, using:

$$L_S = 10 \ \log(4\pi D^2) \quad \text{(dB m}^2) \tag{12.1}$$

where D is the path distance between receive site and satellite (m). (Equation 5.5 in Chapter 5 gives D in kilometres, so multiply the result by 1000 to convert to metres.)

Step 3: calculate the effective area of the dish, A_E, using:

$$A_E = 10 \ \log[\pi(d/2)^2\eta] \quad \text{(dB m}^2) \tag{12.2}$$

where: d = diameter of antenna (m)
  h = fractional efficiency of dish (i.e. $h = 0.67$ for a 67% efficient dish).

Step 4: calculate the carrier power, C, at the LNB output using:

$$C = \text{EIRP} - L_S + A_E + G_{\text{LNB}} - A_{\text{rain}} - A_{\text{atm}} - \alpha - \beta \quad \text{(dBW)} \tag{12.3}$$

where: EIRP   = the effective isotropic radiated power at receive site (dBW)

A_E = effective area of antenna (dB m^2)

L_S = spreading loss (dB m^2)

G = gain of LNB (dB)

A_{atm} = gaseous attenuation due to atmospheric absorption (dB)

A_{rain} = rain attenuation for a given percentage of the time (dB)

α = coupling loss (dB) by waveguide components (e.g. OMT)

β = losses due to antenna pointing errors, polarization errors and ageing (dB).

Note: A_{rain} can be neglected in clear-sky calculations.
The combined attenuation from rain and gaseous absorption can be taken as around 2–2.5 dB for most of Europe in the 12 GHz band.

Step 5: calculate the LNB signal output level using:

$$V = \sqrt{10^{0.1C} \times R} \quad \text{(V)} \tag{12.4}$$

where R is the system impedance (ohms) (e.g. 75 ohms).

Step 6: Convert V to dBuV using Equation 12.5 (or dBmV using Equation 12.6):

$$\text{Signal level} = 20 \ \log\left(\frac{V}{10^{-6}}\right) \quad \text{(dBuV)} \tag{12.5}$$

$$\text{Signal level} = 20 \ \log\left(\frac{V}{10^{-3}}\right) \quad \text{(dBmV)} \tag{12.6}$$

Example
Calculate the LNB signal output level in dBuV for a 90 cm dish targeted at the Astra cluster (19.2°E) in Brighton, UK, given that the path distance to the satellite is 38 696 km, the EIRP is 52 dBW and both 1.3 dB LNB gains are 55 dB. Allow 0.3 dB for antenna pointing and polarization errors and 0.2 dB for coupling losses (OMT).

Step 1: performing the detailed link budget calculation outlined in Chapter 5 for 99.7% signal availability (average year) results in the following values:

Atmospheric absorption = 0.14 dB
Rain attenuation = 0.90 dB
Noise increase due to rain =1.13 dB
Downlink degradation = 2.04 dB
C/N (clear sky) = 17.68 dB
C/N (degraded sky) = 15.35 dB
S/N (clear sky) = 49.91 dB
S/N (degraded sky) = 47.58 dB (CCIR grade 5)

Step 2:

$$L_S = 10 \log [4 \times 3.142 \times (38696 \times 10^3)^2]$$
$$= 10 \log (1.88 \times 10^{16})$$
$$= 162.74 \text{ dB m}$$

Step 3:

$$A_E = 10 \log [3.142 \times (0.90/2)^2 \times 0.65]$$
$$= 10 \log 0.41$$
$$= -3.83 \text{ dB m}^2$$

Step 4:

$$C = 52 - 162.74 + (-3.83) + 55 - 0.90 - 0.14 - 0.2 - 0.3$$
$$= -61.1 \text{dBW}$$

Step 5:

$$V = \sqrt{10^{0.1(-61.1)} \times 75}$$
$$= 7.6 \times 10^{-3} \text{ V}$$

Step 6:

$$\text{Signal level} = 20 \log \left(\frac{7.6 \times 10^{-3}}{10^{-6}}\right)$$
$$= 20 \log 7600$$
$$= 77.6 \text{ dBuV}$$

This is close to the target value of 78.5 dBuV, and so should be adequate for the purposes of IF distribution.

Appendix 1

Glossary of terms

Absolute zero: The temperature at which all molecular movement ceases; zero degrees kelvin (0 K) or −273 degrees Celsius (−273°C).

Active device: Any distribution component that provides amplification. That is to say provides a net gain rather than a loss. Some components such as active splitters only provide sufficient gain to compensate for their inherent splitting loss, so are deemed lossless.

ADC: Analogue to digital conversion.

AFC: Automatic frequency control used to lock onto, and maintain, frequency of the selected signal.

Alignment: Fine tuning to maximize sensitivity in a selected channel.

AM: Amplitude modulation.

Amplifier: A circuit that increases the power or voltage of a signal.

Apogee: The highest point (maximum altitude) of a geocentric orbit.

Aperture: The microwave collection area of a dish.

Aspect ratio: The ratio of screen width to height. HDTV is 16:9 and conventional TV is 4:3.

Attenuation: A signal level loss.

Attenuator: A circuit that decreases the power of a signal.

AZ: Azimuth.

Azimuth: The angle between an antenna beam and the meridian plane (horizontal plane).

Bandpass filter: A circuit which passes a restricted band of frequencies. Unwanted lower and upper frequencies are attenuated.

Bandwidth: The total range of frequencies occupied by a signal; window of receiver, etc.

Baseband: The band of frequencies containing the information before modulation; general term for audio and video signals in AV links.

Baud: The accepted unit of digital data-transmission rate.

BBS: Business band service.

Beamwidth: The antenna acceptance angle measured between half power points (3 dB down points).

BER: Bit error rate – a measure of digital demodulation or decoding accuracy.

Bird: A quaint Americanism for satellite.

Bit: A binary digit (a '1' or a '0').

Bit rate: The number of digital bits transmitted per second.

Block: An 8 × 8 matrix of picture elements, or 64 DCT coefficients.

Boresight: Central axis of symmetry in a paraboloid dish – the beam centre.

Branch: (See *Tap line*).

Carrier: The radio frequency wave upon which the baseband signal is modulated.

Cassegrain: A dish using a convex subreflector and a paraboloid main reflector.

CATV: Community (cable) antenna television.

C-band: Satellite frequencies in the 4/6 GHz band.

CCI: Co-channel interference.

CCIR: International Radio Consultative Committee.

CCIR-601: Specifies the image format, acquisition semantic and parts of the coding for digital 'standard' television signals (PAL, NTSC and SECAM). This is now renamed ITU-R BT601.

Chrominance: The colour information in a composite video signal.

C/I: Ratio of carrier signal to interference.

Clamp: Video processing circuit designed to remove unwanted low frequencies.

Clark orbit: A geosynchronous equatorial orbit in which a satellite appears stationary north-to-south with respect to an observer at the earth's surface.

CNR or C/N: Carrier to noise ratio.

Composite baseband: The raw demodulator output before filtering, clamping and decoding.

Composite video: A complete video signal, including luminance, sync and colour information but not audio or data subcarriers.

Coupling loss: (See *Tap loss*).

CP: Circular polarization.

Cross polarized: Of the opposite polarization.

Crosstalk: Interference between cross polarized or adjacent channels.

DAC: Digital to analogue conversion.

dB: Decibels; a logarithmic ratio normally used to express the difference between two powers (dB = 10 log P2/P1).

dBi: The gain of an antenna relative to an isotropic source.

dBm: dB power relative to a 1 milliwatt standard.

dBw: dB power relative to one watt.

DBS: Direct broadcasting by satellite or direct broadcasting service.

DCT: Discrete cosine transform. DVB/MPEG-2 uses DCT coefficients to organize the video information for compression.

Declination: The angle between the equatorial plane and antenna beam.

Decoding: The process that reads an input coded bitstream and produces decoded pictures or audio samples.

Decoupling: (See *Isolation*).

De-emphasis: The reversal of pre-emphasis by reducing the amplitude of high frequency components and noise.

Demodulation: The recovery of baseband information from a modulated carrier.

Deviation: The maximum amount the carrier frequency is shifted by the modulating message or baseband signal.

Discrimination: The ability of a circuit to separate wanted from unwanted signals.

Discriminator: One type of circuit used to demodulate an FM signal.

Down-conversion: Reducing a band of high frequencies to a lower band.

DSO: Dielectric stabilized oscillator used in LNB design.

DTH: Direct to home.

EBU: European Broadcasting Union.

Eclipse protected: A satellite which continues transmission in spite of a solar eclipse.

ECS: European communications satellite.

EDTV: Extra definition television.

EIRP: Equivalent isotropically radiated power, combining the transmitter (or transponder) rf power and the transmitting antenna gain.

EL: Elevation.

Elevation: The angle between the antenna beam and the horizontal. Measured in the vertical plane.

EL/AZ: An antenna with independent steering both in azimuth and elevation.

Energy dispersal: A low frequency signal added to the baseband signal before modulation. Used to reduce the peak power per unit bandwidth of an FM signal in order to reduce its interference potential.

EOL: End of life of a satellite.

ERP: Effective radiated power.

ESA: European Space Agency.

Eutelsat: European Telecommunications Satellite Organization.

F/D: The ratio of focal length to diameter of a dish.

FEC: Forward error correction.

Feedhorn: A small wide-beam antenna system (usually horn shaped) that collects the energy reflected from the dish.

FET: Field effect transistor. A type of low noise transistor relying on electric fields, rather than simple pn junctions.

Field: For an interlaced video signal, a field is the assembly of alternate lines of a frame. Therefore an interlaced frame is composed of two fields, an odd field and an even field.

Field period: The reciprocal of the field rate.

FM: A transmission system in which the modulating waveform is made to vary the carrier frequency.

Footprint: The area on the earth's surface, for a given dish size, within which a signal from a satellite is judged to be of acceptable quality.

Fps: Frames per second.

Frame: Lines of spatial information containing samples starting from the top left of a picture and continuing, through following lines, to the bottom right of the picture. For interlaced video a frame consists of two fields, an even field and an odd field. One of these fields will commence one field period later than the other.

Frame rate: The rate at which frames are output from a decoding process. MPEG output is 30 fps.

Free space loss: The attenuation between transmitter and receiver.

Frequency: Number of cycles per second measured in Hz.

FSS: Fixed satellite service.

Gain: The ratio of output power to input power, normally expressed in dB form.

GaAsFET: Gallium arsenide field effect transistor (used in LNBs).

GEO: Geosynchronous equatorial orbit. (See Clark orbit).

Geosynchronous: An orbit having a period equal to that of the earth's rotation but not necessarily geostationary.

Ground noise: Spurious microwave signals generated from ground temperature.

G/T: Gain to noise-temperature ratio (a figure of merit of a satellite receiving system).

Giga (G): One billion.

GHz: A frequency of 1000 million cycles per second (1000 MHz).

Global beam: Satellite footprint which covers the entire visible earth's surface.

Half-transponder: A compromise method of sending two TV signals through one transponder.

HEMT: High electron mobility transistor (low noise device used in LNBs).

HDTV: High definition television.

Head unit: Combination of LNB, polarizer and feedhorn.

High pass filter: A circuit which only passes signals above a designed frequency.

High-power satellite: A loose expression, normally taken to mean greater than 100 watts transponder power.

Hz: Hertz (1 cycle per second).

IF: Intermediate frequency.

IMD: Intermodulation distortion.

ITU: International Telecommunications Union.

Insertion loss: (see *through loss*).

Isolation: Passive devices are normally directional; in that signals are allowed to pass easily in one direction only. The ratio of forward transmission to reverse expressed in decibels is known as the isolation or decoupling.

Isolator: A device with high signal loss in one direction but low in the other.

Isotropic: Ideally, a point source which transmits signals of equal power in all directions.

Kelvin scale: An absolute temperature scale in which the zero of the scale is −273 degrees Celsius (−273°C).

Ku-band: Frequencies within the range 10.7 to 18 GHz.

Link budget: Overall calculation of power gains and losses from transmission to reception.

Link margin: Amount in dBs by which *C/N* ratio exceeds the receiver's demodulator threshold *C/N* value.

LHCP: Left hand circular polarization.

LNA: Low noise amplifier.

LNB: Low noise block (down-converter). One type of LNC which down-converts a block of frequencies in one go.

LNC: Low noise converter (term often includes the LNB).

Low pass filter: A circuit which only passes signals lower than a designed frequency.

Low power satellite: Transponder power less than 30 watts.

LPF: Low pass filter.

Luminance: Light and shade information in a TV signal (contrast).

MAC: Multiplexed analogue components colour system.

Macroblock: The four, 8 × 8, blocks of data coming from a 16 × 16 section of the luminance component of the picture.

Magnetic variation: Difference between true north and that indicated by a compass (also called magnetic declination).

MATV: Master antenna television system.

Medium power satellite: Loose term for power between 30 and 100 watts.

Modulation index: The ratio of peak deviation to the highest modulating frequency.

Motion compensation: Motion vectors are used to improve efficiency in the prediction of pel values and uses motion vectors.

Motion estimation: The process of estimating motion vectors during the encoding process.

Motion vector: A two-dimensional vector used for motion compensation that provides an offset from the coordinate position in the current picture to the coordinates in a reference picture.

Mount: The structure which supports the antenna.

Multiplexing: A single transmission channel carrying two or more independent signals.

Multiplexing (MUXing): The combining of two or more streams of audio and video information.

Noise figure (NF): Ratio of the noise contributed by a practical amplifier to an ideal noise-free amplifier measured at some reference temperature. Usually expressed in dB.

Noise temperature (NT): Noise measurement of an amplifying system expressed as the absolute temperature of a resistance delivering the same noise power.

NTSC: National Television Standards Commission (USA).

Offset-fed antenna: An antenna with a reflector that forms only part of a true paraboloid in order to minimize blockage caused by the feed and its support structure.

OMT: Orthogonal mode transducer, a waveguide device which separates (or combines) two orthogonally polarized signals.

Orthogonal: At right angles to each other.

Outdoor unit: See Head Unit.

PAL: Phase alternate line colour system employing a delay line.

Paraboloid: Classical shape of the antenna reflector.

Passive device: Any distribution component that does not provide extra amplification to compensate for its inherent loss (e.g. capacitors, inductors and resistors but not semi-conductors).

Path distance: See Slant Range.

Pel: Picture element.

Perigree: Lowest point or minimum altitude in a geocentric orbit.

Period (1): Time taken for a satellite to complete one orbital revolution.

Period (2): The time taken for one cycle of a sinusoidal waveform.

PFD: Power flux density (related to field strength).

Phase distortion: Due to a non-linear relationship between amplifier phase shift and frequency.

PLL: Type of demodulator relying on a phase locked loop.

Polar mount: Antenna mechanism allowing tracking of the geo-arc.

Polarization: The plane or direction of one of the fields (usually the E field) in an electromagnetic wave.

p-p: peak to peak.

Pre-emphasis: The procedure for improving the signal to noise ratio of a transmission by emphasizing the higher baseband frequencies.

Prime focus: The focus of a paraboloid dish.

PSK: Phase shift keying.

QPSK: Quarternary(Quadrature) phase shift keying.

Rain outage: Loss of Ku-band signal caused by heavy rain absorption.

Reed Solomon FEC: A powerful error correction code providing forward error correction with little overhead.

Reference signal: Highly stable signal used to compare other signals.

Return loss: Any device placed in a transmission line will reflect a certain amount of signal back toward the source. The ratio of the signal reflected back to that incident, expressed in decibels, is known as the return loss.

RF: Radio frequency.

RGB: The three primary colours, red green and blue.

RHCP: Right hand circular polarization.

RRO: Radio receive only.

Saturation: The colour intensity parameter in a video signal.

SAW: Surface acoustic wave filter. A device designed to shape the frequency response of a signal. Can replace many tuned circuits.

S-band: 2.6 GHz band.

Scaler feed: Wide corrugated horn feed.

SECAM: Sequence colour à Mémoire (a French TV standard).

Semi-DBS: Popular term for medium power satellite providing DTH programming.

Sidelobe: Response of a dish to signals off the central axis.

Side loss: (see *tap loss*).

SIF: The standard interchange format is used for exchanging video images between NTSC (240 lines) and PAL/SECAM (288 lines). At the field rates of 60 and 50 fields per second, respectively, the two formats have the same data rate. Each line is composed of 352 pixels in all the above systems.

Signal availability: The percentage of time which the performance of a satellite system exceeds some pre-determined design value of the C/N or S/N ratio. It can be quoted either as a percentage of an 'average year' or a percentage of the 'worst month'.

Skew: The twist difference in polarization angle between satellites.

Slant range: Total path length between satellite and a receiver on earth.

SMATV: Satellite master antenna television.

SNR or S/N: Signal to noise ratio.

Solar outage: Signal loss due to sun's position, relative to receiving station.

Sparklies: Popular term for impulse noise spikes visible as annoying black or white flashes over the screen.

Spillover: Usable signal, normally outside expected range.

Spot beam: Circular or elliptical beam covering some defined region of the earth's surface.

Stag's head: Term used to describe a wall mounted antenna.

Star wired: Single cables extending outwards from a common point.

Subcarrier: An information carrying wave which modulates the main carrier.

Sync: Synchronization.

Tap line: The cable branch from the trunk to the subscriber's wall plate.

Tap loss: The attenuation of the signal in decibels incurred between a trunk level input and a tap line.

TDM: Tuner/demodulator in the indoor unit or receiver.

Thermal noise: Random electric variations caused by molecular motion which increases with temperature.

TEM: Transverse electromagnetic wave.

Threshold: Term used in an FM signal where the normally linear relationship between C/N and demodulated signal S/N no longer holds.

Threshold extension: A technique for lowering the C/N value at which non-linear demodulator effects start.

Through loss: The attenuation of the signal in decibels due to the presence of a passive (non-amplified) tap in the trunk.

Transponder: Powered equipment on a satellite used to re-broadcast the uplink signal down to earth-located receivers (the downlink).

Trap: Jargon for any device which attenuates a selective frequency band.

Tree wired: Cables forming a central trunk with branches at various points along its length.

Truncation: Loss of the outer side-frequencies of an FM signal due to filters, displayed on the screen as 'tearing' on video transients.

Trunk: The main distribution cable of a tree network which is tapped off at various points to subscribers' dwellings (akin to a water main in a plumbing system).

TVRO: Television receive only.

TWT: Travelling wave tube.

UHF: 300 MHz to 3 GHz band.

VCO: Voltage controlled oscillator sometimes called VTO (voltage tuned oscillator).

Viterbi decoder: Performs deconvolution of the bitstream.

VSWR: Voltage standing wave ratio; a measurement of impedance mismatch conditions on an antenna, waveguide or transmission line system.

WARC: World Administrative Radio Conference. The International Telecommunications Union meetings for determining radio communication standards.

Weighting: Correction of S/N measurements after allowing for subjective annoyance factors.

Appendix 2

Computer software tools

It can be a problem obtaining up-to-date information about satellites and their EIRP coverage. It is also tedious working out link budgets, particularly if rain attenuation and atmospheric absorption are to be taken into account on a specific space to Earth path. Fortunately there are a number of regularly updated computer software packages that help in this task. This appendix covers some of the better known packages that may be useful to those engaged in satellite installation.

Satmaster Pro for Windows

Satmaster Pro is designed to be of use to broadcasters, SNG operatives, antenna installation companies and sole traders engaged in the satellite TV industry. Antenna aiming, link budget analysis and solar outage prediction provide the backbone of the package. The link budget facilities employ rain attenuation and atmospheric absorption prediction modelling even for low elevations. This and other useful features, such as integral magnetic variation calculation and tens of thousands of stored town/city coordinates, combine to enable a system to be either designed and/or installed at any global location. Satmaster Pro integrates all necessary reference, system design and analysis tools into a single easy-to-use package.

Features
- Calculates look angles and polar mount set-up angles including compass bearing and optional dual feed spacing. Editable database of over 25 000 town and city records arranged in over 200 separate country data files. Dual feed calculations.
- Calculates short form analogue and digital downlink budgets for TV broadcasts. Minimum dish size optimizer for any selected target value of C/N, C/N_0, S/N, CCIR Grade, or E_b/N_0 including LNB output carrier level for IF distribution systems.
- SCPC/MCPC digital link budget calculator. Finds minimum uplink power requirements, transponder bandwidth and power usage per carrier with interference taken into account. Information rates from

56 kbit/s to well beyond 40 Mbit/s. Handles common modulation schemes with any FEC code rate.

- Displays maps of footprints, rain-climatic zones, water vapour density and surface temperatures.
- Solar outage prediction, presenting annual lists of date and time windows, for any global satellite/ground station combination. Locates north/south from the Sun's position for polar mount set-up.
- Plots graphs and generates a large selection of tables useful to broadcasters and installation companies.
- Context sensitive help on all input data fields, and a large reference help file.
- Generates graphs and tables including digital modulation, rain attenuation, atmospheric absorption.
- Performs calculations and conversions and includes an editable mathematical expression evaluator for more complex needs.
- Minimum system requirements: Windows 3.1/WFWG 3.11/Windows 95/NT, IBM 386 or better with 4 Mb RAM, 1.44 Mb 3.5 in diskette drive, 10 Mb free hard disk space. Also runs on Power MAC and UNIX platforms with Insignia SoftWindows.

For more information, and details of your nearest stockist, contact the author:

D. J. Stephenson, 58 Forest Road, Heswall, Merseyside L60 5SW, England
Tel or fax: +44 (0)151 342 4846
E-mail: djsteph@arrowe.u-net.com
Web Site: http://www.arrowe.u-net.com/

Satfinder

This is a huge database, supplied exclusively on CD, for the satellite professional. The database, updated four times a year, can be run on Windows or Macintosh platforms and includes the following topics:

Company information lists over 10 000 companies involved in the satellite communications industry. There are 31 categories of companies or you can look for a particular company. Search any product or service and display full details including contact names and numbers.

Satellite information lists over 450 planned and operational satellites. Details on owner/operator, manufacturer, launch information, bandwidth and operating frequencies, power levels, and many other technical details.

Map information   gives the EIRP, G/T and SFD of most operational satellites. There are over 500 maps – decide on which satellite or country you wish to see coverage and display details.

What's on satellite   lists the video, data and voice activity on over 100 satellites. Frequencies, bandwidth, EIRP, video format and many other details.

Standards and statistics   gives the broadcast standards and basic statistical information on most countries of the world.

Uplink information   gives the equipment specifications and contact information necessary to perform a satellite uplink from anywhere in the world.

More details on Satfinder are available from:

Design Publishers, 800 Siesta Way, Sonoma, CA 95476, USA
Tel: +1 (707) 939-9306 or Fax: +1 (707) 939-9235

The Satellite's Encyclopedia

The Satellite's Encyclopedia (TSE) is a comprehensive, yet low cost, satellite database in the form of a large Windows help file (with a bookmark and annotation system). It is compiled in France by J. P. Donnio and is updated approximately every three months.

The software includes over 3000 hypertext pages, multiple thematic listings of satellites, a large number of high quality EIRP maps, listings of all satellites in the Clarke Belt, details of past, present and future launches with most technical details such as transponder frequencies, etc. Registration of the shareware brings numerous additional sections and EIRP maps not provided in the shareware version. Registration also provides free access to the on-line WWW version.

TSE runs on Windows 3.1x, Windows 95 and Windows NT and requires a 256 colour display, 65k colours recommended. You can download the shareware version from the following Web site: *http://www.tele-satellit.com/tse/*

Both the shareware and registered versions can also be obtained on disks from:

WinShare, 46-48 route de Thionville, 57140 Woippy, France
Tel: +33 3 8730 8557 Fax: +33 3 8732 3775
E-mail: 100031.3257@compuserve.com

Appendix 3

Conditions for geostationary satellite orbit

The orbital radius (r) of a satellite depends on its velocity (v) – the faster it travels around the earth, the smaller will be its radius. This means that, for a given orbital radius, there is one, and only one, corresponding velocity.

For a satellite always to appear stationary to a ground observer, it must satisfy two conditions:

1 It must orbit the equator.
2 The radius of orbit, measured from the Earth's centre, must be such that it travels one revolution in 24 hours.

Physical constants and symbols:

G = constant of gravitation = 6.67×10^{-11} N m^2/kg^2
M = mass of the Earth = 5.976×10^{34} kg
m = mass of satellite
t = time for one satellite revolution = 24 hours = 8.64×10^4 seconds
Radius of Earth at equator = 6378 km
Permittivity of free space = 8.85×10^{-12} F/m.

There are two forces at work, the gravitational force tending to pull the satellite down to Earth and the centripetal force tending to propel the satellite out of orbit and into space. As far as the gravitational force is concerned, we are indebted to Newton who worked out that: 'The force of attraction between two bodies is proportional to the product of their masses and inversely proportional to the square of the distance between them.' In modern terms and using the symbols and gravitational constant given above, the equation can be written as

$$F = \frac{GmM}{r^2}$$

The equation for the centripetal force acting on the satellite is

$$F = \frac{Mv^2}{r}$$

To maintain a stable orbit, these two forces must balance, so

$$\frac{GmM}{r^2} = \frac{mv^2}{r}$$

which simplifies to

$$r = \frac{GM}{v^2} \qquad\qquad\qquad\qquad (A3.1)$$

The velocity v must be eliminated before r can be established. This is not difficult because we know that the satellite must revolve round the Earth in 24 hours and the distance it has to travel in this time is equal to its orbital circumference, so

$$v = \frac{2\pi r}{t} \qquad\qquad\qquad\qquad (A3.2)$$

Substituting (A3.2) in (A3.1) and simplifying gives the final equation for radius of orbit:

$$r = \left[\frac{GMt^2}{4\pi^2} \right]^{1/3}$$

Substituting the values and known constants given above, we have

$$r = \left| \frac{6.67 \times 10^{-11} \times 5.976 \times 10^{24} \times 7.46 \times 10^9}{4\pi^2} \right|^{1/3} = 42\,000 \text{ km approximately}$$

But this is the total orbital radius measured from the Earth's centre. The radius of the Earth at the equator is 6378 km, so this must be subtracted from 42 000, and therefore the height above ground of a geostationary satellite is approximately 35 622 km.

Note: The mass of the satellite is unimportant – it vanished from the equations at an early stage. This may seem strange but fortunate for the satellite industry because the hardware they throw up in space can vary from a few pounds to a ton or more and yet they all occupy the same orbital radius! It is well to remember that the time of swing of a simple pendulum depends only on g and the length of string – the weight on the end of the string has no effect on the time of swing.

Appendix 4

Fixings (catalogue numbers for Rawlplug company products)

Table A4.1   *Rawlplug fixings*

Types and sizes
Boxes of 100 plastic plugs (in clusters of 10)

Plug colour	Ref.	Suitable screw size	Plug length mm		Drill size	Cat. no.
Green	48	Nos. 4, 6, 8 3–4 mm	20	($^3/_4$")	No. 8, 4.5 mm $^3/_{16}$"	**67–008**
Pink	68				No. 10, 5.5 mm $^7/_{32}$"	**67–012**
Orange	610	Nos. 6, 8, 10 3.5–5 mm	25	(1")	No. 12, 6.5 mm $^1/_4$"	**67–016**
Grey	810				No. 10, 5.5 $^7/_{32}$"	**67–020**
White	812	Nos. 8, 10, 12, 4.5-5 mm	35	(1$^3/_8$")	No. 12, 6.5 mm $^1/_4$"	**67–024**
Blue	1014	Nos. 10, 12, 14 5–6 mm			8 mm $^5/_{16}$"	**67–028**
Yellow	1620	Nos. 16, 18, 20	50	(2")	11 mm $^7/_{16}$"	**67–032**

Packed in shrink wrapped outers of 10 boxes.
Source: The Rawlplug Company

Table A4.2   *Rawlbolt fixings*

Types and sizes

Ref.	Bolt size	Shield length 'A' (mm)	Fixing thickness* mm 'B' Max.	Min.	Hole dia. (mm)	Min. hole depth (mm)	Application torque for concrete (Nm)	Box qty	Cat. no.
M6 10L	M6	45	10	0	12	50	6.5	50	**44–015**
M6 25L			25						44–020
M6 40L			40						44–025
M8 10L	M8	50	10	0	14	55	15	50	**44–055**
M8 25L			25						44–060
M8 40L			40						44–065
M10 10L	M10	60	10	0	16	65	27	50	**44–105**
M10 25L			25						44–110
M10 50L			50						44–115
M10 75L			75						44–120
M12 10L	M12	75	10	0	20	85	50	25	**44–155**
M12 25L			25						44–160
M12 40L			40						44–165
M12 60L			60						44–170
M16 15L	M16	115	15	0	25	125	120	10	**44–205**
M16 30L			30	10					44–210
M16 60L			60	30					44–215
M20 60L	M20	130	60	25	32	140	230	10	**44–255**
M20 100L			100	60					44–260
M24 100L	M24	150	100	25	38	160	400	5	**44–305**
M24 150L			150	100				2	44–310

*If the fixing thickness is less than the stated maximum, increase the hole depth by the difference between actual and maximum thickness.
Source: The Rawling Company

Table A4.3 *Rawlok fixings*

Types and sizes
Boxes: hex nut

Ref.	Thread size	Anchor/ hole dia. (mm)	Anchor length 'A' (mm)	Min. hole depth (mm)	Max. fixture thickness (mm)	Rec. tightening torque (Nm)	Box qty	Cat. no.
R6026			26	22	5			**69–502**
R6038	M4.5	6	38		9	2.5	100	69–504
R6058			58	30	27			69–506
R8042			42		9		100	**69–508**
R8042SS								69–
R8066	M6	8	66	35	35	6.0	50	308*
R8092			92		60			69–510
								69–512
R10048			48		9			**69–514**
R10048SS	M8			40		11.0	50	69–
R10075		10	75		36			314*
R10100			100		60		25	69–516
								69–518
R12058			58		9			**69–520**
R12070			70		22		25	69–522
R12070SS	M10	12		50		22.0		69–
R12098			98		50		10	322*
R12126			126		80			69–524
								69–525
R16064			64		13		20	**69–526**
R16108	M12	16	108	55	55	38.0		69–528
R16142			142		90		10	69–530
R20082			84		25		10	**69–533**
R20114	M16	20	114	60	57	95.0		69–534
R20158			158		100		5	69–536

*Indicates stainless steel Rawlok
Source: The Rawlplug Company

Table A4.4   *Rawlbor long life masonry drills*

Types and sizes STS plus shank

Dia. (mm)	Overall length (mm)	Working length (mm)	For Rawlbolt size	Cat no.
5	110	50		28–202
5.5	110	50		28–206
	160	100		28–208
6	110	50		28–210
	160	100		28–214
6.5	110	50		28–218
	160	100		28–222
7	160	100		28–226
8	110	50		28–230
	160	100		28–234
	210	150		28–236
10	160	100		28–238
	260	200		28–242
	450	384		28–244
12	160	100	M6	28–246
	260	200		28–250
	450	400		28–251
13	160	100		28–252
	260	200		28–253
14	160	100	M8	28–254
	260	200		28–256
15	160	100		28–258
	260	200		28–260
16	210	150	M10	28–262
	450	400		28–267
18	210	150		28–270
	450	400		28–272
20	200	150	M12	28–274
	450	400		28–279
22	250	200		28–282
24	250	200		28–286
	450	400		28–290
25	250	200	M16	28–294
	450	400		28–296

Source: The Rawlplug Company

Appendix 5

List of geostationary satellites

This list was compiled in the autumn of 1996.

Key

S	S-Band (2.5–2.6 GHz)
C	C-Band (3.4–4.2 GHz)
Ku	Ku-Band (10.7–12.75 GHz)
Ka	Ka-Band (19.1–20.2 GHz)

Satellite	Longitude	Bands
Intelsat 513	177.00W	C, Ku
TDRS 5	174.00W	S, C, Ku
TDRS 7	171.00W	S, C, Ku
Raduga 21	169.20W	C
Aurora 2	139.00W	C
Satcom C1	137.00W	C
Satcom C4	135.00W	C
Galaxy 1R	133.00W	C
Satcom C3	131.00W	C
Galaxy 5	125.00W	C
Gstar 2	124.90W	Ku
SBS 5	123.00W	Ku
Galaxy 9	123.00W	C, Ku
Telstar 303	120.00W	C
USA 099	120.00W	Ka
Echostar 1	119.00W	Ku
Morelos 2	116.80W	C, Ku
Anik C3	114.90W	Ku
Solidaridad 2	113.00W	C, Ku
Anik E1	111.00W	C, Ku
Solidaridad 1	109.20W	C
Anik E2	107.30W	C, Ku
Marisat 1	106.00W	C
Aurora 1	105.40W	C
G-Star 4	105.00W	Ku
G-Star 1	103.00W	Ku
DBS 1	101.20W	Ku

ASC 2	101.00W	C, Ku
AMSC 1	101.00W	Ku
DBS 3	101.00W	Ku
DBS 2	100.80W	Ku
ACTS	100.00W	Ka
Galaxy 4	99.00W	C, Ku
Telstar 401	97.00W	C, Ku
Galaxy 3R	95.00W	C, Ku
G-Star 3	93.00W	Ku
Brazilsat A2	92.00W	C
Galaxy 7	91.00W	C, Ku
Telstar 402R	89.00W	C, Ku
Spacenet 3R	87.00W	C, Ku
Satcom K1	85.00W	C, Ku
Telstar 302	85.00W	C
Satcom K2	80.90W	C, Ku
Brazilsat A1	79.00W	C
SBS 4	77.00W	Ku
Comstar 4	76.00W	C
Anik C2	75.90W	Ku
Galaxy 6	74.10W	C
SBS 6	74.00W	Ku
Anik C1	71.80W	Ku
SBS 2	71.00W	Ku
Brazilsat B1	70.00W	C
Spacenet 2	69.00W	C, Ku
Brazilsat B2	65.00W	C
Inmarsat 2F4	54.00W	C
Intelsat 706	53.00W	C, Ku
Intelsat 709	50.00W	C, Ku
TDRS 1	49.00W	S, C, Ku
TDRS 6	47.00W	S, C, Ku
PAS 1	45.00W	C, Ku
PAS 3R	43.00W	C, Ku
TDRS 4	41.00W	S, C, Ku
Intelsat 502	40.30W	C, Ku
Orion 1	37.50W	Ku
Intelsat 603	34.50W	C, Ku
Intelsat 505	33.00W	C, Ku
Intelsat 506	31.50W	C, Ku
Hispasat 1B	30.00W	Ku
Hispasat 1A	30.00W	Ku
Intelsat 601	27.50W	C, Ku
Intelsat 605	24.50W	C, Ku
Intelsat K	21.50W	Ku
Intelsat 512	21.30W	C, Ku
TDF 2	18.80W	Ku

TDF 1	18.80W	Ku
Intelsat 705	18.00W	C, Ku
Cosmos 2172	17.00W	C
Cosmos 2054	16.00W	Ku
Inmarsat 2F2	15.80W	C
Marecs B2	15.00W	C
Express 1	14.00W	C, Ku
Cosmos 2291	13.50W	C
Gorizont 26	10.70W	C, Ku
Meteosat 6	10.00W	S
Telecom 2A	8.00W	C, Ku
Telecom 2B	5.00W	C, Ku
Intelsat 707	1.00W	C, Ku
Thor	0.80W	Ku
TVSat 2	0.60W	Ku
Meteosat 5	0.00E	S
Telecom 2C	3.00E	C, Ku
USA 115	4.00E	Ka
TELE-X	5.00E	Ku
Sirius 1	5.20E	Ku
Eutelsat 2F4	7.00E	Ku
Eutelsat 2F2	10.00E	Ku
Raduga 22	11.50E	C
Raduga 29	11.50E	C
Cosmos 2224	12.50E	S
Eutelsat 2F1	13.00E	Ku
Eutelsat Hotbird 1	13.00E	Ku
Italsat 1	13.20E	Ka
Eutelsat 2F3	16.00E	Ku
Astra 1E	19.20E	Ku
Astra 1C	19.20E	Ku
Astra 1A	19.20E	Ku
Astra 1F	19.20E	Ku
Astra 1D	19.20E	Ku
Astra 1B	19.20E	Ku
Telstar 301	20.00E	C
Eutelsat 1F5	21.50E	Ku
Kopernikus 3	23.50E	Ku, Ka
Eutelsat 1F4	25.50E	Ku
Arabsat 2A	26.00E	C, Ku
Gorizont 20	26.00E	C, Ku
Kopernikus 2	28.50E	Ku, Ka
Arabsat 1C	31.00E	S, C
Turksat 1C	31.00E	Ku
Gorizont 17	34.00E	C, Ku
Raduga 28	35.00E	C
Gals 2	36.00E	Ku

Gals 1	36.00E	Ku
Eutelsat1F1	36.00E	Ku
Gorizont 31	40.00E	C, Ku
Turksat 1B	42.00E	Ku
Raduga 1-3	49.00E	C
Raduga 1-2	49.00E	C
Gorizont 32	53.00E	C, Ku
Intelsat 703	57.00E	C, Ku
Intelsat 604	60.00E	C, Ku
Intelsat 602	62.90E	C, Ku
Inmarsat 3F1	63.50E	C
Inmarsat 2F1	64.30E	C
DSCS 2-15	65.00E	C
Intelsat 704	66.00E	C, Ku
PAS 4	68.50E	C, Ku
Raduga 32	70.00E	C
Intelsat 501	72.00E	C, Ku
Marisat 2	72.50E	C
Insat 2A	74.00E	S, C
Luch 1	76.80E	Ku
Thaicom 2	78.50E	C, Ku
Thaicom 1	78.50E	C, Ku
Cosmos 2319	80.00E	C
Gorizont 24	80.00E	C, Ku
Insat 1D	83.00E	S, C
Raduga 30	85.00E	C
TDRS 3	85.00E	S, C, Ku
PRC 22	87.80E	C, Ku
Gorizont 28	90.00E	C, Ku
Measat 1	91.50E	C, Ku
Intelsat 501	91.50E	C, Ku
Insat 2C	93.50E	S, C
Insat 2B	93.50E	S, C
Luch 0	95.00E	Ku
Gorizont 19	96.50E	C, Ku
Gorizont 27	96.50E	C, Ku
PRC 26	98.00E	C
Asiasat 2	100.50E	C, Ku
Gorizont 25	103.50E	C, Ku
Raduga 31	105.00E	C
Asiasat 1	105.50E	C
Palapa B2R	108.00E	C
Yuri 3A	109.50E	Ku
Yuri 3N	109.80E	Ku
Yuri 3B	110.00E	Ku
PRC 25	110.50E	C
Palapa C2	113.00E	C, Ku

Spacenet 1	115.50E	C, Ku
Koreasat 2	116.00E	Ku
Koreasat 1	116.00E	Ku
Palapa B4	118.00E	C
GMS 4	120.00E	S
Raduga 27	128.00E	C
JC-Sat 3	128.00E	Ku
Rimsat G1	130.00E	C, Ku
Gorizont 29	130.00E	C, Ku
Sakura 3A	132.00E	C, Ka
N Star A	132.00E	S, C, Ku, Ka
Apstar 1A	134.00E	C, Ku
N Star B	136.00E	S, C, Ku, Ka
Sakura 3B	136.00E	C, Ka
Apstar 1	138.00E	C
Gorizont 22	140.00E	C, Ku
GMS 5	140.00E	S
Gorizont 18	140.00E	C, Ku
Rimsat G2	142.50E	C, Ku
Palapa B2P	144.00E	C
Gorizont 21	145.00E	C, Ku
JC-Sat 1	150.00E	Ku
ETS 5	150.30E	C
Palapa C1	150.50E	C, Ku
Optus A3	152.00E	Ku
JC-Sat 2	154.00E	Ku
Optus B3	156.00E	Ku
Intelsat 503	157.00E	C, Ku
Superbird A2	158.00E	Ku, Ka
Optus B1	160.00E	Ku
Superbird B1	162.00E	Ku, Ka
Optus A2	164.00E	Ku
PAS 2	169.00E	C, Ku
Intelsat 701	174.00E	C, Ku
Intelsat 702	177.00E	C, Ku
Intelsat 511	180.00E	C, Ku

Appendix 6

AZ/EL table for UK and Eire towns and cities

Inclusion does not necessarily imply reception is possible in every case. Some locations may be outside usable footprint area for some satellites. Add local magnetic variation to true azimuth value for compass bearing. All values are decimalized degrees.

Satellite position (degrees longitude)

	26.5E	23.5E	19.2E	16.0E	13.0E	10.0E	1.0W	5.0W	19.0W	27.5W	31.0W
Aberdeen (57.17N,2.07W)											
AZ		147.1	150.3	155.1	158.8	162.2	165.7	178.7	183.5	199.9	209.5	213.3
EL		20.3	21.2	22.3	23.0	23.6	24.1	24.9	24.9	23.3	21.2	20.2
Aberystwyth (52.42N,4.08W)											
AZ		143.3	146.6	151.5	155.2	158.8	162.4	176.1	181.2	198.6	208.7	212.6
EL		23.7	24.8	26.3	27.2	28.0	28.6	30.0	30.1	28.5	26.2	25.1
Bath (51.38N,2.37W)											
AZ		144.8	148.2	153.2	157.0	160.6	164.3	178.3	183.4	200.9	211.0	214.9
EL		25.3	26.4	27.8	28.7	29.4	30.0	31.2	31.1	29.1	26.6	25.4
Belfast (54.58N,5.92W)											
AZ		142.1	145.3	150.1	153.7	157.2	160.6	174.0	178.9	195.9	205.9	209.9
EL		21.2	22.3	23.7	24.6	25.4	26.1	27.6	27.7	26.6	24.7	23.7
Birmingham (52.47N,1.92W)											
AZ		145.7	149.1	154.0	157.8	161.4	165.1	178.8	183.9	201.2	211.1	215.0
EL		24.5	25.5	26.9	27.7	28.4	29.0	30.0	29.9	27.9	25.5	24.2
Blackpool (53.83N,3.05W)											
AZ		144.9	148.2	153.1	156.8	160.4	164.0	177.5	182.4	199.5	209.4	213.3
EL		22.9	23.9	25.2	26.1	26.8	27.4	28.5	28.5	26.8	24.6	23.4
Bournemouth (50.72N,1.90W)											
AZ		145.1	148.5	153.5	157.3	161.0	164.8	178.8	184.0	201.7	211.8	215.7
EL		26.0	27.2	28.6	29.5	30.2	30.8	31.9	31.8	29.7	27.1	25.8
Brighton (50.83N,0.13E)											
AZ		147.4	150.9	156.0	159.9	163.6	167.4	181.5	186.6	204.1	214.0	217.9
EL		26.7	27.7	29.1	29.9	30.5	31.0	31.8	31.6	29.0	26.2	24.8
Bristol (51.45N,2.58W)											
AZ		144.6	148.0	152.9	156.7	160.4	164.1	178.0	183.1	200.6	210.7	214.7
EL		25.1	26.2	27.7	28.6	29.3	29.9	31.1	31.1	29.1	26.6	25.4
Cambridge (52.22N,0.13E)											
AZ		147.9	151.3	156.4	160.2	163.9	167.6	181.4	186.5	203.7	213.5	217.4
EL		25.4	26.4	27.7	28.5	29.1	29.6	30.3	30.1	27.7	25.0	23.7
Cardiff (51.48N,3.22W)											
AZ		143.9	147.2	152.2	156.0	159.6	163.3	177.2	182.3	199.9	210.0	214.0
EL		24.9	26.0	27.4	28.4	29.1	29.8	31.0	31.0	29.2	26.8	25.6

Chester	(53.20N,2.90W)										
AZ		144.9	148.2	153.1	156.8	160.4	164.0	177.6	182.6	199.8	209.8	213.7
EL		23.5	24.5	25.9	26.8	27.5	28.1	29.2	29.2	27.4	25.1	24.0

Cork	(51.90N,8.47W)										
AZ		138.4	141.6	146.3	150.0	153.4	157.0	170.5	175.6	193.3	203.7	207.8
EL		22.3	23.6	25.3	26.4	27.3	28.2	30.2	30.5	29.8	28.0	27.0

Derby	(52.92N,1.48W)										
AZ		146.3	149.7	154.7	158.5	162.1	165.7	179.4	184.4	201.6	211.5	215.4
EL		24.2	25.3	26.6	27.4	28.0	28.6	29.5	29.4	27.4	24.9	23.7

Douglas	(54.15N,4.47W)										
AZ		143.5	146.8	151.6	155.3	158.8	162.3	175.7	180.7	197.7	207.7	211.6
EL		22.1	23.2	24.5	25.4	26.2	26.8	28.1	28.2	26.8	24.7	23.6

Dover	(51.13N,1.32E)										
AZ		148.9	152.4	157.5	161.4	165.1	168.9	183.0	188.1	205.4	3215.2	219.1
EL		26.8	27.8	29.1	29.8	30.4	30.9	31.4	31.1	28.4	25.5	24.1

Dublin	(53.33N,6.25W)										
AZ		141.3	144.5	149.3	153.0	156.5	160.0	173.5	178.4	195.8	205.9	209.9
EL		2.1	23.2	24.7	25.7	26.5	27.3	28.9	29.1	27.9	26.0	25.0

Dundee	(56.47N,2.97W)										
AZ		145.9	149.2	154.0	157.6	161.1	164.6	177.6	182.4	199.0	208.7	212.6
EL		20.6	21.6	22.8	23.5	24.1	24.7	25.7	25.7	24.1	22.1	21.1

Edinburgh	(55.95N,3.22W)										
AZ		145.4	148.7	153.5	157.2	160.7	164.2	177.3	182.2	198.8	208.6	212.5
EL		21.0	21.9	23.2	24.0	24.6	25.2	26.2	26.2	24.7	22.7	21.6

Exeter	(50.72N,3.52W)										
AZ		143.3	146.6	151.6	155.4	159.0	162.7	176.7	181.9	199.7	209.9	213.9
EL		25.4	26.6	28.1	29.0	29.8	30.5	31.9	31.9	30.1	27.6	26.4

Fort William	(56.82N,5.12W)										
AZ		143.7	146.9	151.6	155.2	158.6	162.1	175.1	179.9	196.5	206.2	210.1
EL		19.6	20.6	21.8	22.7	23.4	23.9	25.2	25.3	24.2	22.4	21.4

Galway	(53.27N,9.05W)										
AZ		138.3	141.5	146.2	149.8	153.2	156.7	170.0	175.0	192.3	202.6	206.7
EL		21.0	22.2	23.8	24.9	25.8	26.7	28.7	29.0	28.4	26.8	25.9

Glasgow	(55,88N,4.25W)										
AZ		144.3	147.6	152.3	156.0	159.4	162.9	176.1	180.9	197.6	207.4	211.3
EL		20.7	21.7	23.0	23.8	24.5	25.1	26.3	26.3	25.0	23.0	22.0

Gloucester	(51.88N,2.23W)										
AZ		145.1	148.5	153.5	157.3	160.9	164.6	178.4	183.5	201.0	211.0	214.9
EL		24.9	26.0	27.3	28.2	28.9	29.5	30.6	30.6	28.6	26.1	24.9

Grimsby	(53.58N,0.08W)										
AZ		148.1	151.5	156.5	160.3	163.9	167.5	181.1	186.1	203.1	212.8	216.7
EL		24.1	25.1	26.3	27.0	27.6	28.1	28.8	28.6	26.4	23.8	22.6

Harwich	(51.95N,1.28E)										
AZ		149.1	152.6	157.7	161.6	165.2	169.0	182.9	188.0	205.1	214.9	218.7
EL		26.1	27.0	28.2	29.0	29.6	30.0	30.5	30.3	27.6	24.8	23.4

Holyhead	(53.32N,4.63W)										
AZ		143.0	146.3	151.2	154.8	158.4	162.0	175.5	180.5	197.7	207.7	211.7
EL		22.7	23.8	25.3	26.2	27.0	27.6	29.0	29.1	27.7	25.6	24.4

Inverness	(57.45N,4.25W)										
AZ		144.8	148.0	152.8	156.4	159.8	163.2	176.1	180.9	197.3	207.0	210.9
EL		19.3	20.3	21.5	22.3	22.9	23.5	24.6	24.6	23.4	21.5	20.6

Jersey   (49.25N,2.17W)

AZ	144.2	147.6	152.7	156.6	160.3	164.1	178.5	183.7	201.8	212.0	216.0
EL	27.2	28.4	29.9	30.9	31.7	32.3	33.5	33.4	31.2	28.5	27.1

Kirkwall   (58.98N,2.97W)

AZ	146.6	149.8	154.6	158.1	161.5	165.0	177.7	182.4	198.5	208.0	211.9
EL	18.4	19.3	20.3	21.0	21.6	22.1	23.0	23.0	21.6	19.8	18.8

Lands End   (50.05N,5.73W)

AZ	140.6	143.9	148.8	152.5	156.1	159.8	173.8	179.0	197.1	207.5	211.6
EL	25.0	26.3	27.9	29.0	29.9	30.7	32.5	32.6	31.3	29.0	27.8

Leicester   (52.63N, 1.08W)

AZ	146.7	150.1	155.1	158.9	162.5	166.2	179.9	184.9	202.1	212.0	215.9
EL	24.6	25.7	27.0	27.8	28.4	28.9	29.8	29.7	27.6	25.0	23.8

Lerwick   (60.15N,1.15W)

AZ	148.9	152.1	156.8	160.4	163.8	167.2	179.8	184.4	200.4	209.7	213.5
EL	17.9	18.7	19.6	20.2	20.7	21.1	21.8	21.7	20.1	18.2	17.3

Limerick   (52.67N,8.63W)

AZ	138.5	141.7	146.4	150.0	153.5	157.0	170.4	175.4	193.0	203.3	207.4
EL	21.7	22.9	24.5	25.6	26.5	27.4	29.4	29.7	29.0	27.3	26.3

Liverpool   (53.42N,2.92W)

AZ	144.9	148.3	153.2	156.9	160.4	164.1	177.6	182.6	199.8	209.7	213.6
EL	23.3	24.3	25.7	26.5	27.2	27.8	29.0	28.9	27.2	24.9	23.8

London   (51.50N,0.17W)

AZ	147.3	150.8	155.8	159.7	163.4	167.1	181.1	186.2	203.5	213.4	217.3
EL	26.0	27.0	28.3	29.1	29.8	30.3	31.0	30.9	28.5	25.7	24.4

Londonderry   (54.98N7.33W)

AZ	140.7	143.9	148.6	152.2	155.7	159.1	172.3	177.2	194.2	204.2	208.2
EL	20.3	21.4	22.9	23.8	24.6	25.4	27.0	27.3	26.4	24.7	23.7

Manchester   (53.47N,2.25W)

AZ	145.7	149.0	153.9	157.7	161.3	164.9	178.4	183.4	200.5	210.4	214.3
EL	23.5	24.5	25.8	26.7	27.3	27.9	28.9	28.9	27.0	24.7	23.5

Middlesborough   (54.58N,1.23W)

AZ	147.2	150.5	155.4	159.2	162.7	166.3	179.7	184.6	201.5	211.2	215.1
EL	22.8	23.8	25.0	25.8	26.4	26.9	27.7	27.6	25.7	23.3	22.1

Newcastle on Tyne   (52.43N,3.10W)

AZ	144.4	147.7	152.6	156.4	160.0	163.6	177.4	182.4	199.8	209.8	213.7
EL	24.1	25.2	26.6	27.5	28.2	28.8	30.0	30.0	28.2	25.9	24.7

Northampton (52.23N,0.90W)

AZ	146.7	150.2	155.2	159.0	162.6	166.3	180.1	185.2	202.5	212.4	216.3
EL	25.1	26.1	27.4	28.2	28.9	29.4	30.3	30.1	27.9	25.3	24.1

Norwich   (52.63N,1.30E)

AZ	149.4	152.8	157.9	161.7	165.4	169.1	182.9	187.9	205.0	214.7	218.5
EL	25.5	26.4	27.6	28.3	28.8	29.3	29.8	29.5	26.9	24.2	22.9

Nottingham   (52.97N,1.17W)

AZ	146.7	150.1	155.1	158.8	162.5	166.1	179.8	184.8	201.9	211.8	215.7
EL	24.3	25.3	26.6	27.4	28.1	28.6	29.5	29.4	27.3	24.8	23.5

Oban   (56.42N,5.48W)

AZ	143.1	146.4	151.1	154.7	158.1	161.6	174.6	179.4	196.1	205.9	209.8
EL	19.8	20.8	22.1	23.0	23.7	24.3	25.6	25.8	24.6	22.8	21.9

Oxford   (51.77N,1.25W)

AZ	146.2	149.6	154.6	158.4	162.1	165.8	179.7	184.8	202.2	212.2	216.0
EL	25.4	26.4	27.7	28.6	29.3	29.8	30.8	30.7	28.5	25.9	24.6

Plymouth (50.38N,4.17W)

AZ	142.4	145.8	150.7	154.7	158.1	161.9	175.9	181.1	199.0	209.2	213.3
EL	25.4	26.6	28.2	29.2	30.0	30.7	32.2	32.3	30.6	28.2	26.9

Portsmouth (50.80N,1.08W)

AZ	146.0	149.4	154.5	158.4	162.1	165.8	179.9	185.0	202.6	212.7	216.6
EL	26.3	27.4	28.7	29.6	30.3	30.9	31.8	31.7	29.4	26.7	25.4

Reading (51.47N,0.98W)

AZ	146.4	149.8	154.8	158.7	162.3	166.1	180.0	185.1	202.6	212.5	216.4
EL	25.7	26.8	28.1	29.0	29.6	30.2	31.1	31.0	28.7	26.1	24.7

Southampton (50.92N,1.42W)

AZ	145.7	149.1	154.1	158.0	161.7	165.4	179.5	184.6	202.2	212.2	216.2
EL	26.0	27.1	28.5	29.4	30.1	30.7	31.7	31.6	29.4	26.7	25.4

Stoke on Trent (53.00N,2.17W)

AZ	145.6	149.0	153.9	157.7	161.3	164.9	178.5	183.5	200.7	210.7	214.6
EL	23.9	25.0	26.3	27.1	27.8	28.4	29.4	29.4	27.5	25.1	23.9

Stornaway (58.20N,6.38W)

AZ	142.7	145.9	150.6	154.1	157.5	160.9	173.7	178.4	194.8	204.4	208.3
EL	18.0	19.0	20.2	21.0	21.7	22.3	23.7	23.8	22.9	21.3	20.5

Sunderland (54.92N,1.38W)

AZ	147.1	150.5	155.3	159.1	162.6	166.2	179.5	184.4	201.2	210.9	214.8
EL	22.5	23.4	24.7	25.4	26.0	26.5	27.4	27.3	25.4	23.1	21.9

Swansea (51.63N,3.95W)

AZ	143.1	146.5	151.4	155.2	158.8	162.4	176.2	181.3	198.9	209.1	213.1
EL	24.4	25.6	27.1	28.0	28.8	29.5	30.8	30.9	29.2	26.9	25.7

Telford (52.67N,2.47W)

AZ	145.2	148.5	153.5	157.2	160.8	164.5	178.2	183.2	200.5	210.4	214.4
EL	24.1	25.2	26.5	27.4	28.1	28.7	29.8	29.7	27.9	25.5	24.3

Wick (58.43N,3.10W)

AZ	146.3	149.6	154.3	157.9	161.3	164.7	177.5	182.2	198.5	208.0	211.9
EL	18.8	19.7	20.8	21.6	22.1	22.6	23.6	23.6	22.2	20.3	19.4

York (53.97N,1.08W)

AZ	147.1	150.5	155.4	159.2	162.8	166.4	179.9	184.8	201.8	211.6	215.4
EL	23.4	24.4	25.7	26.4	27.0	27.6	28.4	28.3	26.2	23.8	22.6

Appendix 7

Global azimuth and elevation tables

As an alternative to using tedious trigonometrical equations, the following tables may be used to obtain values for dish azimuth and elevation. Use in the following way:

Let: LS = the longitude of the satellite in degrees east
         LR = the longitude of the receive site in degrees east (e.g.
              $3°W = 357°E$)
    (LS−LR) = net longitudinal difference (degrees).

Step 1: Calculate longitudinal difference between satellite and receive site (LS−LR).
Step 2: Look up the azimuth table with site latitude and (LS−LR).
Step 3: Look up the elevation table with latitude and (LS−LR) value.

Interpretation rules
Rule 1: If (LS−LR) lies between −180° and −360°, adjust by adding 360°.
Rule 2: If (LS−LR) lies between +180° and +360°, adjust by subtracting 360°.
Rule 3: If the final result of (LS−LR) is positive (after correction if necessary), the satellite is east of south, otherwise it is west of south (east or west of north in the southern hemisphere).

Note 1: Rules 1 and 2 only need application when straddling the Greenwich meridian (0°).
Note 2: If either the final value of (LS−LR), or elevation, exceeds the range of the tables then the satellite is over the horizon.
Note 3: Fractional degrees may be linearly estimated between table entries.
Note 4: If using a compass to set the azimuth, remember to make allowance for the magnetic variation.

Example 1: satellite 13°E, receive site 53°N, 3°W (357°E)
$(LS−LR) = 13 − 357 = −344°$.
Adjust by adding 360° (Rule 1), $(−344° + 360°) = 16°$.
$(LS−LR) = 16°$
Looking up the azimuth table returns a value of 19.8°.

(LS−LR) is positive so azimuth = 19.2° east of south.
Looking up the elevation table returns a value of 27.6°.

Example 2: satellite 160°E, receive site 30°S, 150°E
(LS−LR) = 160 − 150 = 10°.
Looking up the azimuth table returns a value of 19.4°.
(LS−LR) is positive so azimuth = 19.4° east of north.
Looking up the elevation table returns a value of 53.3°.

Example 3: satellite 27.5°W (332.5°E), receive site 50°N, 15°E
(LS−LR) = 332.5 − 15 = 317.5°.
Adjust by subtracting 360° (Rule 2), (317.5° − 360°) = −42.5°.
Looking up the azimuth table and estimating fraction gives 50.1°.
(LS−LR) is negative so azimuth is 50.1° west of south.
Looking up the elevation table returns a value of 20.2°.

Example 4: satellite 125°W (235°E), receive site 25.78°N, 80.1°W (279.9°E)
(LS−LR) = 235° − 279.9° = −44.9°.
Looking up the azimuth table returns a value of about 66.4°.
(LS−LR) is negative so azimuth = 66.4° west of south.
Looking up the elevation table returns a value of 32.3°.

Table A7.1 *Azimuth angles*

| Latitude (°) | Longitude difference LS − LR(°) | | | | | | | | | | |
	0.0	1.0	2.0	3.0	4.0	5.0	6.0	7.0	8.0	9.0	10.0
00	0.0	90.0	90.0	90.0	90.0	90.0	90.0	90.0	90.0	90.0	90.0
1.0	0.0	45.0	63.4	71.6	76.0	78.7	80.6	81.9	82.9	83.7	84.3
2.0	0.0	26.6	45.0	56.3	63.5	68.3	71.6	74.1	76.1	77.6	78.8
3.0	0.0	18.4	33.7	45.0	53.2	59.1	63.5	66.9	69.6	71.7	73.5
4.0	0.0	14.0	26.6	36.9	45.1	51.4	56.4	60.4	63.6	66.2	68.4
5.0	0.0	11.3	21.8	31.0	38.7	45.1	50.3	54.6	58.2	61.2	63.7
6.0	0.0	9.5	18.5	26.6	33.8	39.9	45.2	49.6	53.4	56.6	59.3
7.0	0.0	8.2	46.0	23.3	29.8	35.7	40.8	45.2	49.1	52.4	55.3
8.0	0.0	7.1	14.1	20.6	26.7	32.2	37.1	41.4	45.3	48.7	51.7
9.0	0.0	6.4	12.6	18.5	24.1	29.2	33.9	38.1	41.9	45.4	48.4
10.0	0.0	5.7	11.4	16.8	21.9	26.7	31.2	35.3	39.0	42.4	45.4
11.0	0.0	5.2	10.4	15.4	20.1	24.6	28.8	32.8	36.4	39.7	42.7
12.0	0.0	4.8	9.5	14.1	18.6	22.8	26.8	30.6	34.1	37.3	40.3
13.0	0.0	4.4	8.8	13.1	17.3	21.3	25.0	28.6	32.0	35.1	38.1
14.0	0.0	4.1	8.2	12.2	16.1	19.9	23.5	26.9	30.2	33.2	36.1
15.0	0.0	3.9	7.7	11.4	15.1	18.7	22.1	25.4	28.5	31.5	34.3
16.0	0.0	3.6	7.2	10.8	14.2	17.6	20.9	24.0	27.0	29.9	32.6
17.0	0.0	3.4	6.8	10.2	13.5	16.7	19.8	22.8	25.7	28.4	31.1
18.0	0.0	3.2	6.4	9.6	12.8	15.8	18.8	21.7	24.5	27.1	29.7
19.0	0.0	3.1	6.1	9.1	12.1	15.0	17.9	20.7	23.3	25.9	28.4
20.0	0.0	2.9	5.8	8.7	11.6	14.3	17.1	19.7	22.3	24.8	27.3

Table A7.1   *Azimuth angles*

Latitude (°)	Longitude difference LS − LR(°)										
	0.0	1.0	2.0	3.0	4.0	5.0	6.0	7.0	8.0	9.0	10.0
21.0	0.0	2.8	5.6	8.3	11.0	13.7	16.3	18.9	21.4	23.8	26.2
22.0	0.0	2.7	5.3	8.0	10.6	13.1	15.7	18.1	20.6	22.9	25.2
23.0	0.0	2.6	5.1	7.6	10.1	12.6	15.1	17.4	19.8	22.1	24.3
24.0	0.0	2.5	4.9	7.3	9.8	12.1	14.5	16.8	19.1	21.3	23.4
25.0	0.0	2.4	4.7	7.1	9.4	11.7	14.0	16.2	18.4	20.5	22.6
26.0	0.0	2.3	4.6	6.8	9.1	11.3	13.5	15.6	17.8	19.9	21.9
27.0	0.0	2.2	4.4	6.6	8.8	10.9	13.0	15.1	17.2	19.2	21.2
28.0	0.0	2.1	4.3	6.4	8.5	10.6	12.6	14.7	16.7	18.6	20.6
29.0	0.0	2.1	4.1	6.2	8.2	10.2	12.2	14.2	16.2	18.1	20.0
30.0	0.0	2.0	4.0	6.0	8.0	9.9	11.9	13.8	15.7	17.6	19.4
31.0	0.0	1.9	3.9	5.8	7.7	9.6	11.5	13.4	15.3	17.1	18.9
32.0	0.0	1.9	3.8	5.6	7.5	9.4	11.2	13.0	14.9	16.6	18.4
33.0	0.0	1.8	3.7	5.5	7.3	9.1	10.9	12.7	14.5	16.2	17.9
34.0	0.0	1.8	3.6	5.4	7.1	8.9	10.6	12.4	14.1	15.8	17.5
35.0	0.0	1.7	3.5	5.2	7.0	8.7	10.4	12.1	13.8	15.4	17.1
36.0	0.0	1.7	3.4	5.1	6.8	8.5	10.1	11.8	13.4	15.1	16.7
37.0	0.0	1.7	3.3	5.0	6.6	8.3	9.9	11.5	13.1	14.7	16.3
38.0	0.0	1.6	3.2	4.9	6.5	8.1	9.7	11.3	12.9	14.4	16.0
39.0	0.0	1.6	3.2	4.8	6.3	7.9	9.5	11.0	12.6	14.1	15.7
40.0	0.0	1.6	3.1	4.7	6.2	7.8	9.3	10.8	12.3	13.8	15.3
41.0	0.0	1.5	3.0	4.6	6.1	7.6	9.1	10.6	12.1	13.6	15.0
42.0	0.0	1.5	3.0	4.5	6.0	7.4	8.9	10.4	11.9	13.3	14.8
43.0	0.0	1.5	2.9	4.4	5.9	7.3	8.8	10.2	11.6	13.1	14.5
44.0	0.0	1.4	2.9	4.3	5.7	7.2	8.6	10.0	11.4	12.8	14.2
45.0	0.0	1.4	2.8	4.2	5.6	7.1	8.5	9.9	11.2	12.6	14.0
46.0	0.0	1.4	2.8	4.2	5.6	6.9	8.3	9.7	11.1	12.4	13.8
47.0	0.0	1.4	2.7	4.1	5.5	6.8	8.2	9.5	10.9	12.2	13.6
48.0	0.0	1.3	2.7	4.0	5.4	6.7	8.1	9.4	10.7	12.0	13.3
49.0	0.0	1.3	2.6	4.0	5.3	6.6	7.9	9.2	10.5	11.9	13.2
50.0	0.0	1.3	2.6	3.6	5.2	6.5	7.8	9.1	10.4	11.7	13.0
51.0	0.0	1.3	2.6	3.9	5.1	6.4	7.7	9.0	10.3	11.5	12.8
52.0	0.0	1.3	2.5	3.8	5.1	6.3	7.6	8.9	10.1	11.4	12.6
53.0	0.0	1.3	2.5	3.8	5.0	6.3	7.5	8.7	10.0	11.2	12.5
54.0	0.0	1.2	2.5	3.7	4.9	6.2	7.4	8.6	9.9	11.1	12.3
55.0	0.0	1.2	2.4	3.7	4.9	6.1	7.3	8.5	9.7	10.9	12.1
56.0	0.0	1.2	2.4	3.6	4.8	6.0	7.2	8.4	9.6	10.8	12.0
57.0	0.0	1.2	2.4	3.6	4.8	6.0	7.1	8.3	9.5	10.7	11.9
58.0	0.0	1.2	2.4	3.5	4.7	5.9	7.1	8.2	9.4	10.6	11.7
59.0	0.0	1.2	2.3	3.5	4.7	5.8	7.0	8.2	9.3	10.5	11.6
60.0	0.0	1.2	2.3	3.5	4.6	5.8	6.9	8.1	9.2	10.4	11.5
61.0	0.0	1.1	2.3	3.4	4.6	5.7	6.9	8.0	9.1	10.3	11.4
62.0	0.0	1.1	2.3	3.4	4.5	5.7	6.8	7.9	9.0	10.2	11.3
63.0	0.0	1.1	2.2	3.4	4.5	5.6	6.7	7.8	9.0	10.1	11.2
64.0	0.0	1.1	2.2	3.3	4.4	5.6	6.7	7.8	8.9	10.0	11.1
65.0	0.0	1.1	2.2	3.3	4.4	5.5	6.6	7.7	8.8	9.9	11.0
66.0	0.0	1.1	2.2	3.3	4.4	5.5	6.6	7.7	8.7	9.8	10.9
67.0	0.0	1.1	2.2	3.3	4.3	5.4	6.5	7.6	8.7	9.8	10.8
68.0	0.0	1.1	2.2	3.2	4.3	5.4	6.5	7.5	8.6	9.7	10.8
69.0	0.0	1.1	2.1	3.2	4.3	5.4	6.4	7.5	8.6	9.6	10.7
70.0	0.0	1.1	2.1	3.2	4.3	5.3	6.4	7.4	8.5	9.6	10.6
71.0	0.0	1.1	2.1	3.2	4.2	5.3	6.3	7.4	8.5	9.5	10.6
72.0	0.0	1.1	2.1	3.2	4.2	5.3	6.3	7.4	8.4	9.5	10.5
73.0	0.0	1.0	2.1	3.1	4.2	5.2	6.3	7.3	8.4	9.4	10.4
74.0	0.0	1.0	2.1	3.1	4.2	5.2	6.2	7.3	8.3	9.4	10.4
75.0	0.0	1.0	2.1	3.1	4.1	5.2	6.2	7.2	8.3	9.3	10.3
76.0	0.0	1.0	2.1	3.1	4.1	5.2	6.2	7.2	8.2	9.3	10.3

Table A7.1 *Azimuth angles*

Latitude (°)	Longitude difference LS − LR(°)										
	0.0	1.0	2.0	3.0	4.0	5.0	6.0	7.0	8.0	9.0	10.0
77.0	0.0	1.0	2.1	3.1	4.1	5.1	6.2	7.2	8.2	9.2	10.3
78.0	0.0	1.0	2.0	3.1	4.1	5.1	6.1	7.2	8.2	9.2	10.2
79.0	0.0	1.0	2.0	3.1	4.1	5.1	6.1	7.1	8.1	9.2	10.2
80.0	0.0	1.0	2.0	3.0	4.1	5.1	6.1	7.1	8.1	9.1	10.2

Latitude (°)	Longitude difference LS − LR(°)									
	11.0	12.0	13.0	14.0	15.0	16.0	17.0	18.0	19.0	20.0
0.0	90.0	90.0	90.0	90.0	90.0	90.0	90.0	90.0	90.0	90.0
1.0	84.9	85.3	85.7	86.0	86.3	86.5	86.7	86.9	87.1	87.3
2.0	79.8	80.7	81.4	82.0	82.6	83.1	83.5	83.9	84.2	84.5
3.0	74.9	76.2	77.2	78.1	78.9	79.7	80.3	80.8	81.4	81.8
4.0	70.3	71.8	73.2	74.4	75.4	76.3	77.1	77.9	78.5	79.2
5.0	65.8	67.7	69.3	70.7	72.0	73.1	74.1	75.0	75.8	76.5
6.0	61.7	63.8	65.6	67.3	68.7	70.0	71.1	72.2	73.1	74.0
7.0	57.9	60.2	62.2	64.0	65.5	67.0	68.3	69.4	70.5	71.5
8.0	54.4	56.8	58.9	60.8	62.6	64.1	65.5	66.8	68.0	69.1
9.0	51.2	53.6	55.9	57.9	59.7	61.4	62.9	64.3	65.6	66.7
10.0	48.2	50.8	53.1	55.1	57.1	58.8	60.4	61.9	63.2	64.5
11.0	45.5	48.1	50.4	52.6	54.5	56.4	58.0	59.6	61.0	62.3
12.0	43.1	45.6	48.0	50.2	52.2	54.1	55.8	57.4	58.9	60.3
13.0	40.8	43.4	45.7	47.9	50.0	51.9	53.7	55.3	56.8	58.3
14.0	38.8	41.3	43.7	45.9	47.9	49.8	51.6	53.3	54.9	56.4
15.0	36.9	39.4	41.7	43.9	46.0	47.9	49.8	51.5	53.1	54.6
16.0	35.2	37.6	39.9	42.1	44.2	46.1	48.0	49.7	51.3	52.9
17.0	33.6	36.0	38.3	40.5	42.5	44.4	46.3	48.0	49.7	51.2
18.0	32.2	34.5	36.8	38.9	40.9	42.9	44.7	46.4	48.1	49.7
19.0	30.8	33.1	35.3	37.4	39.5	41.4	43.2	44.9	46.6	48.2
20.0	29.6	31.9	34.0	36.1	38.1	40.0	41.8	43.5	45.2	46.8
21.0	28.5	30.7	32.8	34.8	36.8	38.7	40.5	42.2	43.9	45.4
22.0	27.4	29.6	31.6	33.6	35.6	37.4	39.2	40.9	42.6	44.2
23.0	26.4	28.5	30.6	32.5	34.4	36.3	38.0	39.7	41.4	43.0
24.0	25.5	27.6	29.6	31.5	33.4	35.2	36.9	38.6	40.2	41.8
25.0	24.7	26.7	28.6	30.5	32.4	34.2	35.9	37.6	39.2	40.7
26.0	23.9	25.9	27.8	29.6	31.4	33.2	34.9	36.5	38.1	39.7
27.0	23.2	25.1	27.0	28.8	30.5	32.3	34.0	35.6	37.2	38.7
28.0	22.5	24.4	26.2	28.0	29.7	31.4	33.1	34.7	36.3	37.8
29.0	21.8	23.7	25.5	27.2	28.9	30.6	32.2	33.8	35.4	36.9
30.0	21.2	23.0	24.8	26.5	28.2	29.8	31.4	33.0	34.6	36.1
31.0	20.7	22.4	24.1	25.8	27.5	29.1	30.7	32.2	33.8	35.2
32.0	20.1	21.9	23.5	25.2	26.8	28.4	30.0	31.5	33.0	34.5
33.0	19.6	21.3	23.0	24.6	26.2	27.8	29.3	30.8	32.3	33.8
34.0	19.2	20.8	22.4	24.0	25.6	27.1	28.7	30.2	31.6	33.1
35.0	18.7	20.3	21.9	23.5	25.0	26.6	28.1	29.5	31.0	32.4
36.0	18.3	19.9	21.4	23.0	24.5	26.0	27.5	28.9	30.4	31.8
37.0	17.9	19.5	21.0	22.5	24.0	25.5	26.9	28.4	29.8	31.2
38.0	17.5	19.0	20.6	22.0	23.5	25.0	26.4	27.8	29.2	30.6
39.0	17.2	18.7	20.1	21.6	23.1	24.5	25.9	27.3	28.7	30.0
40.0	16.8	18.3	19.8	21.2	22.6	24.0	25.4	26.8	28.2	29.5
41.0	16.5	18.0	19.4	20.8	22.2	23.6	25.0	26.3	27.7	29.0
42.0	16.2	17.6	19.0	20.4	21.8	23.2	24.6	25.9	27.2	28.5
43.0	15.9	17.3	18.7	20.1	21.4	22.8	24.1	25.5	26.8	28.1
44.0	15.6	17.0	18.4	19.7	21.1	22.4	23.8	25.1	26.4	27.7
45.0	15.4	16.7	18.1	19.4	20.8	22.1	23.4	24.7	26.0	27.2
46.0	15.1	16.5	17.8	19.1	20.4	21.7	23.0	24.3	25.6	26.8

Table A7.1   *Azimuth angles*

Latitude (°)	Longitude difference LS − LR(°)									
	11.0	12.0	13.0	14.0	15.0	16.0	17.0	18.0	19.0	20.0
47.0	20.3	16.2	17.5	18.8	20.1	21.4	22.7	24.0	25.2	26.5
48.0	14.9	16.0	17.3	18.5	19.8	21.1	22.4	23.6	24.9	26.1
49.0	14.7	15.7	17.0	18.3	19.5	20.8	22.1	23.3	24.5	25.7
50.0	14.4	15.5	16.8	18.0	19.3	20.5	21.8	23.0	24.2	25.4
51.0	14.2	15.3	16.5	17.8	19.0	20.3	21.5	22.7	23.9	25.1
52.0	14.0	15.1	16.3	17.6	18.8	20.0	21.2	22.4	23.6	24.8
53.0	13.9	14.9	16.1	17.3	18.5	19.8	20.9	22.1	23.3	24.5
54.0	13.7	14.7	15.9	17.1	18.3	19.5	20.7	21.9	23.1	24.2
55.0	13.5	14.5	15.7	16.9	18.1	19.3	20.5	21.6	22.8	24.0
56.0	13.3	14.4	15.6	16.7	17.9	19.1	20.2	21.4	22.6	23.7
57.0	13.2	14.2	15.4	16.6	17.7	18.9	20.0	21.2	22.3	23.5
58.0	13.0	14.1	15.2	16.4	17.5	18.7	19.8	21.0	22.1	23.2
59.0	12.9	13.9	15.1	16.2	17.4	18.5	19.6	20.8	21.9	23.0
60.0	12.8	13.8	14.8	16.1	17.2	18.3	19.4	20.6	21.7	22.8
61.0	12.7	13.7	14.8	15.9	17.0	18.2	19.3	20.4	21.5	22.6
62.0	12.5	13.5	14.7	15.8	16.9	18.0	19.1	20.2	21.3	22.4
63.0	12.4	13.4	14.5	15.6	16.7	17.8	18.9	20.0	21.1	22.2
64.0	12.3	13.3	14.4	15.5	16.6	17.7	18.8	19.9	21.0	22.0
65.0	12.2	13.2	14.3	15.4	16.5	17.6	18.6	19.7	20.8	21.9
66.0	12.1	13.1	14.2	15.3	16.3	17.4	18.5	19.6	20.7	21.7
67.0	12.0	13.0	14.1	15.2	16.2	17.3	18.4	19.4	20.5	21.6
68.0	11.9	12.9	14.0	15.1	16.1	17.2	18.2	19.3	20.4	21.4
69.0	11.8	12.8	13.9	15.0	16.0	17.1	18.1	19.2	20.2	21.3
70.0	11.8	12.7	13.8	14.9	15.9	17.0	18.0	19.1	20.1	21.2
71.0	11.7	12.7	13.7	14.8	15.8	16.9	17.9	19.0	20.0	21.1
72.0	11.6	12.6	13.6	14.7	15.7	16.8	17.8	18.9	19.9	20.9
73.0	11.6	12.5	13.6	14.6	15.7	16.7	17.7	18.8	19.8	20.8
74.0	11.5	12.5	13.5	14.5	15.6	16.6	17.6	18.7	19.7	20.7
75.0	11.4	12.4	13.4	14.5	15.5	16.5	17.6	18.6	19.6	20.6
76.0	11.4	12.4	13.4	14.4	15.4	16.5	17.5	18.5	19.5	20.6
77.0	11.3	12.3	13.3	14.4	15.4	16.4	17.4	18.4	19.5	20.5
78.0	11.3	12.3	13.3	14.3	15.3	16.3	17.4	18.4	19.4	20.4
79.0	11.2	12.2	13.2	14.3	15.3	16.3	17.3	18.3	19.3	20.3
80.0	11.2	11.2	12.2	13.2	14.2	15.2	16.2	17.2	18.3	19.3

Latitude (°)	Longitude difference LS − LR(°)									
	21.0	22.0	23.0	24.0	25.0	26.0	27.0	28.0	29.0	30.0
0.0	90.0	90.0	90.0	90.0	90.0	90.0	90.0	90.0	90.0	90.0
1.0	87.4	87.5	87.6	87.8	87.9	88.0	88.0	88.1	88.2	88.3
2.0	84.8	85.1	85.3	85.5	85.7	85.9	86.0	86.2	86.4	86.5
3.0	82.2	82.6	83.0	83.3	83.6	83.9	84.1	84.4	84.6	84.8
4.0	79.7	80.2	80.7	81.1	81.5	81.9	82.2	82.5	82.8	83.1
5.0	77.2	77.8	78.4	78.9	79.4	79.9	80.3	80.7	81.1	81.4
6.0	74.8	75.5	76.2	76.8	77.4	77.9	78.4	78.9	79.3	79.7
7.0	72.4	73.2	74.0	74.7	75.4	76.0	76.5	77.1	77.6	78.1
8.0	70.1	71.0	71.8	72.6	73.4	74.1	74.7	75.3	75.9	76.4
9.0	67.8	68.8	69.8	70.6	71.5	72.2	72.9	73.6	74.2	74.8
10.0	65.7	66.7	67.8	68.7	69.6	70.4	71.2	71.9	72.6	73.3
11.0	63.6	64.7	65.8	66.8	67.7	68.6	69.5	70.3	71.0	71.7
12.0	61.6	62.8	63.9	65.0	66.0	66.9	67.8	68.6	69.4	70.2
13.0	59.6	60.9	62.1	63.2	64.2	65.2	66.2	67.1	67.9	68.7
14.0	57.8	59.1	60.3	61.5	62.6	63.6	64.6	65.5	66.4	67.3
15.0	56.0	57.4	58.6	59.8	61.0	62.0	63.1	64.0	65.0	65.9

Table A7.1 *Azimuth angles*

Latitude (°)	Longitude difference LS − LR(°)									
	21.0	*22.0*	*23.0*	*24.0*	*25.0*	*26.0*	*27.0*	*28.0*	*29.0*	*30.0*
16.0	54.3	55.7	57.0	58.2	59.4	60.5	61.6	62.6	63.6	64.5
17.0	52.7	54.1	55.4	56.7	57.9	59.1	60.2	61.2	62.2	63.1
18.0	51.2	52.6	53.9	55.2	56.5	57.6	58.8	59.8	60.9	61.8
19.0	49.7	51.1	52.5	53.8	55.1	56.3	57.4	58.5	59.6	60.6
20.0	48.3	49.8	51.1	52.5	53.7	55.0	56.1	57.2	58.3	59.4
21.0	47.0	48.4	49.8	51.2	52.5	53.7	54.9	56.0	57.1	58.2
22.0	45.7	47.2	48.6	49.9	51.2	52.5	53.7	54.8	55.9	57.0
23.0	44.5	46.0	47.4	48.7	50.0	51.3	52.5	53.7	54.8	55.9
24.0	43.3	44.8	46.2	47.6	48.9	50.2	51.4	52.6	53.7	54.8
25.0	42.2	43.7	45.1	46.5	47.8	49.1	50.3	51.5	52.7	53.8
26.0	41.2	42.7	44.1	45.4	46.8	48.1	49.3	50.5	51.7	52.8
27.0	40.2	41.7	43.1	44.4	45.8	47.1	48.3	49.5	50.7	51.8
28.0	39.3	40.7	42.1	43.5	44.8	46.1	47.3	48.6	49.7	50.9
29.0	38.4	39.8	41.2	42.6	43.9	45.2	46.4	47.6	48.8	50.0
30.0	37.5	38.9	40.3	41.7	43.0	44.3	45.5	46.8	47.9	49.1
31.0	36.7	38.1	39.5	40.8	42.2	43.4	44.7	45.9	47.1	48.3
32.0	35.9	37.3	38.7	40.0	41.3	42.6	43.9	45.1	46.3	47.5
33.0	35.2	36.6	37.9	39.3	40.6	41.8	43.1	44.3	45.5	46.7
34.0	34.5	35.8	37.2	38.5	39.8	41.1	42.3	43.6	44.7	45.9
35.0	33.8	35.2	36.5	37.8	39.1	40.4	41.6	42.8	44.0	45.2
36.0	33.1	34.5	35.8	37.1	38.4	39.7	40.9	42.1	43.3	44.5
37.0	32.5	33.9	35.2	36.5	37.8	39.0	40.3	41.5	42.6	43.8
38.0	31.9	33.3	34.6	35.9	37.1	38.4	39.6	40.8	42.0	43.2
39.0	31.4	32.7	34.0	35.3	36.5	37.8	39.0	40.2	41.4	42.5
40.0	30.8	32.2	33.4	34.7	36.0	37.2	38.4	39.6	40.8	41.9
41.0	30.3	31.6	32.9	34.2	35.4	36.6	37.8	39.0	40.2	41.3
42.0	29.8	31.1	32.4	33.6	34.9	36.1	37.3	38.5	39.6	40.8
43.0	29.4	30.6	31.9	33.1	34.4	35.6	36.8	37.9	39.1	40.2
44.0	28.9	30.2	31.4	32.7	33.9	35.1	36.3	37.4	38.6	39.7
45.0	28.5	29.7	31.0	32.2	33.4	34.6	35.8	36.9	38.1	39.2
46.0	28.1	29.3	30.5	31.8	33.0	34.1	35.3	36.5	37.6	38.8
47.0	27.7	28.9	30.1	31.3	32.5	33.7	34.9	36.0	37.2	38.3
48.0	27.3	28.5	29.7	30.9	32.1	33.3	34.4	35.6	36.7	37.8
49.0	27.0	28.2	29.4	30.5	31.7	32.9	34.0	35.2	36.3	37.4
50.0	26.6	27.8	29.0	30.2	31.3	32.5	33.6	34.8	35.9	37.0
51.0	26.3	27.5	28.6	29.8	31.0	32.1	33.3	34.4	35.5	36.6
52.0	26.0	27.1	28.3	29.5	30.6	31.8	32.9	34.0	35.1	36.2
53.0	25.7	26.8	28.0	29.1	30.3	31.4	32.5	33.7	34.8	35.9
54.0	25.4	26.5	27.7	28.8	30.0	31.1	32.2	33.3	34.4	35.5
55.0	25.1	26.3	27.4	28.5	29.7	30.8	31.9	33.0	34.1	35.2
56.0	24.8	26.0	27.1	28.2	29.4	30.5	31.6	32.7	33.8	34.9
57.0	24.6	25.7	26.8	28.0	29.1	30.2	31.3	32.4	33.5	34.5
58.0	24.4	25.5	26.6	27.7	28.8	29.9	31.0	32.1	33.2	34.2
59.0	24.1	25.2	26.3	27.4	28.5	29.6	30.7	31.8	32.9	34.0
60.0	23.9	25.0	26.1	27.2	28.3	29.4	30.5	31.5	32.6	33.7
61.0	23.7	24.8	25.9	27.0	28.1	29.1	30.2	31.3	32.4	33.4
62.0	23.5	24.6	25.7	26.8	27.8	28.9	30.0	31.1	32.1	33.2
63.0	23.3	24.4	25.5	26.6	27.6	28.7	29.8	30.8	31.9	32.9
64.0	23.1	24.2	25.3	26.4	27.4	28.5	29.5	30.6	31.7	32.7
65.0	23.0	24.0	25.1	26.2	27.2	28.3	29.3	30.4	31.5	32.5
66.0	22.8	23.9	24.9	26.0	27.0	28.1	29.2	30.2	31.2	32.3
67.0	22.6	23.7	24.8	25.8	26.9	27.9	29.0	30.0	31.1	32.1
68.0	22.5	23.5	24.6	25.7	26.7	27.7	28.8	29.8	30.9	31.9
69.0	22.4	23.4	24.5	25.5	26.5	27.6	28.6	29.7	30.7	31.7
70.0	22.2	23.3	24.3	25.4	26.4	27.4	28.5	29.5	30.5	31.6
71.0	22.1	23.1	24.2	25.2	26.3	27.3	28.3	29.4	30.4	31.4

Table A7.1   *Azimuth angles*

Latitude (°)	Longitude difference LS − LR(°)									
	21.0	22.0	23.0	24.0	25.0	26.0	27.0	28.0	29.0	30.0
72.0	22.0	23.0	24.1	25.1	26.1	27.2	28.2	29.2	30.2	31.3
73.0	21.9	22.9	23.9	25.0	26.0	27.0	28.0	29.1	30.1	31.1
74.0	21.8	22.8	23.8	24.9	25.9	26.9	27.9	28.9	30.0	31.0
75.0	21.7	22.7	23.7	24.7	25.8	26.8	27.8	28.8	29.8	30.9
76.0	21.6	22.6	23.6	24.6	25.7	26.7	27.7	28.7	29.7	30.8
77.0	21.5	22.5	23.5	24.6	25.6	26.6	27.6	28.6	29.6	30.6
78.0	21.4	22.4	23.5	24.5	25.5	26.5	27.5	28.5	29.5	30.6
79.0	21.4	22.4	23.4	24.4	25.4	26.4	27.4	28.4	29.5	30.5
80.0	21.3	22.3	23.3	24.3	25.3	26.3	27.4	28.4	29.4	30.4

Latitude (°)	Longitude difference LS − LR(°)									
	31.0	32.0	33.0	34.0	35.0	36.0	37.0	38.0	39.0	40.0
0.0	90.0	90.0	90.0	90.0	90.0	90.0	90.0	90.0	90.0	90.0
1.0	88.3	88.4	88.5	88.5	88.6	88.6	88.7	88.7	88.8	88.8
2.0	86.7	86.8	86.9	87.0	87.1	87.2	87.3	87.4	87.5	87.6
3.0	85.0	85.2	85.4	85.6	85.7	85.9	86.0	86.2	86.3	86.4
4.0	83.4	83.6	83.9	84.1	84.3	84.5	84.7	84.9	85.1	85.2
5.0	81.7	82.1	82.4	82.6	82.9	83.2	83.4	83.6	83.9	84.1
6.0	80.1	80.5	80.9	81.2	81.5	81.8	82.1	82.4	82.6	82.9
7.0	78.5	79.0	79.4	79.8	80.1	80.5	80.8	81.1	81.4	81.7
8.0	77.0	77.4	77.9	78.3	78.8	79.2	79.5	79.9	80.2	80.6
9.0	75.4	75.9	76.5	76.9	77.4	77.8	78.3	78.7	79.1	79.4
10.0	73.9	74.5	75.0	75.6	76.1	76.6	77.0	77.5	77.9	78.3
11.0	72.4	73.0	73.6	74.2	74.8	75.3	75.8	76.3	76.7	77.2
12.0	70.9	71.6	72.2	72.9	73.5	74.0	74.6	75.1	75.6	76.1
13.0	69.5	70.2	70.9	71.6	72.2	72.8	73.4	73.9	74.5	75.0
14.0	68.1	68.8	69.6	70.3	70.9	71.6	72.2	72.8	73.4	73.9
15.0	66.7	67.5	68.3	69.0	69.7	70.4	71.0	71.7	72.3	72.9
16.0	65.4	66.2	67.0	67.8	68.5	69.2	69.9	70.6	71.2	71.8
17.0	64.1	64.9	65.8	66.6	67.3	68.1	68.8	69.5	70.1	70.8
18.0	62.8	63.7	64.6	65.4	66.2	67.0	67.7	68.4	69.1	69.8
19.0	61.5	62.5	63.4	64.2	65.1	65.9	66.6	67.4	68.1	68.8
20.0	60.4	61.3	62.2	63.1	64.0	64.8	65.6	66.4	67.1	67.8
21.0	59.2	60.2	61.1	62.0	62.9	63.7	64.6	65.4	66.1	66.9
22.0	58.1	59.1	60.0	61.0	61.9	62.7	63.6	64.4	65.2	65.9
23.0	57.0	58.0	59.0	59.9	60.8	61.7	62.6	63.4	64.2	65.0
24.0	55.9	56.9	57.9	58.9	59.8	60.8	61.6	62.5	63.3	64.1
25.0	54.9	55.9	56.9	57.9	58.9	59.8	60.7	61.6	62.4	63.3
26.0	53.9	54.9	56.0	57.0	58.0	58.9	59.8	60.7	61.6	62.4
27.0	52.9	54.0	55.0	56.1	57.0	58.0	58.9	59.8	60.7	61.6
28.0	52.0	53.1	54.1	55.2	56.2	57.1	58.1	59.0	59.9	60.8
29.0	51.1	52.2	53.3	54.3	55.3	56.3	57.2	58.2	59.1	60.0
30.0	50.2	51.3	52.4	53.5	54.5	55.5	56.4	57.4	58.3	59.2
31.0	49.4	50.5	51.6	52.6	53.7	54.7	55.6	56.6	57.5	58.5
32.0	48.6	49.7	50.8	51.8	52.9	53.9	54.9	55.9	56.8	57.7
33.0	47.8	48.9	50.0	51.1	52.1	53.1	54.1	55.1	56.1	57.0
34.0	47.1	48.2	49.3	50.3	51.4	52.4	53.4	54.4	55.4	56.3
35.0	46.3	47.5	48.5	49.6	50.7	51.7	52.7	53.7	54.7	55.6
36.0	45.6	46.8	47.9	48.9	50.0	51.0	52.0	53.0	54.0	55.0
37.0	45.0	46.1	47.2	48.3	49.3	50.4	51.4	52.4	53.4	54.4
38.0	44.3	45.4	46.5	47.6	48.7	49.7	50.8	51.8	52.8	53.7
39.0	43.7	44.8	45.9	47.0	48.1	49.1	50.1	51.1	52.1	53.1
40.0	43.1	44.2	45.3	46.4	47.4	48.5	49.5	50.6	51.6	52.5
41.0	42.5	43.6	44.7	45.8	46.9	47.9	49.0	50.0	51.0	52.0

Table A7.1 *Azmith angles*

Latitude (°)	Longitude difference LS − LR(°)									
	31.0	32.0	33.0	34.0	35.0	36.0	37.0	38.0	39.0	40.0
42.0	41.9	43.0	44.1	45.2	46.3	47.4	48.4	49.4	50.4	51.4
43.0	41.4	42.5	43.6	44.7	45.8	46.8	47.9	48.9	49.9	50.9
44.0	40.9	42.0	43.1	44.2	45.2	46.3	47.3	48.4	49.4	50.4
45.0	40.4	41.5	42.6	43.6	44.7	45.8	46.8	47.9	48.9	49.9
46.0	39.9	41.0	42.1	43.2	44.2	45.3	46.3	47.4	48.4	49.4
47.0	39.4	40.5	41.6	42.7	43.8	44.8	45.9	46.9	47.9	48.9
48.0	39.0	40.1	41.1	42.2	43.3	44.4	45.4	46.4	47.5	48.5
49.0	38.5	39.6	40.7	41.8	42.9	43.9	45.0	46.0	47.0	48.0
50.0	38.1	39.2	40.3	41.4	42.4	43.5	44.5	45.6	46.6	47.6
51.0	37.7	38.8	39.9	41.0	42.0	43.1	44.1	45.2	46.2	47.2
52.0	37.3	38.4	39.5	40.6	41.6	42.7	43.7	44.8	45.8	46.8
53.0	37.0	38.0	39.1	40.2	41.2	42.3	43.3	44.4	45.4	46.4
54.0	36.6	37.7	38.8	39.8	40.9	41.9	43.0	44.0	45.0	46.0
55.0	36.3	37.3	38.4	39.5	40.5	41.6	42.6	43.6	44.7	45.7
56.0	35.9	37.0	38.1	39.1	40.2	41.2	42.3	43.3	44.3	45.3
57.0	35.6	36.7	37.8	38.8	39.9	40.9	41.9	43.0	44.0	45.0
58.0	35.3	36.4	37.4	38.5	39.5	40.6	41.6	42.7	43.7	44.7
59.0	35.0	36.1	37.1	38.2	39.2	40.3	41.3	42.3	43.4	44.4
60.0	34.8	35.8	36.9	37.9	39.0	40.0	41.0	42.1	43.1	44.1
61.0	34.5	35.5	36.6	37.6	38.7	39.7	40.7	41.8	42.8	43.8
62.0	34.2	35.3	36.3	37.4	38.4	39.4	40.5	41.5	42.5	43.5
63.0	34.0	35.0	36.1	37.1	38.2	39.2	40.2	41.2	42.3	43.3
64.0	33.8	34.8	35.8	36.9	37.9	39.0	40.0	41.0	42.0	43.0
65.0	33.5	34.6	35.6	36.7	37.7	38.7	39.7	40.8	41.8	42.8
66.0	33.3	34.4	35.4	36.4	37.5	38.5	39.5	40.5	41.6	42.6
67.0	33.1	34.2	35.2	36.2	37.3	38.3	39.3	40.3	41.3	42.4
68.0	32.9	34.0	35.0	36.0	37.1	38.1	39.1	40.1	41.1	42.1
69.0	32.8	33.8	34.8	35.8	36.9	37.9	38.9	39.9	40.9	41.9
70.0	32.6	33.6	34.6	35.7	36.7	37.7	38.7	39.7	40.8	41.8
71.0	32.4	33.5	34.5	35.5	36.5	37.5	38.6	39.6	40.6	41.6
72.0	32.3	33.3	34.3	35.3	36.4	37.4	38.4	39.4	40.4	41.4
73.0	32.1	33.2	34.2	35.2	36.2	37.2	38.2	39.2	40.3	41.3
74.0	32.0	33.0	34.0	35.1	36.1	37.1	38.1	39.1	40.1	41.1
75.0	31.9	32.9	33.9	34.9	35.9	36.9	38.0	39.0	40.0	41.0
76.0	31.8	32.8	33.8	34.8	35.8	36.8	37.8	38.8	39.8	40.9
77.0	31.7	32.7	33.7	34.7	35.7	36.7	37.7	38.7	39.7	40.7
78.0	31.6	32.6	33.6	34.6	35.6	36.6	37.6	38.6	39.6	40.6
79.0	31.5	32.5	33.5	34.5	35.5	36.5	37.5	38.5	39.5	40.5
80.0	31.4	32.4	33.4	34.4	35.4	36.4	37.4	38.4	39.4	40.4

Latitude (°)	Longitude difference LS − LR(°)									
	41.0	42.0	43.0	44.0	45.0	46.0	47.0	48.0	49.0	50.0
0.0	90.0	90.0	90.0	90.0	90.0	90.0	90.0	90.0	90.0	90.0
1.0	88.8	88.9	88.9	89.0	89.0	89.0	89.1	89.1	89.1	89.2
2.0	87.7	87.8	87.9	87.9	88.0	88.1	88.1	88.2	88.3	88.3
3.0	86.6	86.7	86.8	86.9	87.0	87.1	87.2	87.3	87.4	87.5
4.0	85.4	85.6	85.7	85.9	86.0	86.1	86.3	86.4	86.5	86.7
5.0	84.3	84.5	84.7	84.8	85.0	85.2	85.4	85.5	85.7	85.8
6.0	83.1	83.4	83.6	83.8	84.0	84.2	84.4	84.6	84.8	85.0
7.0	82.0	82.3	82.6	82.8	83.1	83.3	83.5	83.7	84.0	84.2
8.0	80.9	81.2	81.5	81.8	82.1	82.3	82.6	82.9	83.1	83.3
9.0	79.8	80.1	80.5	80.8	81.1	81.4	81.7	82.0	82.3	82.5
10.0	78.7	79.1	79.5	79.8	80.1	80.5	80.8	81.1	81.4	81.7
11.0	77.6	78.0	78.4	78.8	79.2	79.6	79.9	80.3	80.6	80.9
12.0	76.5	77.0	77.4	77.8	78.3	78.6	79.0	79.4	79.8	80.1

Table A7.1   Azimuth angles

| Latitude (°) | Longitude difference LS − LR(°) | | | | | | | | | |
	41.0	42.0	43.0	44.0	45.0	46.0	47.0	48.0	49.0	50.0
13.0	75.5	76.0	76.4	76.9	77.3	77.7	78.2	78.5	78.9	79.3
14.0	74.4	75.0	75.5	75.9	76.4	76.9	77.3	77.7	78.1	78.5
15.0	73.4	74.0	74.5	75.0	75.5	76.0	76.4	76.9	77.3	77.7
16.0	72.4	73.0	73.5	74.1	74.6	75.1	75.6	76.1	76.5	77.0
17.0	71.4	72.0	72.6	73.2	73.7	74.2	74.7	75.3	75.7	76.2
18.0	70.4	71.1	71.7	72.3	72.8	73.4	73.9	74.5	75.0	75.5
19.0	69.5	70.1	70.8	71.4	72.0	72.5	73.1	73.7	74.2	74.7
20.0	68.5	69.2	69.9	70.5	71.1	71.7	72.3	72.9	73.4	74.0
21.0	67.6	68.3	69.0	69.6	70.3	70.9	71.5	72.1	72.7	73.3
22.0	66.7	67.4	68.1	68.8	69.5	70.1	70.7	71.4	72.0	72.6
23.0	65.8	66.5	67.3	68.0	68.7	69.3	70.0	70.6	71.2	71.8
24.0	64.9	65.7	66.4	67.2	67.9	68.6	69.2	69.9	70.5	71.2
25.0	64.1	64.9	65.6	66.4	67.1	67.8	68.5	69.2	69.8	70.5
26.0	63.2	64.0	64.8	65.6	66.3	67.1	67.8	68.5	69.1	69.8
27.0	62.4	63.2	64.0	64.8	65.6	66.3	67.1	67.8	68.5	69.1
28.0	61.6	62.5	63.3	64.1	64.9	65.6	66.4	67.1	67.8	68.5
29.0	60.9	61.7	62.5	63.3	64.1	64.9	65.7	66.4	67.1	67.9
30.0	60.1	61.0	61.8	62.6	63.4	64.2	65.0	65.8	66.5	67.2
31.0	59.4	60.2	61.1	61.9	62.7	63.6	64.3	65.1	65.9	66.6
32.0	58.6	59.5	60.4	61.2	62.1	62.9	63.7	64.5	65.3	66.0
33.0	57.9	58.8	59.7	60.6	61.4	62.3	63.1	63.9	64.7	65.4
34.0	57.2	58.2	59.1	59.9	60.8	61.6	62.5	63.3	64.1	64.9
35.0	56.6	57.5	58.4	59.3	60.2	61.0	61.9	62.7	63.5	64.3
36.0	55.9	56.9	57.8	58.7	59.6	60.4	61.3	62.1	62.9	63.7
37.0	55.3	56.2	57.2	58.1	59.0	59.8	60.7	61.5	62.4	63.2
38.0	54.7	55.6	56.6	57.5	58.4	59.3	60.1	61.0	61.8	62.7
39.0	54.1	55.0	56.0	56.9	57.8	58.7	59.6	60.5	61.3	62.2
40.0	53.5	54.5	55.4	56.4	57.3	58.2	59.1	59.9	60.8	61.7
41.0	53.0	53.9	54.9	55.8	56.7	57.6	58.5	59.4	60.3	61.2
42.0	52.4	53.4	54.3	55.3	56.2	57.1	58.0	58.9	59.8	60.7
43.0	51.9	52.9	53.8	54.8	55.7	56.6	57.5	58.4	59.3	60.2
44.0	51.4	52.3	53.3	54.3	55.2	56.1	57.1	58.0	58.9	59.8
45.0	50.9	51.9	52.8	53.8	54.7	55.7	56.6	57.5	58.4	59.3
46.0	50.4	51.4	52.4	53.3	54.3	55.2	56.1	57.1	58.0	58.9
47.0	49.9	50.9	51.9	52.9	53.8	54.8	55.7	56.6	57.6	58.5
48.0	49.5	50.5	51.4	52.4	53.4	54.3	55.3	56.2	57.1	58.1
49.0	49.0	50.0	51.0	52.0	53.0	53.9	54.9	55.8	56.7	57.7
50.0	48.6	49.6	50.6	51.6	52.5	53.5	54.5	55.4	56.3	57.3
51.0	48.2	49.2	50.2	51.2	52.1	53.1	54.1	55.0	56.0	56.9
52.0	47.8	48.8	49.8	50.8	51.8	52.7	53.7	54.6	55.6	56.5
53.0	47.4	48.4	49.4	50.4	51.4	52.4	53.3	54.3	55.2	56.2
54.0	47.1	48.1	49.1	50.0	51.0	52.0	53.0	53.9	54.9	55.8
55.0	46.7	47.7	48.7	49.7	50.7	51.7	52.6	53.6	54.5	55.5
56.0	46.4	47.4	48.4	49.4	50.3	51.3	52.3	53.3	54.2	55.2
57.0	46.0	47.0	48.0	49.0	50.0	51.0	52.0	52.9	53.9	54.9
58.0	45.7	46.7	47.7	48.7	49.7	50.7	51.7	52.6	53.6	54.6
59.0	45.4	46.4	47.4	48.4	49.4	50.4	51.4	52.3	53.3	54.3
60.0	45.1	46.1	47.1	48.1	49.1	50.1	51.1	52.1	53.0	54.0
61.0	44.8	45.8	46.8	47.8	48.8	49.8	50.8	51.8	52.8	53.7
62.0	44.6	45.6	46.6	47.6	48.6	49.5	50.5	51.5	52.5	53.5
63.0	44.3	45.3	46.3	47.3	48.3	49.3	50.3	51.3	52.2	53.2
64.0	44.0	45.1	46.1	47.1	48.1	49.0	50.0	51.0	52.0	53.0
65.0	43.8	44.8	45.8	46.8	47.8	48.8	49.8	50.8	51.8	52.7
66.0	43.6	44.6	45.6	46.6	47.6	48.6	49.6	50.6	51.5	52.5
67.0	43.4	44.4	45.4	46.4	47.4	48.4	49.4	50.3	51.3	52.3
68.0	43.2	44.2	45.2	46.2	47.2	48.2	49.2	50.1	51.1	52.1

Table A7.1   *Azimuth angles*

| Latitude (°) | Longitude difference LS − LR(°) | | | | | | | | | |
	41.0	42.0	43.0	44.0	45.0	46.0	47.0	48.0	49.0	50.0
69.0	43.0	44.0	45.0	46.0	47.0	48.0	49.0	49.9	50.9	51.9
70.0	42.8	43.8	44.8	45.8	46.8	47.8	48.8	49.8	50.8	51.7
71.0	42.6	43.6	44.6	45.6	46.6	47.6	48.6	49.6	50.6	51.6
72.0	42.4	43.4	44.4	45.4	46.4	47.4	48.4	49.4	50.4	51.4
73.0	42.3	43.3	44.3	45.3	46.3	47.3	48.3	49.3	50.3	51.3
74.0	42.1	43.1	44.1	45.1	46.1	47.1	48.1	49.1	50.1	51.1
75.0	42.0	43.0	44.0	45.0	46.0	47.0	48.0	49.0	50.0	51.0
76.0	41.9	42.9	43.9	44.9	45.9	46.9	47.9	48.9	49.9	50.8
77.0	41.7	42.7	43.7	44.7	45.7	46.7	47.7	48.7	49.7	50.7
78.0	41.6	42.6	43.6	44.6	45.6	46.6	47.6	48.6	49.6	50.6
79.0	41.5	42.5	43.5	44.5	45.5	46.5	47.5	48.5	49.5	50.5
80.0	41.4	42.4	43.4	44.4	45.4	46.4	47.4	48.4	49.4	50.4

| Latitude (°) | Longitude difference LS − LR(°) | | | | | | | | | |
	51.0	52.0	53.0	54.0	55.0	56.0	57.0	58.0	59.0	60.0
0.0	90.0	90.0	90.0	90.0	90.0	90.0	90.0	90.0	90.0	90.0
1.0	89.2	89.2	89.2	89.3	89.3	89.3	89.4	89.4	89.4	89.4
2.0	88.4	88.4	88.5	88.5	88.6	88.7	88.7	88.8	88.8	88.8
3.0	87.6	87.7	87.7	87.8	87.9	88.0	88.1	88.1	88.2	88.3
4.0	86.8	86.9	87.0	87.1	87.2	87.3	87.4	87.5	87.6	87.7
5.0	86.0	86.1	86.2	86.4	86.5	86.6	86.8	86.9	87.0	87.1
6.0	85.2	85.3	85.5	85.7	85.8	86.0	86.1	86.3	86.4	86.5
7.0	84.4	84.6	84.8	84.9	85.1	85.3	85.5	85.6	85.8	86.0
8.0	83.6	83.8	84.0	84.2	84.4	84.6	84.8	85.0	85.2	85.4
9.0	82.8	83.0	83.3	83.5	83.7	84.0	84.2	84.4	84.6	84.8
10.0	82.0	82.3	82.5	82.8	83.1	83.3	83.6	83.8	84.0	84.3
11.0	81.2	81.5	81.8	82.1	82.4	82.7	82.9	83.2	83.5	83.7
12.0	80.4	80.8	81.1	81.4	81.7	82.0	82.3	82.6	82.9	83.2
13.0	79.7	80.0	80.4	80.7	81.0	81.4	81.7	82.0	82.3	82.6
14.0	78.9	79.3	79.7	80.0	80.4	80.7	81.1	81.4	81.7	82.0
15.0	78.2	78.6	79.0	79.4	79.7	80.1	80.5	80.8	81.2	81.5
16.0	77.4	77.8	78.3	78.7	79.1	79.5	79.9	80.2	80.6	81.0
17.0	76.7	77.1	77.6	78.0	78.4	78.8	79.2	79.6	80.0	80.4
18.0	76.0	76.4	76.9	77.3	77.8	78.2	78.7	79.1	79.5	79.9
19.0	75.2	75.7	76.2	76.7	77.2	77.6	78.1	78.5	78.9	79.4
20.0	74.5	75.0	75.5	76.0	76.5	77.0	77.5	77.9	78.4	78.8
21.0	73.8	74.4	74.9	75.4	75.9	76.4	76.9	77.4	77.8	78.3
22.0	73.1	73.7	74.2	74.8	75.3	75.8	76.3	76.8	77.3	77.8
23.0	72.4	73.0	73.6	74.2	74.7	75.2	75.8	76.3	76.8	77.3
24.0	71.8	72.4	73.0	73.5	74.1	74.7	75.2	75.7	76.3	76.8
25.0	71.1	71.7	72.3	72.9	73.5	74.1	74.7	75.2	75.8	76.3
26.0	70.5	71.1	71.7	72.3	72.9	73.5	74.1	74.7	75.2	75.8
27.0	69.8	70.5	71.1	71.7	72.4	73.0	73.6	74.2	74.7	75.3
28.0	69.2	69.9	70.5	71.2	71.8	72.4	73.0	73.7	74.2	74.8
29.0	68.6	69.3	69.9	70.6	71.2	71.9	72.5	73.1	73.8	74.4
30.0	68.0	68.7	69.4	70.0	70.7	71.4	72.0	72.6	73.3	73.9
31.0	67.4	68.1	68.8	69.5	70.2	70.8	71.5	72.2	72.8	73.4
32.0	66.8	67.5	68.2	68.9	69.6	70.3	71.0	71.7	72.3	73.0
33.0	66.2	66.9	67.7	68.4	69.1	69.8	70.5	71.2	71.9	72.5
34.0	65.6	66.4	67.2	67.9	68.6	69.3	70.0	70.7	71.4	72.1
35.0	65.1	65.9	66.6	67.4	68.1	68.8	69.6	70.3	71.0	71.7
36.0	64.5	65.3	66.1	66.9	67.6	68.4	69.1	69.8	70.5	71.3
37.0	64.0	64.8	65.6	66.4	67.1	67.9	68.7	69.4	70.1	70.8
38.0	63.5	64.3	65.1	65.9	66.7	67.4	68.2	69.0	69.7	70.4
39.0	63.0	63.8	64.6	65.4	66.2	67.0	67.8	68.5	69.3	70.0

Table A7.1   *Azimuth angles*

| Latitude (°) | Longitude difference LS – LR(°) | | | | | | | | | |
	51.0	52.0	53.0	54.0	55.0	56.0	57.0	58.0	59.0	60.0
40.0	62.5	63.3	64.2	65.0	65.8	66.6	67.3	68.1	68.9	69.6
41.0	62.0	62.9	63.7	64.5	65.3	66.1	66.9	67.7	68.5	69.3
42.0	61.5	62.4	63.2	64.1	64.9	65.7	66.5	67.3	68.1	68.9
43.0	61.1	61.9	62.8	63.6	64.5	65.3	66.1	66.9	67.7	68.5
44.0	60.6	61.5	62.4	63.2	64.1	64.9	65.7	66.5	67.3	68.1
45.0	60.2	61.1	61.9	62.8	63.7	64.5	65.3	66.2	67.0	67.8
46.0	59.8	60.7	61.5	62.4	63.3	64.1	65.0	65.8	66.6	67.4
47.0	59.4	60.3	61.1	62.0	62.9	63.7	64.6	65.4	66.3	67.1
48.0	59.0	59.9	60.8	61.6	62.5	63.4	64.2	65.1	65.9	66.8
49.0	58.6	59.5	60.4	61.3	62.1	63.0	63.9	64.8	65.6	66.5
50.0	58.2	59.1	60.0	60.9	61.8	62.7	63.6	64.4	65.3	66.1
51.0	57.8	58.7	59.6	60.5	61.4	62.3	63.2	64.1	65.0	65.8
52.0	57.5	58.4	69.3	60.2	61.1	62.0	62.9	63.8	64.7	65.5
53.0	57.1	58.0	59.0	59.9	60.8	61.7	62.6	63.5	64.4	65.2
54.0	56.8	57.7	58.6	59.6	60.5	61.4	62.3	63.2	64.1	65.0
55.0	56.4	57.4	58.3	59.2	60.2	61.1	62.0	62.9	63.8	64.7
56.0	56.1	57.1	58.0	58.9	59.9	60.8	61.7	62.6	63.5	64.4
57.0	55.8	56.8	57.7	58.6	59.6	60.5	61.4	62.3	63.3	64.2
58.0	55.5	56.5	57.4	58.4	59.3	60.2	61.2	62.1	63.0	63.9
59.0	55.2	56.2	57.1	58.1	59.0	60.0	60.9	61.8	62.7	63.7
60.0	55.0	55.9	56.9	57.8	58.8	59.7	60.6	61.6	62.5	63.4
61.0	54.7	55.7	56.6	57.6	58.5	59.5	60.4	61.3	62.3	63.2
62.0	54.4	55.4	56.4	57.3	58.3	59.2	60.2	61.1	62.1	63.0
63.0	54.2	55.2	56.1	57.1	58.0	59.0	59.9	60.9	61.8	62.8
64.0	54.0	54.9	55.9	56.9	57.8	58.8	59.7	60.7	61.6	62.6
65.0	53.7	54.7	55.7	56.6	57.6	58.6	59.5	60.5	61.4	62.4
66.0	53.5	54.5	55.5	56.4	57.4	58.4	59.3	60.3	61.2	62.2
67.0	53.3	54.3	55.3	56.2	57.2	58.2	59.1	60.1	61.1	62.0
68.0	53.1	54.1	55.1	56.0	57.0	58.0	58.9	59.9	60.9	61.8
69.0	52.9	53.9	54.9	55.9	56.8	57.8	58.8	59.7	60.7	61.7
70.0	52.7	53.7	54.7	55.7	56.7	57.6	58.6	59.6	60.5	61.5
71.0	52.6	53.5	54.5	55.5	56.5	57.5	58.4	59.4	60.4	61.4
72.0	52.4	53.4	54.4	55.4	56.3	57.3	58.3	59.3	60.3	61.2
73.0	52.2	53.2	54.2	55.2	56.2	57.2	58.2	59.1	60.1	61.1
74.0	52.1	53.1	54.1	55.1	56.1	57.0	58.0	59.0	60.0	61.0
75.0	52.0	53.0	53.9	54.9	55.9	56.9	57.9	58.9	59.9	60.9
76.0	51.8	52.8	53.8	54.8	55.8	56.8	57.8	58.8	59.8	60.7
77.0	51.7	82.7	53.7	54.7	55.7	56.7	57.7	58.7	59.7	60.6
78.0	51.6	52.6	53.6	54.6	55.6	56.6	57.6	58.6	59.6	60.5
79.0	51.5	52.5	53.5	54.5	55.5	56.5	57.5	58.5	59.5	60.5
80.0	51.4	52.4	53.4	54.4	55.4	56.4	57.4	58.4	59.4	60.4

| Latitude (°) | Longitude difference LS – LR(°) | | | | | | | | | |
	61.0	62.0	63.0	64.0	65.0	66.0	67.0	68.0	69.0	70.0
0.0	90.0	90.0	90.0	90.0	90.0	90.0	90.0	90.0	90.0	90.0
1.0	89.4	89.5	89.5	89.5	89.5	89.6	89.6	89.6	89.6	89.6
2.0	88.9	88.9	89.0	89.0	89.1	89.1	89.2	89.2	89.2	89.3
3.0	88.3	88.4	88.5	88.5	88.6	88.7	88.7	88.8	88.8	88.9
4.0	87.8	87.9	88.0	88.1	88.1	88.2	88.3	88.4	88.5	88.5
5.0	87.2	87.3	87.5	87.6	87.7	87.8	87.9	88.0	88.1	88.2
6.0	86.7	86.8	87.0	87.1	87.2	87.3	87.5	87.6	87.7	87.8
7.0	86.1	86.3	86.4	86.6	86.7	86.9	87.0	87.2	87.3	87.5
8.0	85.6	85.8	85.9	86.1	86.3	86.5	86.6	86.8	86.9	87.1
9.0	85.0	85.2	85.4	85.6	85.8	86.0	86.2	86.4	86.6	86.7
10.0	84.5	84.7	84.9	85.2	85.4	85.6	85.8	86.0	86.2	86.4

Table A 7.1 *Azimuth angles*

Latitude (°)	Longitude difference LS – LR (°)									
	61.0	62.0	63.0	64.0	65.0	66.0	67.0	68.0	69.0	70.0
11.0	84.0	84.2	84.4	84.7	84.9	85.1	85.4	85.6	85.8	86.0
12.0	83.4	83.7	84.0	84.2	84.5	84.7	85.0	85.2	85.4	85.7
13.0	82.9	83.2	83.5	83.7	84.0	84.3	84.5	84.8	85.1	85.3
14.0	82.4	82.7	83.0	83.3	83.6	83.9	84.1	84.4	84.7	85.0
15.0	81.8	82.2	82.5	82.8	83.1	83.4	83.7	84.0	84.3	84.6
16.0	81.3	81.7	82.0	82.3	82.7	83.0	83.3	83.6	84.0	84.3
17.0	80.8	81.2	81.5	81.9	82.2	82.6	82.9	83.3	83.6	83.9
18.0	80.3	80.7	81.1	81.4	81.8	82.2	82.5	82.9	83.2	83.6
19.0	79.8	80.2	80.6	81.0	81.4	81.8	82.1	82.5	82.9	83.2
20.0	79.3	79.7	80.1	80.5	80.9	81.3	81.7	82.1	82.5	82.9
21.0	78.8	79.2	79.7	80.1	80.5	80.9	81.4	81.8	82.2	82.6
22.0	78.3	78.7	79.2	79.6	~l0.1	80.5	81.0	81.4	81.8	82.2
23.0	77.8	78.3	78.7	79.2	'9.7	80.1	80.6	81.0	81.5	81.9
24.0	77.3	77.8	78.3	78.8	79.3	79.7	80.2	80.7	81.1	81.6
25.0	76.8	77.3	77.8	78.4	78.9	79.3	79.8	80.3	80.8	81.3
26.0	76.3	76.9	77.4	77.9	78.4	79.0	79.5	80.0	80.4	80.9
27.0	75.9	76.4	77.0	77.5	78.0	78.6	79.1	79.6	80.1	80.6
28.0	75.4	76.0	76.5	77.1	77.7	78.2	78.7	79.3	79.8	80.3
29.0	75.0	75.5	76.1	76.7	77.3	77.8	78.4	78.9	79.5	80.0
30.0	74.5	75.1	75.7	76.3	76.9	77.4	78.0	78.6	79.1	79.7
31.0	74.1	74.7	75.3	75.9	76.5	77.1	77.7	78.2	78.8	79.4
32.0	73.6	74.3	74.9	75.5	76.1	76.7	77.3	77.9	78.5	79.1
33.0	73.2	73.8	74.5	75.1	75.7	76.4	77.0	77.6	78.2	78.8
34.0	72.8	73.4	74.1	74.7	75.4	76.0	76.6	77.3	77.9	78.5
35.0	72.4	73.0	73.7	74.4	75.0	75.7	76.3	77.0	77.6	78.2
36.0	72.0	72.6	73.3	74.0	74.7	75.3	76.0	76.6	77.3	77.9
37.0	71.6	72.3	73.0	73.6	74.3	75.0	75.7	76.3	77.0	77.6
38.0	71.2	71.9	72.6	73.3	74.0	74.7	75.4	76.0	76.7	77.4
49.0	70.8	71.5	72.2	72.9	73.6	74.3	75.0	75.7	76.4	77.1
40.0	70.4	71.1	71.9	72.6	73.3	74.0	74.7	75.4	76.1	76.8
41.0	70.0	70.8	71.5	72.3	73.0	73.7	74.4	75.2	75.9	76.6
42.0	69.6	70.4	71.2	71.9	72.7	73.4	74.1	74.9	75.6	76.3
43.0	69.3	70.1	70.8	71.6	72.4	73.1	73.9	74.6	75.3	76.1
44.0	68.9	69.7	70.5	71.3	72.1	72.8	73.6	74.3	75.1	75.8
45.0	68.6	69.4	70.2	71.0	71.8	72.5	73.3	74.1	74.8	75.6
46.0	68.3	69.1	69.9	70.7	71.5	72.2	73.0	73.8	74.6	75.3
47.0	67.9	68.8	69.6	70.4	71.2	72.0	72.8	73.5	74.3	75.1
48.0	67.6	68.4	69.3	70.1	70.9	71.7	72.5	73.3	74.1	74.9
49.0	67.3	68.1	69.0	69.8	70.6	71.4	72.2	73.0	73.8	74.6
50.0	67.0	67.8	68.7	69.5	70.3	71.2	72.0	72.8	73.6	74.4
51.0	66.7	67.5	68.4	69.2	70.1	70.9	71.7	72.6	73.4	74.2
52.0	66.4	67.3	68.1	69.0	69.8	70.7	71.5	72.3	73.2	74.0
53.0	66.1	67.0	67.9	68.7	69.6	70.4	71.3	72.1	73.0	73.8
54.0	65.8	66.7	67.6	68.5	69.3	70.2	71.0	71.9	72.7	73.6
55.0	65.6	66.5	67.3	68.2	69.1	70.0	70.8	71.7	72.5	73.4
56.0	65.3	66.2	67.1	68.0	68.9	69.7	70.6	71.5	72.3	73.2
57.0	65.1	66.0	66.9	67.8	68.6	69.5	70.4	71.3	72.2	73.0
58.0	64.8	65.7	66.6	67.5	68.4	69.3	70.2	71.1	72.0	72.8
59.0	64.6	65.5	66.4	67.3	68.2	69.1	70.0	70.9	71.8	72.7
60.0	64.4	65.3	66.2	67.1	68.0	68.9	69.8	70.7	71.6	72.5
61.0	64.1	65.1	66.0	66.9	67.8	68.7	69.6	70.5	71.4	72.3
62.0	63.9	64.9	65.8	66.7	67.6	68.5	69.5	70.4	71.3	72.2
63.0	63.7	64.7	65.6	66.5	67.4	68.4	69.3	70.2	71.1	72.0
64.0	63.5	64.5	65.4	66.3	67.3	68.2	69.1	70.0	71.0	71.9
65.0	63.3	64.3	65.2	66.2	67.1	68.0	69.0	69.9	70.8	71.7
66.0	63.1	64.1	63.0	66.0	66.9	67.9	68.8	69.7	70.7	71.6

Table A7.1 *Azimuth angles*

Latitude (°)	Longitude difference LS − LR (°)									
	61.0	62.0	63.0	64.0	65.0	66.0	67.0	68.0	69.0	70.0
67.0	63.0	63.9	64.3	65.8	66.8	67.7	68.7	69.6	70.5	71.5
68.0	62.8	63.8	64.7	65.7	66.6	67.6	68.5	69.5	70.4	71.4
69.0	62.6	63.6	64.6	65.5	66.5	67.4	68.4	69.3	70.3	71.2
70.0	62.5	63.5	64.4	65.4	66.3	67.3	68.3	69.2	70.2	71.1
71.0	62.3	63.3	64.3	65.2	66.2	67.2	68.1	69.1	70.1	71.0
72.0	62.2	63.2	64.1	65.1	66.1	67.1	68.0	69.0	69.9	70.9
73.0	62.1	63.0	64.0	65.0	66.0	66.9	67.9	'68.9	69.8	70.8
74.0	61.9	62.9	63.9	64.9	65.9	66.8	67.8	68.8	69.7	70.7
75.0	61.8	62.8	63.8	64.8	65.8	66.7	67.7	68.7	69.7	70.6
76.0	61.7	62.7	63.7	64.7	65.7	66.6	67.6	68.6	69.6	70.5
77.0	61.6	62.6	63.6	64.6	65.6	66.5	67.5	68.5	69.5	70.5
78.0	61.5	62.5	63.5	64.5	65.5	66.5	67.5	68.4	69.4	70.4
79.0	61.4	62.4	63.4	64.4	65.4	66.4	67.4	68.4	69.4	70.3
80.0	61.4	62.4	63.4	64.3	65.3	66.3	67.3	68.3	69.3	70.3

Latitude (°)	Longitude difference LS − LR (°)									
	71.0	72.0	73.0	74.0	75.0	76.0	77.0	78.0	79.0	80.0
0.0	90.0	90.0	90.0	90.0	90.0	90.0	90.0	90.0	90.0	90.0
1.0	89.7	89.7	89.7	89.7	89.7	89.8	89.8	89.8	89.8	89.8
2.0	89.3	89.4	89.4	89.4.	89.5	89.5	89.5	89.6	89.6	89.6
3.0	89.0	89.0	89.1	89.1	89.2	89.3	89.3	89.4	89.4	89.5
4.0	88.6	88.7	88.8	88.9	88.9	89.0	89.1	89.2	89.2	89.3
5.0	88.3	88.4	88.5	88.6	88.7	88.8	88.8	88.9	89.0	89.1
6.0	87.9	88.1	88.2	88.3	88.4	88.5	88.6	88.7	88.8	88.9
7.0	87.6	87.7	87.9	88.0	88.1	88.3	88.4	88.5	88.6	88.8
8.0	87.3	87.4	87.6	87.7	87.9	88.0	88.2	88.3	88.5	88.6
9.0	86.9	87.1	87.3	87.4	87.6	87.8	87.9	88.1	88.3	88.4
10.0	86.6	86.8	87.0	87.1	87.3	87.5	87.7	87.9	88.1	88.2
11.0	86.2	86.5	86.7	86.9	87.1	87.3	87.5	87.7	87.9	88.1
12.0	85.9	86.1	86.4	86.6	86.8	87.0	87.3	87.5	87.7	87.9
13.0	85.6	85.8	86.1	86.3	86.6	86.8	87.0	87.3	87.5	87.7
14.0	85.2	85.5	85.8	86.0	86.3	86.5	86.8	87.1	87.3	87.6
15.0	84.9	85.2	85.5	85.8	86.0	86.3	86.6	86.9	87.1	87.4
16.0	84.6	84.9	85.2	85.5	85.8	86.1	86.4	86.6	86.9	87.2
17.0	84.3	84.6	84.9	85.2	85.5	85.8	86.1	86.4	86.7	87.0
18.0	83.9	84.3	84.6	84.9	85.3	85.6	85.9	86.2	86.6	86.9
19.0	83.6	84.0	84.3	84.7	85.0	85.4	85.7	86.0	86.4	86.7
20.0	83.3	83.7	84.0	84.4	84.8	85.1	85.5	85.8	86.2	86.5
21.0	83.0	83.4	83.7	84.1	84.5	84.9	85.3	85.6	86.0	86.4
22.0	82.7	83.1	83.5	83.9	84.3	84.7	85.1	85.4	85.8	86.2
23.0	82.3	82.8	83.2	83.6	84.0	84.4	84.8	85.3	85.7	86.1
24.0	82.0	82.5	82.9	83.3	83.8	84.2	84.6	85.1	85.5	85.9
25.0	81.7	82.2	82.6	83.1	83.5	84.0	84.4	84.9	85.3	85.7
26.0	81.4	81.9	82.4	82.8	83.3	83.8	84.2	84.7	85.1	85.6
27.0	81.1	81.6	82.1	82.6	83.1	83.5	84.0	84.5	85.0	85.4
28.0	80.8	81.3	81.8	82.3	82.8	83.3	83.8	84.3	84.8	85.3
29.0	80.5	81.0	81.6	82.1	82.6	83.1	83.6	84.1	84.6	85.1
30.0	80.2	80.8	81.3	81.8	82.4	82.9	83.4	83.9	84.4	85.0
31.0	79.9	80.5	81.1	81.6	82.1	82.7	83.2	83.8	84.3	84.8
32.0	79.7	80.2	80.8	81.4	81.9	82.5	83.0	83.6	84.1	84.7
33.0	79.4	80.0	80.5	81.1	81.7	82.3	82.8	83.4	84.0	84.5
34.0	79.1	79.7	80.3	80.9	81.5	82.1	82.6	83.2	83.8	84.4
35.0	78.8	79.4	80.1	80.7	81.3	81.9	82.5	83.0	83.6	84.2
36.0	78.6	79.2	79.8	80.4	81.0	81.7	82.3	82.9	83.5	84.1
37.0	78.3	78.9	79.6	80.2	80.8	81.5	82.1	82.7	83.3	83.9

Table A7.1 *Azimuth angles*

Latitude (°)	Longitude difference LS − LR (°)									
	71.0	72.0	73.0	74.0	75.0	76.0	77.0	78.0	79.0	80.0
38.0	78.0	78.7	79.3	80.0	80.6	81.3	81.9	82.5	83.2	83.8
39.0	77.8	78.4	79.1	79.8	80.4	81.1	81.7	82.4	83.0	83.7
40.0	77.5	78.2	78.9	79.6	80.2	80.9	81.6	82.2	82.9	83.5
41.0	77.3	78.0	78.7	79.3	80.0	80.7	81.4	82.1	82.7	83.4
42.0	77.0	77.7	78.4	79.1	79.8	80.5	81.2	81.9	82.6	83.3
43.0	76.8	77.5	78.2	78.9	79.6	80.3	81.1	81.8	82.4	83.1
44.0	76.5	77.3	78.0	78.7	79.5	80.2	80.9	81.6	82.3	83.0
45.0	76.3	77.1	77.8	78.5	79.3	80.0	80.7	81.5	82.2	82.9
46.0	76.1	76.8	77.6	78.3	79.1	79.8	80.6	81.3	82.0	82.8
47.0	75.9	76.6	77.4	78.2	78.9	79.7	80.4	81.2	81.9	82.7
48.0	75.6	76.4	77.2	78.0	78.7	79.5	80.3	81.0	81.8	82.5
49.0	75.4	76.2	77.0	77.8	78.6	79.3	80.1	80.9	81.7	82.4
50.0	75.2	76.0	76.8	77.6	78.4	79.2	80.0	80.8	81.5	82.3
51.0	75.0	75.8	76.6	77.4	78.2	79.0	79.8	80.6	81.4	82.2
52.0	74.8	75.6	76.5	77.3	78.1	78.9	79.7	80.5	81.3	82.1
53.0	74.6	75.5	76.3	77.1	77.9	78.7	79.6	80.4	81.2	82.0
54.0	74.4	75.3	76.1	76.9	77.8	78.6	79.4	80.2	81.1	81.9
55.0	74.2	75.1	75.9	76.8	77.6	78.5	79.3	80.1	81.0	81.8
56.0	74.1	74.9	75.8	76.6	77.5	78.3	79.2	80.0	80.8	81.7
57.0	73.9	74.8	75.6	76.5	77.3	78.2	79.0	79.9	80.7	81.6
58.0	73.7	74.6	75.5	76.3	7~.2	78.1	78.9	79.8	80.6	81.5
59.0	73.6	74.4	75.3	76.2	77.1	77.9	78.8	79.7	80.5	81.4
60.0	73.4	74.3	75.2	76.1	76.9	77.8	78.7	79.6	80.4	81.3
61.0	73.2	74.1	75.0	75.9	76.8	77.7	78.6	79.5	80.4	81.2
62.0	73.1	74.0	74.9	75.8	76.7	77.6	78.5	79.4	80.3	81.2
63.0	72.9	73.9	74.8	75.7	76.6	77.5	78.4	79.3	80.2	81.1
64.0	72.8	73.7	74.6	75.5	76.5	77.4	78.3	79.2	80.1	81.0
65.0	72.7	73.6	74.5	75.4	76.4	77.3	78.2	79.1	80.0	80.9
66.0	72.5	73.5	74.4	75.3	76.2	77.2	78.1	79.0	79.9	80.8
67.0	72.4	73.3	74.3	75.2	76.1	77.1	78.0	78.9	79.9	80.8
68.0	72.3	73.2	74.2	75.1	76.0	77.0	77.9	78.9	79.8	80.7
69.0	72.2	73.1	74.1	75.0	76.0	76.9	77.8	78.8	79.7	80.7
70.0	72.1	73.0	74.0	74.9	75.9	76.8	77.8	78.7	79.6	80.6
71.0	72.0	72.9	73.9	74.8	75.8	76.7	77.7	78.6	79.6	80.5
72.0	71.9	72.8	73.8	74.7	75.7	76.7	77.6	78.6	79.5	80.5
73.0	71.8	72.7	73.7	74.7	75.6	76.6	77.5	78.5	79.5	80.4
74.0	71.7	72.7	73.6	74.6	75.6	76.5	77.5	78.5	79.4	80.4
75.0	71.6	72.6	73.5	74.5	75.5	76.5	77.4	78.4	79.4	80.3
76.0	71.5	72.5	73.5	74.5	75.4	76.4	77.4	78.3	79.3	80.3
77.0	71.5	72.4	73.4	74.4	75.4	76.3	77.3	78.3	79.3	80.3
78.0	71.4	72.4	73.4	74.3	75.3	76.3	77.3	78.3	79.2	80.2
79.0	71.3	72.3	73.3	74.3	75.3	76.2	77.2	78.2	79.2	80.2
80.0	71.3	72.3	73.2	74.2	75.2	76.2	77.2	78.2	79.2	80.1

Table A7.2 *Elevation angles*

Latitude (°)	Longitude difference LS − LR (°)										
	0.0	1.0	2.0	3.0	4.0	5.0	6.0	7.0	8.0	9.0	10.0
0.0	90.0	88.8	87.6	86.5	85.3	84.1	82.9	81.8	80.6	79.4	78.2
1.0	88.8	88.3	87.4	86.3	85.1	84.0	82.8	81.7	80.5	79.3	78.2
2.0	87.6	87.4	86.7	85.8	84.7	83.7	82.6	81.4	80.3	79.2	78.0
3.0	86.5	86.3	85.8	85.0	84.1	83.1	82.1	81.0	79.9	78.8	77.7
4.0	85.3	85.1	84.7	84.1	83.3	82.5	81.5	80.5	79.5	78.4	77.3
5.0	84.1	84.0	83.7	83.1	82.5	81.7	80.8	79.9	78.9	77.9	76.9
6.0	82.9	82.8	82.6	82.1	81.5	80.8	80.0	79.2	78.2	77.3	76.3

Table A7.2 Elevation angles

| Latitude (°) | Longitude difference LS − LR (°) | | | | | | | | | | |
	0.0	1.0	2.0	3.0	4.0	5.0	6.0	7.0	8.0	9.0	10.0
7.0	81.8	81.7	81.4	81.0	80.5	79.9	79.2	78.4	77.5	76.6	75.7
8.0	80.6	80.5	80.3	79.9	79.5	78.9	78.2	77.5	76.7	75.9	75.0
9.0	79.4	79.3	79.2	78.8	78.4	77.9	77.3	76.6	75.9	75.1	74.2
10.0	78.2	78.2	78.0	77.7	77.3	76.9	76.3	75.7	75.0	74.2	73.4
11.0	77.1	77.0	76.8	76.6	76.2	75.8	75.3	74.7	74.0	73.3	72.6
12.0	75.9	75.8	75.7	75.5	75.1	74.7	74.3	73.7	73.1	72.4	71.7
13.0	74.7	74.7	74.5	74.3	74.0	73.6	73.2	72.7	72.1	71.5	70.8
14.0	73.5	73.5	73.4	73.2	72.9	72.6	72.1	71.6	71.1	70.5	69.9
15.0	72.4	72.3	72.2	72.0	71.8	71.4	71.1	70.6	70.1	69.5	68.9
16.0	71.2	71.2	71.1	70.9	70.6	70.3	70.0	69.5	69.1	68.5	67.9
17.0	70.0	70.0	69.9	69.7	69.5	69.2	68.9	68.5	68.0	67.5	67.0
18.0	68.9	68.8	68.8	68.6	68.4	68.1	67.8	67.4	67.0	66.5	66.0
19.0	67.7	67.7	67.6	67.4	67.2	67.0	66.7	66.3	65.9	65.4	64.9
20.0	66.5	66.5	66.4	66.3	66.1	65.9	65.6	65.2	64.8	64.4	63.9
21.0	65.4	65.4	65.3	65.2	65.0	64.7	64.5	64.1	63.8	63.3	62.9
22.0	64.2	64.2	64.1	64.C	63.8	63.6	63.3	63.0	62.7	62.3	61.8
23.0	63.1	63.0	63.0	62.9	62.7	62.5	62.2	61.9	61.6	61.2	60.8
24.0	61.9	61.9	61.8	61.7	61.6	61.4	61.1	60.8	60.5	60.1	59.7
25.0	60.8	60.7	60.7	60.6	60.4	60.2	60.0	59.7	59.4	59.1	58.7
26.0	59.6	59.6	59.5	59.4	59.3	59.1	58.9	58.6	58.3	58.0	57.6
27.0	58.5	58.4	58.4	58.3	58.1	58.0	57.8	57.5	57.2	56.9	56.6
28.0	57.3	57.3	57.2	57.1	57.0	56.8	56.6	56.4	56.1	55.8	55.5
29.0	56.2	56.2	56.1	56.0	55.9	55.7	55.5	55.3	55.0	54.7	54.4
30.0	55.0	55.0	55.0	54.9	54.8	54.6	54.4	54.2	53.9	53.7	53.3
31.0	53.9	53.9	53.8	53.7	53.6	53.5	53.3	53.1	52.8	52.6	52.3
32.0	52.7	52.7	52.7	52.6	52.5	52.4	52.2	52.0	51.7	51.5	51.2
33.0	51.6	51.6	51.5	51.5	51.4	51.2	51.1	50.9	50.6	50.4	50.1
34.0	50.5	50.5	50.4	50.3	50.2	50.1	50.0	49.8	49.5	49.3	49.0
35.0	49.3	49.3	49.3	49.2	49.1	49.0	48.8	48.7	48.5	48.2	48.0
36.0	48.2	48.2	48.2	48.1	48.0	47.9	47.7	47.6	47.4	47.1	46.9
37.0	47.1	47.1	47.0	47.0	46.9	46.8	46.6	46.5	46.3	46.0	45.8
38.0	46.0	46.0	45.9	45.9	45.8	45.7	45.5	45.4	45.2	45.0	44.7
39.0	44.8	44.8	44.8	44.7	44.7	44.5	44.4	44.3	44.1	43.9	43.7
40.0	43.7	43.7	43.7	43.6	43.5	43.4	43.3	43.2	43.0	42.8	42.6
41.0	42.6	42.6	42.6	42.5	42.4	42.3	42.2	42.1	41.9	41.7	41.5
42.0	41.5	41.5	41.5	41.4	41.3	41.2	41.1	41.0	40.8	40.6	40.4
43.0	40.4	40.4	40.3	40.3	40.2	40.1	40.0	39.9	39.7	39.6	39.4
44.0	39.3	39.3	39.2	39.2	39.1	39.0	38.9	38.8	38.6	38.5	38.3
45.0	38.2	38.2	38.1	38.1	38.0	37.9	37.8	37.7	37.6	37.4	37.2
46.0	37.1	37.1	37.0	37.0	36.9	36.8	36.7	36.6	36.5	36.3	36.1
47.0	36.0	36.0	35.9	35.9	35.8	35.7	35.6	35.5	35.4	35.2	35.1
48.0	34.9	34.9	34.8	34.8	34.7	34.7	34.6	34.5	34.3	34.2	34.0
49.0	33.8	33.8	33.7	33.7	33.6	33.6	33.5	33.4	33.2	33.1	33.0
50.0	32.7	32.7	32.7	32.6	32.6	32.5	32.4	32.3	32.2	32.0	31.9
51.0	31.6	31.6	31.6	31.5	31.5	31.4	31.3	31.2	31.1	31.0	30.8
52.0	30.5	30.5	30.5	30.4	30.4	30.3	30.2	30.2	30.0	29.9	29.8
53.0	29.4	29.4	29.4	29.4	29.3	29.3	29.2	29.1	29.0	28.9	28.7
54.0	28.3	28.3	28.3	28.3	28.2	28.2	28.1	28.0	27.9	27.8	27.7
55.0	27.3	27.3	27.2	27.2	27.2	27.1	27.0	27.0	26.9	26.7	26.6
56.0	26.2	26.2	26.2	26.1	26.1	26.0	26.0	25.9	25.8	25.7	25.6
57.0	25.1	25.1	25.1	25.1	25.0	25.0	24.9	24.8	24.7	24.6	24.5
58.0	24.1	24.1	24.0	24.0	24.0	23.9	23.9	23.8	23.7	23.6	23.5
59.0	23.0	23.0	23.0	22.9	22.9	22.9	22.8	22.7	22.6	22.5	22.4
60.0	21.9	21.9	21.9	21.9	21.8	21.8	21.7	21.7	21.6	21.5	21.4
61.0	20.9	20.9	20.9	20.8	20.8	20.7	20.7	20.6	20.5	20.5	20.4
62.0	19.8	19.8	19.8	19.8	19.7	19.7	19.6	19.6	19.5	19.4	19.3

Table A7.2 *Elevation angles*

Latitude (°)	Longitude difference LS − LR (°)										
	0.0	1.0	2.0	3.0	4.0	5.0	6.0	7.0	8.0	9.0	10.0
63.0	18.8	18.8	18.7	18.7	18.7	18.6	18.6	18.5	18.5	18.4	18.3
64.0	17.7	17.7	17.7	17.7	17.6	17.6	17.6	17.5	17.4	17.4	17.3
65.0	16.7	16.7	16.7	16.6	16.6	16.6	16.5	16.5	16.4	16.3	16.2
66.0	15.6	15.6	15.6	15.6	15.6	15.5	15.5	15.4	15.4	15.3	15.2
67.0	14.6	14.6	14.6	14.5	14.5	14.5	14.4	14.4	14.3	14.3	14.2
68.0	13.5	13.5	13.5	13.5	13.5	13.5	13.4	13.4	13.3	13.2	13.2
69.0	12.5	12.5	12.5	12.5	12.5	12.4	12.4	12.3	12.3	12.2	12.2
70.0	11.5	11.5	11.5	11.4	11.4	11.4	11.4	11.3	11.3	11.2	11.1
71.0	10.4	10.4	10.4	10.4	10.4	10.4	10.3	10.3	10.2	10.2	10.1
72.0	9.4	9.4	9.4	9.4	9.4	9.3	9.3	9.3	9.2	9.2	9.1
73.0	8.4	8.4	8.4	8.4	8.3	8.3	8.3	8.3	8.2	8.2	8.1
74.0	7.4	7.4	7.4	7.3	7.3	7.3	7.3	7.2	7.2	7.2	7.1
75.0	6.4	6.4	6.3	6.3	6.3	6.3	6.3	6.2	6.2	6.2	6.1
76.0	5.3	5.3	5.3	5.3	5.3	5.3	5.3	5.2	5.2	5.2	5.1
77.0	4.3	4.3	4.3	4.3	4.3	4.3	4.3	4.2	4.2	4.2	4.1
78.0	3.3	3.3	3.3	3.3	3.3	3.3	3.2	3.2	3.2	3.2	3.1
79.0	2.3	2.3	2.3	2.3	2.3	2.3	2.2	2.2	2.2	2.2	2.1
80.0	1.3	1.3	1.3	1.3	1.3	1.3	1.2	1.2	1.2	1.2	1.1

Latitude (°)	Longitude difference LS − LR (°)									
	11.0	12.0	13.0	14.0	15.0	16.0	17.0	18.0	19.0	20.0
0.0	77.1	75.9	74.7	73.5	72.4	71.2	70.0	68.9	67.7	66.5
1.0	77.0	75.8	74.7	73.5	72.3	71.2	70.0	68.8	67.7	66.5
2.0	76.8	75.7	74.5	73.4	72.2	71.1	69.9	68.8	67.6	66.4
3.0	76.6	75.5	74.3	73.2	72.0	70.9	69.7	68.6	67.4	66.3
4.0	76.2	75.1	74.0	72.9	71.8	70.6	69.5	68.4	67.2	66.1
5.0	75.8	74.7	73.6	72.6	71.4	70.3	69.2	68.1	67.0	65.9
6.0	75.3	74.3	73.2	72.1	71.1	70.0	68.9	67.8	66.7	65.6
7.0	74.7	73.7	72.7	71.6	70.6	69.5	68.5	67.4	66.3	65.2
8.0	74.0	73.1	72.1	71.1	70.1	69.1	68.0	67.0	65.9	64.8
9.0	73.3	72.4	71.5	70.5	59.5	68.5	67.5	66.5	65.4	64.4
10.0	72.6	71.7	70.8	69.9	68.9	67.9	67.0	66.0	64.9	63.9
11.0	71.8	70.9	70.1	69.2	68.3	67.3	66.4	65.4	64.4	63.4
12.0	70.9	70.1	69.3	68.5	67.6	66.7	65.7	64.8	63.8	62.8
13.0	70.1	69.3	68.5	67.7	66.8	66.0	65.1	64.1	63.2	62.3
14.0	69.2	68.5	67.7	66.9	66.1	65.2	64.4	63.5	62.6	61.6
15.0	68.3	67.6	66.8	66.1	65.3	64.5	63.6	62.8	61.9	61.0
16.0	67.3	66.7	66.0	65.2	64.5	63.7	62.9	62.0	61.2	60.3
17.0	66.4	65.7	65.1	64.4	63.6	62.9	62.1	61.3	60.4	59.6
18.0	65.4	64.8	64.1	63.5	62.8	62.0	61.3	60.5	59.7	58.9
19.0	64.4	63.8	63.2	62.6	61.9	61.2	60.4	59.7	58.9	58.1
20.0	63.4	62.8	62.3	61.6	61.0	60.3	59.6	58.9	58.1	57.3
21.0	62.4	61.9	61.3	60.7	60.1	59.4	58.7	58.0	57.3	56.5
22.0	61.4	60.9	60.3	59.7	59.1	58.5	57.8	57.2	56.4	55.7
23.0	60.3	59.9	59.3	58.8	58.2	57.6	56.9	56.3	55.6	54.9
24.0	59.3	58.8	58.3	57.8	57.2	56.7	56.0	55.4	54.7	54.0
25.0	58.3	57.8	57.3	5~.8	56.3	55.7	55.1	54.5	53.9	53.2
26.0	57.2	56.8	56.3	55.8	55.3	54.8	54.2	53.6	53.0	52.3
27.0	56.2	55.8	55.3	54.8	54.3	53.8	53.2	52.7	52.1	51.4
28.0	55.1	54.7	54.3	53.8	53.3	52.8	52.3	51.7	51.2	50.6
29.0	54.1	53.7	53.3	52.8	52.4	51.9	51.3	50.8	50.2	49.7
30.0	53.0	52.6	52.2	51.8	51.4	50.9	50.4	49.9	49.3	48.7
31.0	51.9	51.6	51.2	50.8	50.4	49.9	49.4	48.9	48.4	47.8
32.0	50.9	50.5	50.2	49.8	49.3	48.9	48.4	47.9	47.4	46.9
33.0	49.8	49.5	49.1	48.7	48.3	47.9	47.5	47.0	46.5	46.0

Table A7.2 *Elevation angles*

Latitude (°)	Longitude difference LS − LR (°)									
	11.0	12.0	13.0	14.0	15.0	16.0	17.0	18.0	19.0	20.0
34.0	48.7	48.4	48.1	47.7	47.3	46.9	46.5	46.0	45.5	45.0
35.0	47.7	47.4	47.0	46.7	46.3	45.9	45.5	45.0	44.6	44.1
36.0	46.6	46.3	46.0	45.6	45.3	44.9	44.5	44.1	43.6	43.1
37.0	45.5	45.3	44.9	44.6	44.3	43.9	43.5	43.1	42.6	42.2
38.0	44.5	44.2	43.9	43.6	43.2	42.9	42.5	42.1	41.7	41.2
39.0	43.4	43.1	42,9	42.5	42.2	41.9	41.5	41.1	40.7	40.3
40.0	42.3	42.1	41.8	41.5	41.2	40.9	40.5	40.1	39.7	39.3
41.0	41.3	41.0	40.8	40.5	40.2	39.8	39.5	39.1	38.8	38.4
42.0	40.2	40.0	39.7	39.4	39.1	38.8	38.5	38.1	37.8	37.4
43.0	39.1	38.9	38.7	38.4	38.1	37.8	37.5	37.1	36.8	36.4
44.0	38.1	37.9	37.6	37.4	37.1	36.8	36.5	36.2	35.8	35.5
45.0	37.0	36.8	36.6	36.3	36.1	35.8	35.5	35.2	34.8	34.5
46.0	36.0	35.8	35.5	35.3	35.0	34.8	34.5	34.2	33.8	33.5
47.0	34.9	34.7	34.5	34.2	34.0	33.7	33.5	33.2	32.9	32.5
48.0	33.8	33.6	33.4	33.2	33.0	32.7	32.5	32.2	31.9	31.6
49.0	32.8	32.6	32.4	32.2	31.9	31.7	31.4	31.2	30.9	30.6
50.0	31.7	31.6	31.4	31.1	30.9	30.7	30.4	30.2	29.9	29.6
51.0	30.7	30.5	30.3	30.1	29.9	29.7	29.4	29.2	28.9	28.6
52.0	29.6	29.5	29.3	29.1	28.9	28.7	28.4	28.2	27.9	27.6
53.0	28.6	28.4	28.2	28.1	27.9	27.6	27.4	27.2	26.9	26.7
54.0	27.5	27.4	27.2	27.0	26.8	26.6	26.4	26.2	25.9	25.7
55.0	26.5	26.3	26.2	26.0	25.8	25.6	25.4	25.2	25.0	24.7
56.0	25.4	25.3	25.1	25.0	24.8	24.6	24.4	24.2	24.0	23.7
57.0	24.4	24.3	24.1	24.0	23.8	23.6	23.4	23.2	23.0	22.8
58.0	23.4	23.2	23.1	22.9	22.8	22.6	22.4	22.2	22.0	21.8
59.0	22.3	22.2	22.1	21.9	21.8	21.6	21.4	21.2	21.0	20.8
60.0	21.3	21.2	21.0	20.9	20.7	20.6	20.4	20.2	20.0	19.8
61.0	20.3	20.1	20.0	19.9	19.7	19.6	19.4	19.2	19.1	18.9
62.0	19.2	19.1	19.0	18.9	18.7	18.6	18.4	18.3	18.1	17.9
63.0	18.2	18.1	18.0	17.9	17.7	17.6	17.4	17.3	17.1	16.9
64.0	17.2	17.1	17.0	16.8	16.7	16.6	16.4	16.3	16.1	16.0
65.0	16.2	16.1	16.0	15.8	15.7	15.6	15.5	15.3	15.2	15.0
66.0	15.1	15.0	14.9	14.8	14.7	14.6	14.5	14.3	14.2	14.0
67.0	14.1	14.0	13.9	13.8	13.7	13.6	13.5	13.4	13.2	13.1
68.0	13.1	13.0	12.9	12.8	12.7	12.6	12.5	12.4	12.2	12.1
69.0	12.1	12.0	11.9	11.8	11.7	11.6	11.5	11.4	11.3	11.1
70.0	11.1	11.0	10.9	10.8	10.7	10.6	10.5	10.4	10.3	10.2
71.0	10.1	10.0	9.9	9.8	9.8	9.7	9.6	9.5	9.3	9.2
72.0	9.1	9.0	8.9	8.9	8.8	8.7	8.6	8.5	8.4	8.3
73.0	8.1	8.0	7.9	7.9	7.8	7.7	7.6	7.5	7.4	7.3
74.0	7.1	7.0	6.9	6.9	6.8	6.7	6.6	6.6	6.5	6.4
75.0	6.1	6.0	6.0	5.9	5.8	5.7	5.7	5.6	5.5	5.4
76.0	5.1	5.0	5.0	4.9	4.8	4.8	4.7	4.6	4.6	4.5
77.0	4.1	4.0	4.0	3.9	3.9	3.8	3.7	3.7	3.6	3.5
78.0	3.1	3.0	3.0	2.9	2.9	2.8	2.8	2.7	2.6	2.6
79.0	2.1	2.1	2.0	2.0	1.9	1.9	1.8	1.8	1.7	1.6
80.0	1.1	1.1	1.0	1.0	1.0	0.9	0.9	0.8	0.8	0.7

Latitude (°)	Longitude difference LS − LR (°)									
	21.0	22.0	23.0	24.0	25.0	26.0	27.0	28.0	29.0	30.0
0.0	65.4	64.2	63.1	61.9	60.8	59.6	58.5	57.3	56.2	55.0
1.0	65.4	64.2	63.0	61.9	60.7	59.6	58.4	57.3	56.2	55.0
2.0	65.3	64.1	63.0	61.8	60.7	59.5	58.4	57.2	56.1	55.0
3.0	65.2	64.0	62.9	61.7	60.6	59.4	58.3	57.1	56.0	54.9
4.0	65.0	63.8	62.7	61.6	60.4	59.3	58.1	57.0	55.9	54.8

Table A7.2 *Elevation angles*

| Latitude (°) | Longitude difference LS − LR (°) | | | | | | | | | |
	21.0	22.0	23.0	24.0	25.0	26.0	27.0	28.0	29.0	30.0
5.0	64.7	63.6	62.5	61.4	60.2	59.1	58.0	56.8	55.7	54.6
6.0	64.5	63.3	62.2	61.1	60.0	58.9	57.8	56.6	55.5	54.4
7.0	64.1	63.0	61.9	60.8	59.7	58.6	57.5	56.4	55.3	54.2
8.0	63.8	62.7	61.6	60.5	59.4	58.3	57.2	56.1	55.0	53.9
9.0	63.3	62.3	61.2	60.1	59.1	58.0	56.9	55.8	54.7	53.7
10.0	62.9	61.8	60.8	59.7	58.7	57.6	56.6	55.5	54.4	53.3
11.0	62.4	61.4	60.3	59.3	58.3	57.2	56.2	55.1	54.1	53.0
12.0	61.9	60.9	59.9	58.8	57.8	56.8	55.8	54.7	53.7	52.6
13.0	61.3	60.3	59.3	58.3	57.3	56.3	55.3	54.3	53.3	52.2
14.0	60.7	59.7	58.8	57.8	56.8	55.8	54.8	53.8	52.8	51.8
15.0	60.1	59.1	58.2	57.2	56.3	55.3	54.3	53.3	52.4	51.4
16.0	59.4	58.5	57.6	56.7	55.7	54.8	53.8	52.8	51.9	50.9
17.0	58.7	57.8	56.9	56.0	55.1	54.2	53.2	52.3	51.3	50.4
18.0	58.0	57.2	56.3	55.4	54.5	53.6	52.7	51.7	50.8	49.9
19.0	57.3	56.4	55.6	54.7	53.9	53.0	52.1	51.2	50.2	49.3
20.0	56.5	55.7	54.9	54.0	53.2	52.3	51.4	50.6	49.7	48.7
21.0	55.8	55.0	54.2	53.3	52.5	51.7	50.8	49.9	49.0	48.2
22.0	55.0	54.2	53.4	52.6	51.8	51.0	50.1	49.3	48.4	47.6
23.0	54.2	53.4	52.7	51.9	51.1	50.3	49.5	48.6	47.8	46.9
24.0	53.3	52.6	51.9	51.1	50.4	49.6	48.8	48.0	47.1	46.3
25.0	52.5	51.8	51.1	50.4	49.6	48.8	48.1	47.3	46.5	45.6
26.0	51.7	51.0	50.3	49.6	48.8	48.1	47.3	46.6	45.8	45.0
27.0	50.8	50.1	49.5	48.8	48.1	47.3	46.6	45.8	45.1	44.3
28.0	49.9	49.3	48.6	48.0	47.3	46.6	45.8	45.1	44.3	43.6
29.0	49.0	48.4	47.8	47.1	46.5	45.8	45.1	44.3	43.6	42.9
30.0	48.2	47.6	46.9	46.3	45.6	45.0	44.3	43.6	42.9	42.2
31.0	47.3	46.7	46.1	45.5	44.8	44.2	43.5	42.8	42.1	41.4
32.0	46.4	45.8	45.2	44.6	44.0	43.3	42.7	42.0	41.4	40.7
33.0	45.4	44.9	44.3	43.7	43.1	42.5	41.9	41.2	40.6	39.9
34.0	44.5	44.0	43.4	42.9	42.3	41.7	41.1	40.4	39.8	39.2
35.0	43.6	43.1	42.5	42.0	41.4	40.8	40.2	39.6	39.0	38.4
36.0	42.7	42.2	41.6	41.1	40.6	40.0	39.4	38.8	38.2	37.6
37.0	41.7	41.2	40.7	40.2	39.7	39.1	38.6	38.0	37.4	36.8
38.0	40.8	40.3	39.8	3~.3	38.8	38.3	37.7	37.2	36.6	36.0
39.0	39.8	39.4	38.9	3~.4	37.9	37.4	36.9	36.3	35.8	35.2
40.0	38.9	38.5	38.0	37.5	37.0	36.5	36.0	35.5	34.9	34.4
41.0	37.9	37.5	37.1	36.6	36.1	35.7	35.2	34.6	34.1	33.6
42.0	37.0	36.6	36.1	35.7	35.2	34.8	34.3	33.8	33.3	32.8
43.0	36.0	35.6	35.2	34.8	34.3	33.9	33.4	32.9	32.4	31.9
44.0	35.1	34.7	34.3	33.9	33.4	33.0	32.5	32.1	31.6	31.1
45.0	34.1	33.7	33.3	32.9	32.5	32.1	31.7	31.2	30.7	30.3
46.0	33.2	32.8	32.4	32.0	31.6	31.2	30.8	30.3	29.9	29.4
47.0	32.2	31.8	31.5	31.1	30.7	30.3	29.9	29.5	29.0	28.6
48.0	31.2	30.9	30.5	30.2	29.8	29.4	29.0	28.6	28.2	27.7
49.0	30.3	29.9	29.6	29.2	28.9	28.5	28.1	27.7	27.3	26.9
50.0	29.3	29.0	28.6	28.3	28.0	27.6	27.2	26.8	26.4	26.0
51.0	28.3	28.0	27.7	27.4	27.0	26.7	26.3	25.9	25.6	25.2
52.0	27.4	27.1	26.8	26.4	26.1	25.8	25.4	25.1	24.7	24.3
53.0	26.4	26.1	25.8	25.5	25.2	24.9	24.5	24.2	23.8	23.4
54.0	25.4	25.2	24.9	24.6	24.3	23.9	23.6	23.3	22.9	22.6
55.0	24.5	24.2	23.9	23.6	23.3	23.0	22.7	22.4	22.1	21.7
56.0	23.5	23.2	23.0	22.7	22.4	22.1	21.8	21.5	21.2	20.8
57.0	22.5	22.3	22.0	21.8	21.5	21.2	20.9	20.6	20.3	20.0
58.0	21.6	21.3	21.1	20.8	20.6	20.3	20.0	19.7	19.4	19.1
59.0	20.6	20.4	20.1	19.9	19.6	19.4	19.1	18.8	18.5	18.2
60.0	19.6	19.4	19.2	19.0	18.7	18.5	18.2	17.9	17.6	17.4

Table A7.2 Elevation angles

Latitude (°)	Longitude difference LS − LR (°)									
	21.0	22.0	23.0	24.0	25.0	26.0	27.0	28.0	29.0	30.0
61.0	18.7	18.5	18.2	18.0	17.8	17.5	17.3	17.0	16.8	16.5
62.0	17.7	17.5	17.3	17.1	16.9	16.6	16.4	16.1	15.9	15.6
63.0	16.7	16.6	16.4	16.1	15.9	15.7	15.5	15.2	15.0	14.7
64.0	15.8	15.6	15.4	15.2	15.0	14.8	14.6	14.3	14.1	13.9
65.0	14.8	14.7	14.5	14.3	14.1	13.9	13.7	13.4	13.2	13.0
66.0	13.9	13.7	13.5	13.3	13.2	13.0	12.8	12.6	12.3	12.1
67.0	12.9	12.8	12.6	12.4	12.2	12.1	11.9	11.7	11.5	11.2
68.0	12.0	11.8	11.7	11.5	11.3	11.1	11.0	10.8	10.6	10.4
69.0	11.0	10.9	10.7	10.6	10.4	10.2	10.1	9.9	9.7	9.5
70.0	10.1	9.9	9.8	9.6	9.5	9.3	9.2	9.0	8.8	8.6
71.0	9.1	9.0	8.8	8.7	8.6	8.4	8.3	8.1	7.9	7.8
72.0	8.2	8.0	7.9	7.8	7.6	7.5	7.4	7.2	7.0	6.9
73.0	7.2	7.1	7.0	6.9	6.7	6.6	6.5	6.3	6.2	6.0
74.0	6.3	6.2	6.0	5.9	5.8	5.7	5.6	5.4	5.3	5.1
75.0	5.3	5.2	5.1	5.0	4.9	4.8	4.7	4.5	4.4	4.3
76.0	4.4	4.3	4.2	4.1	4.0	3.9	3.8	3.7	3.5	3.4
77.0	3.4	3.4	3.3	3.2	3.1	3.0	2.9	2.8	2.7	2.5
78.0	2.5	2.4	2.3	2.3	2.2	2.1	2.0	1.9	1.8	1.7
79.0	1.6	1.5	1.4	1.3	1.3	1.2	1.1	1.0	0.9	0.8
80.0	0.6	0.6	0.5	0.4	0.4	0.3	0.2	0.1	0.0	–

Latitude (°)	Longitude difference LS − LR (°)									
	31.0	32.0	33.0	34.0	35.0	36.0	37.0	38.0	39.0	40.0
0.0	53.9	52.7	51.6	50.5	49.3	48.2	47.1	46.0	44.8	43.7
1.0	53.9	52.7	51.6	50.5	49.3	48.2	47.1	46.0	44.8	43.7
2.0	53.8	52.7	51.5	50.4	49.3	48.2	47.0	45.9	44.8	43.7
3.0	53.7	52.6	51.5	50.3	49.2	48.1	47.0	45.9	44.7	43.6
4.0	53.6	52.5	51.4	50.2	49.1	48.0	46.9	45.8	44.7	43.5
5.0	53.5	52.4	51.2	50.1	49.0	47.9	46.8	45.7	44.5	43.4
6.0	53.3	52.2	51.1	50.0	48.8	47.7	46.6	45.5	44.4	43.3
7.0	53.1	52.0	50.9	49.8	48.7	47.6	46.5	45.4	44.3	43.2
8.0	52.8	51.7	50.6	49.5	48.5	47.4	46.3	45.2	44.1	43.0
9.0	52.6	51.5	50.4	49.3	48.2	47.1	46.0	45.0	43.9	42.8
10.0	52.3	51.2	50.1	49.0	48.0	46.9	45.8	44.7	43.7	42.6
11.0	51.9	50.9	49.8	48.7	47.7	46.6	45.5	44.5	43.3	42.3
12.0	51.6	50.5	49.5	48.4	47.4	46.3	45.3	44.2	43.1	42.1
13.0	51.2	50.2	49.1	48.1	47.0	46.0	44.9	43.9	42.9	41.8
14.0	50.8	49.8	48.7	47.7	46.7	45.6	44.6	43.6	42.5	41.5
15.0	50.4	49.3	48.3	47.3	46.3	45.3	44.3	43.2	42.2	41.2
16.0	49.9	48.9	47.9	46.9	45.9	44.9	43.9	42.9	41.9	40.9
17.0	49.4	48.4	47.5	46.5	45.5	44.5	43.5	42.5	41.5	40.5
18.0	48.9	.17.6	47.0	46.0	45.0	44.1	43.1	42.1	41.1	40.1
19.0	48.4	47.4	46.5	45.5	44.6	43.6	42.6	41.7	40.7	39.7
20.0	47.8	46.9	46.0	45.0	44.1	43.1	42.2	41.2	40.3	39.3
21.0	47.3	46.4	45.4	44.5	43.6	42.7	41.7	40.8	39.8	38.9
22.0	46.7	45.8	44.9	44.0	43.1	42.2	41.2	40.3	39.4	38.5
23.0	46.1	45.2	44.3	43.4	42.5	41.6	40.7	39.8	38.9	38.0
24.0	45.5	44.6	43.7	42.9	42.0	41.1	40.2	39.3	38.4	37.5
25.0	44.8	44.0	43.1	42.3	41.4	40.6	39.7	38.8	37.9	37.0
26.0	44.2	43.3	42.5	41.7	40.8	40.0	39.1	38.3	37.4	36.5
27.0	43.5	42.7	41.9	41.1	40.2	39.4	38.6	37.7	36.9	36.0
28.0	42.8	42.0	41.2	40.4	39.6	38.8	38.0	37.2	36.3	35.5
29.0	42.1	41.4	40.6	39.8	39.0	38.2	37.4	36.6	35.8	34.9
30.0	41.4	40.7	39.9	39.2	38.4	37.6	36.8	36.0	35.2	34.4
31.0	40.7	40.0	39.2	38.5	37.7	37.0	36.2	35.4	34.6	33.8

Table A7.2 *Elevation angles*

Latitude (°)	Longitude difference LS − LR (°)									
	31.0	*32.0*	*33.0*	*34.0*	*35.0*	*36.0*	*37.0*	*38.0*	*39.0*	*40.0*
32.0	40.0	39.3	38.5	37.8	37.1	36.3	35.6	34.8	34.0	33.2
33.0	39.2	38.5	37.8	37.1	36.4	35.7	34.9	34.2	33.4	32.7
34.0	38.5	37.8	37.1	36.4	35.7	35.0	34.3	33.5	32.8	32.1
35.0	37.7	37.1	36.4	35.7	35.0	34.3	33.6	32.9	32.2	31.5
36.0	37.0	36.3	35.7	35.0	34.3	33.6	33.0	32.3	31.5	30.8
37.0	36.2	35.6	34.9	34.3	33.6	33.0	32.3	31.6	30.9	30.2
38.0	35.4	34.8	34.2	33.5	32.9	32.3	31.6	30.9	30.3	29.6
39.0	34.6	34.0	33.4	32.8	32.2	31.5	30.9	30.3	29.6	28.9
40.0	33.8	33.2	32.7	32.1	31.5	30.8	30.2	29.6	28.9	28.3
41.0	33.0	32.5	31.9	31.3	30.7	30.1	29.5	28.9	28.3	27.6
42.0	32.2	31.7	31.1	30.5	30.0	29.4	28.8	28.2	27.6	27.0
43.0	31.4	30.9	30.3	29.8	29.2	28.6	28.1	27.5	26.9	26.3
44.0	30.6	30.1	29.5	29.0	28.5	27.9	27.3	26.8	26.2	25.6
45.0	28.8	29.3	28.8	28.2	27.7	27.2	26.6	26.1	25.5	24.9
46.0	28.9	28.5	28.0	27.4	26.9	26.4	25.9	25.3	24.8	24.2
47.0	28.1	27.6	27.2	26.7	26.2	25.6	25.1	24.6	24.1	23.5
48.0	27.3	26.8	26.3	25.9	25.4	24.9	24.4	23.9	23.4	22.8
49.0	26.4	26.0	25.5	25.1	24.6	24.1	23.6	23.1	22.6	22.1
50.0	25.6	25.2	24.7	24.3	23.8	23.4	22.9	22.4	21.9	21.4
51.0	24.7	24.3	23.9	23.5	23.0	22.6	22.1	21.6	21.2	20.7
52.0	23.9	23.5	23.1	22.7	22.2	21.8	21.4	20.9	20.4	20.0
53.0	23.1	22.7	22.3	21.9	21.4	21.0	20.6	20.1	19.7	19.2
54.0	22.2	21.8	21.4	21.0	20.6	20.2	19.8	19.4	19.0	18.5
55.0	21.3	21.0	20.6	20.2	19.8	19.4	19.0	18.6	18.2	17.8
56.0	20.5	20.1	19.8	19.4	19.0	18.7	18.3	17.9	17.5	17.0
57.0	19.6	19.3	19.0	18.6	18.2	17.9	17.5	17.1	16.7	16.3
58.0	18.8	18.5	18.1	17.8	17.4	17.1	16.7	16.3	16.0	15.6
59.0	17.9	17.6	17.3	17.0	16.6	16.3	15.9	15.6	15.2	14.8
60.0	17.1	16.8	16.5	16.1	15.8	15.5	15.1	14.8	14.4	14.1
61.0	16.2	15.9	15.6	15.3	15.0	14.7	14.4	14.0	13.7	13.3
62.0	15.3	15.1	14.8	14.5	14.2	13.9	13.6	13.2	12.9	12.6
63.0	14.5	14.2	13.9	13.7	13.4	13.1	12.8	12.5	12.2	11.8
64.0	13.6	13.4	13.1	12.8	12.6	12.3	12.0	11.7	11.4	11.1
65.0	12.8	12.5	12.3	12.0	11.7	11.5	11.2	10.9	10.6	10.3
66.0	11.9	11.7	11.4	11.2	10.9	10.7	10.4	10.1	9.9	9.6
67.0	11.0	10.8	10.6	10.3	10.1	9.9	9.6	9.3	9.1	8.8
68.0	10.2	10.0	9.7	9.5	9.3	9.1	8.8	8.6	8.3	8.1
69.0	9.3	9.1	8.9	8.7	8.5	8.2	8.0	7.8	7.5	7.3
70.0	8.4	8.3	8.1	7.9	7.6	7.4	7.2	7.0	6.8	6.5
71.0	7.6	7.4	7.2	7.0	6.8	6.6	6.4	6.2	6.0	5.8
72.0	6.7	6.5	6.4	6.2	6.0	5.8	5.6	5.4	5.2	5.0
73.0	5.9	5.7	5.5	5.4	5.2	5.0	4.8	4.6	4.5	4.3
74.0	5.0	4.8	1.7	4.5	4.4	4.2	4.0	3.9	3.7	3.5
75.0	4.1	4.0	3.9	3.7	3.6	3.4	3.2	3.1	2.9	2.7
76.0	3.3	3.2	3.0	2.9	2.7	2.6	2.4	2.3	2.1	2.0
77.0	2.4	2.3	2.2	2.1	1.9	1.8	1.7	1.5	1.4	1.2
78.0	1.6	1.5	1.3	1.2	1.1	1.0	0.9	0.7	0.6	0.5
79.0	0.7	0.6	0.5	0.4	0.3	0.2	0.1	–	–	–
80.0	–	–	–	–	–	–	–	–	–	–

Latitude (°)	Longitude difference LS − LR (°)									
	41.0	*42.0*	*43.0*	*44.0*	*45.0*	*46.0*	*47.0*	*48.0*	*49.0*	*50.0*
0.0	42.6	41.5	40.4	39.3	38.2	37.1	36.0	34.9	33.8	32.7
1.0	42.6	41.5	40.4	39.3	38.2	37.1	36.0	34.9	33.8	32.7
2.0	42.6	41.5	40.3	39.2	38.1	37.0	35.9	34.8	33.7	32.7

Table A7.2 *Elevation angles*

Latitude (°)	Longitude difference LS − LR (°)									
	41.0	42.0	43.0	44.0	45.0	46.0	47.0	48.0	49.0	50.0
3.0	42.5	41.4	40.3	39.2	38.1	37.0	35.9	34.8	33.7	32.6
4.0	42.4	41.3	40.2	39.1	38.0	36.9	35.8	34.7	33.6	32.6
5.0	42.3	41.2	40.1	39.0	37.9	36.8	35.7	34.7	33.6	32.5
6.0	42.2	41.1	40.0	38.9	37.8	36.7	35.6	34.6	33.5	32.4
7.0	42.1	41.0	39.9	38.8	37.7	36.6	35.5	34.5	33.4	32.3
8.0	41.9	40.8	31.7	38.6	37.6	36.5	35.4	34.3	33.2	32.2
9.0	41.7	40.6	39.6	38.5	37.4	36.3	35.2	34.2	33.1	32.0
10.0	41.5	40.4	39.4	38.3	37.2	36.1	35.1	34.0	33.0	31.9
11.0	41.3	40.2	39.1	38.1	37.0	36.0	34.9	33.8	32.8	31.7
12.0	41.0	40.0	38.9	37.9	36.8	35.8	34.7	33.6	32.6	31.6
13.0	40.8	39.7	38.7	37.6	36.6	35.5	34.5	33.4	32.4	31.4
14.0	40.5	39.4	38.4	37.4	36.3	35.3	34.2	33.2	32.2	31.1
15.0	40.2	39.1	38.1	37.1	36.1	35.0	34.0	33.0	31.9	30.9
16.0	39.8	38.8	37.8	36.8	35.8	34.8	33.7	32.7	31.7	30.7
17.0	39.5	38.5	37.5	36.5	35.5	34.5	33.5	32.5	31.4	30.4
18.0	39.1	38.1	37.1	36.2	35.2	34.2	33.2	32.2	31.2	30.2
19.0	38.8	37.8	36.8	35.8	34.8	33.8	32.9	31.9	30.9	29.9
20.0	38.4	37.4	36.4	35.5	34.5	33.5	32.5	31.6	30.6	29.6
21.0	37.9	37.0	36.0	35.1	34.1	33.2	32.2	31.2	30.3	29.3
22.0	37.5	36.6	35.6	34.7	33.7	32.8	31.8	30.9	29.9	29.0
23.0	37.1	36.1	35.2	34.3	33.3	32.4	31.5	30.5	29.6	28.6
24.0	36.6	35.7	34.8	33.9	32.9	32.0	31.1	30.2	29.2	28.3
25.0	36.1	35.2	34.3	33.4	32.5	31.6	30.7	29.8	28.9	28.0
26.0	35.7	34.8	33.9	33.0	32.1	31.2	30.3	29.4	28.5	27.6
27.0	35.2	34.3	33.4	32.5	31.7	30.8	29.9	29.0	28.1	27.2
28.0	34.6	33.8	32.9	32.1	31.2	30.3	29.5	28.6	27.7	26.8
29.0	34.1	33.3	32.4	31.6	30.7	29.9	29.0	28.2	27.3	26.4
30.0	33.6	32.8	31.9	31.1	30.3	29.4	28.6	27.7	26.9	26.0
31.0	33.0	32.2	31.4	30.6	29.8	28.9	28.1	27.3	26.4	25.6
32.0	32.5	31.7	30.9	30.1	29.3	28.5	27.6	26.8	26.0	25.2
33.0	31.9	31.1	30.3	29.5	28.8	28.0	27.2	26.3	25.5	24.7
34.0	31.3	30.5	29.8	29.0	28.2	27.4	26.7	25.9	25.1	24.3
35.0	30.7	30.0	29.2	28.5	27.7	26.9	26.2	25.4	24.6	23.8
36.0	30.1	29.4	28.6	27.9	27.2	26.4	25.6	24.9	24.1	23.4.
37.0	29.5	28.8	28.1	27.3	26.6	25.9	25.1	24.4	23.6	22.9
38.0	28.9	28.2	27.5	26.8	26.1	25.3	24.6	23.9	23.1	22.4
39.0	28.3	27.6	26.9	26.2	25.5	24.8	24.1	23.4	22.6	21.9
40.0	27.6	27.0	26.3	25.6	24.9	24.2	23.5	22.8	22.1	21.4
41.0	27.0	26.3	25.7	25.0	24.3	23.7	23.0	22.3	21.6	20.9
42.0	26.3	25.7	25.0	24.4	23.7	23.1	22.4	21.7	21.1	20.4
43.0	25.7	25.0	24.4	23.8	23.1	22.5	21.8	21.2	20.5	19.9
44.0	25.0	24.4	23.8	23.2	22.5	21.9	21.3	20.6	20.0	19.3
45.0	24.3	23.7	23.1	22.5	21.9	21.3	20.7	20.1	19.4	18.8
46.0	23.7	23.1	22.5	21.9	21.3	20.7	20.1	19.5	18.9	18.3
47.0	23.0	22.4	21.8	21.3	20.7	20.1	19.5	18.9	18.3	17.7
48.0	22.3	21.7	21.2	20.6	20.1	19.5	18.9	18.3	17.8	17.2
49.0	21.6	21.1	20.5	20.0	19.4	18.9	18.3	17.8	17.2	16.6
50.0	20.9	20.4	19.9	19.3	18.8	18.3	17.7	17.2	16.6	16.0
51.0	20.2	19.7	19.2	18.7	18.2	17.6	17.1	16.6	16.0	15.5
52.0	19.5	19.0	18.5	18.0	17.5	17.0	16.5	16.0	15.4	14.9
53.0	18.8	18.3	17.8	17.4	16.9	16.4	15.9	15.4	14.8	14.3
54.0	18.1	17.6	17.1	16.7	16.2	15.7	15.2	14.7	14.3	13.7
55.0	17.3	16.9	16.5	16.0	15.5	15.1	14.6	14.1	13.7	13.2
56.0	16.6	16.2	15.8	15.3	14.9	14.4	14.0	13.5	13.0	12.6
57.0	15.9	15.5	15.1	14.6	14.2	13.8	13.3	12.9	12.4	12.0
58.0	15.2	14.8	14.4	14.0	13.5	13.1	12.7	12.3	11.8	11.4

Table A7.2 *Elevation angles*

Latitude (°)	Longitude difference LS − LR (°)									
	41.0	42.0	43.0	44.0	45.0	46.0	47.0	48.0	49.0	50.0
59.0	14.5	14.1	13.7	13.3	12.9	12.5	12.1	11.6	11.2	10.8
60.0	13.7	13.3	13.0	12.6	12.2	11.8	11.4	11.0	10.6	10.2
61.0	13.0	12.6	12.3	11.9	11.5	11.1	10.8	10.4	10.0	9.6
62.0	12.3	11.9	11.6	11.2	10.8	10.5	10.1	9.7	9.4	9.0
63.0	11.5	11.2	10.8	10.5	10.2	9.8	9.5	9.1	8.7	8.4
64.0	10.8	10.5	10.1	9.8	9.5	9.1	8.8	8.5	8.1	7.7
65.0	10.0	9.7	9.4	9.1	8.8	8.5	8.1	7.8	7.5	7.1
66.0	9.3	9.0	8.7	8.4	8.1	7.8	7.5	7.2	6.8	6.5
67.0	8.5	8.3	8.0	7.7	7.4	7.1	6.8	6.5	6.2	5.9
68.0	7.8	7.5	7.3	7.0	6.7	6.4	6.2	5.9	5.6	5.3
69.0	7.1	6.8	6.6	6.3	6.0	5.8	5.5	5.2	4.9	4.6
70.0	6.3	6.1	5.8	5.6	5.3	5.1	4.8	4.6	4.3	4.0
71.0	5.6	5.3	5.1	5.9	4.6	4.4	4.2	3.9	3.7	3.4
72.0	4.8	4.6	4.4	4.2	3.9	3.7	3.5	3.2	3.0	2.8
73.0	4.1	3.9	3.7	3.5	3.2	3.0	2.8	2.6	2.4	2.1
74.0	3.3	3.1	2.9	2.7	2.5	2.3	2.1	1.9	1.7	1.5
75.0	2.6	2.4	2.2	2.0	1.8	1.7	1.5	1.3	1.1	0.9
76.0	1.8	1.7	1.5	1.3	1.2	1.0	0.8	0.6	0.4	0.2
77.0	1.1	0.9	0.8	0.6	0.5	0.3	0.1	–	–	–
78.0	0.3	0.2	0.0	–	–	–	–	–	–	–
79.0	–	–	–	–	–	–	–	–	–	–
80.0	–	–	–	–	–	–	–	–	–	–

Latitude (°)	Longitude difference LS − LR (°)									
	51.0	52.0	53.0	54.0	55.0	56.0	57.0	58.0	59.0	60.0
0.0	31.6	30.5	29.4	28.3	27.3	26.2	25.1	24.1	23.0	21.9
1.0	31.6	30.5	29.4	28.3	27.3	26.2	25.1	24.1	23.0	21.9
2.0	31.6	30.5	29.4	28.3	27.2	26.2	25.1	24.0	23.0	21.9
3.0	31.5	30.4	29.4	28.3	27.2	26.1	25.1	24.0	22.9	21.9
4.0	51.0	52.0	53.0	54.0	55.0	56.0	57.0	58.0	59.0	60.0
4.0	31.5	30.4	29.3	28.2	27.2	26.1	25.0	24.0	22.9	21.8
5.0	31.4	30.3	29.3	28.2	27.1	26.0	25.0	23.9	22.9	21.8
6.0	31.3	30.2	29.2	28.1	27.0	26.0	24.9	23.9	22.8	21.7
7.0	31.2	30.2	29.1	28.0	27.0	25.9	24.8	23.8	22.7	21.7
8.0	31.1	30.0	29.0	27.9	26.9	25.8	24.7	23.7	22.6	21.6
9.0	31.0	29.9	28.9	27.8	26.7	25.7	24.6	23.6	22.5	21.5
10.0	30.8	29.8	28.7	27.7	26.6	25.6	24.5	23.5	22.4	21.4
11.0	30.7	29.6	28.6	27.5	26.5	25.4	24.4	23.4	22.3	21.3
12.0	30.5	29.5	28.4	27.4	26.3	25.3	24.3	23.2	22.2	21.2
13.0	30.3	29.3	28.2	27.2	26.2	25.1	24.1	23.1	22.1	21.0
14.0	30.1	29.1	28.1	27.0	26.0	25.0	24.0	22.9	21.9	20.9
15.0	29.9	28.9	27.9	26.8	25.8	24.8	23.8	22.8	21.8	20.7
16.0	29.7	28.7	27.6	26.6	25.6	24.6	23.6	22.6	21.6	20.6
17.0	29.4	28.4	27.4	26.4	25.4	24.4	23.4	22.4	21.4	20.4
18.0	29.2	28.2	27.2	26.2	25.2	24.2	23.2	22.2	21.2	20.2
19.0	28.9	27.9	26.9	25.9	25.0	24.0	23.0	22.0	21.0	20.0
20.0	28.6	27.6	26.7	25.7	24.7	23.7	22.8	21.8	20.8	19.8
21.0	28.3	27.4	26.4	25.4	24.5	23.5	22.5	21.6	20.6	19.6
22.0	28.0	27.1	26.1	25.2	24.2	23.2	22.3	21.3	20.4	19.4
23.0	27.7	26.8	25.8	24.9	23.9	23.0	22.0	21.1	20.1	19.2
24.0	27.4	26.4	25.5	24.6	23.6	22.7	21.8	20.8	19.9	19.0
25.0	27.0	26.1	25.2	24.3	23.3	22.4	21.5	20.6	19.6	18.7
26.0	26.7	25.8	24.9	23.9	23.0	22.1	21.2	20.3	19.4	18.5
27.0	26.3	25.4	24.5	23.6	22.7	21.8	20.9	20.0	19.1	18.2
28.0	25.9	25.1	24.2	23.3	22.4	21.5	20.6	19.7	18.8	17.9

Table A7.2 Elevation angles

Latitude (°)	Longitude difference LS − LR (°)									
	51.0	52.0	53.0	54.0	55.0	56.0	57.0	58.0	59.0	60.0
29.0	25.6	24.7	23.8	22.9	22.1	21.2	20.3	19.4	18.5	17.6
30.0	25.2	24.3	23.4	22.6	21.7	20.8	20.0	19.1	18.2	17.4
31.0	24.7	23.9	23.1	22.2	21.3	20.5	19.6	18.8	17.9	17.1
32.0	24.3	23.5	22.7	21.8	21.0	20.1	19.3	18.5	17.6	16.8
33.0	23.9	23.1	22.3	21.4	20.6	19.8	19.0	18.1	17.3	16.5
34.0	23.5	22.7	21.9	21.0	20.2	19.4	18.6	17.8	17.0	16.1
35.0	23.0	22.2	21.4	20.6	19.8	19.0	18.2	17.4	16.6	15.8
36.0	22.6	21.8	21.0	20.2	19.4	18.7	17.9	17.1	16.3	15.5
37.0	22.1	21.4	20.6	19.8	19.0	18.3	17.5	16.7	15.9	15.1
38.0	21.6	20.9	20.1	19.4	18.6	17.9	17.1	16.3	15.6	14.8
39.0	21.2	20.4	19.7	19.0	18.2	17.5	16.7	16.0	15.2	14.4
40.0	20.7	20.0	19.2	18.5	17.8	17.0	16.3	15.6	14.8	14.1
41.0	20.2	19.5	18.8	18.1	17.3	16.6	15.9	15.2	14.5	13.7
42.0	19.7	19.0	18.3	17.6	16.9	16.2	15.5	14.8	14.1	13.3
43.0	19.2	18.5	17.8	17.1	16.5	15.8	15.1	14.4	13.7	13.0
44.0	18.7	18.0	17.4	16.7	16.0	15.3	14.6	14.0	13.3	12.6
45.0	18.2	17.5	16.9	16.2	15.5	14.9	14.2	13.5	12.9	12.2
46.0	17.6	17.0	16.4	15.7	15.1	14.4	13.8	13.1	12.5	11.8
47.0	17.1	16.5	15.9	15.2	14.6	14.0	13.3	12.7	12.1	11.4
48.0	16.6	16.0	15.4	14.7	14.1	13.5	12.9	12.3	11.6	11.0
49.0	16.0	15.4	14.8	14.3	13.7	13.0	12.4	11.8	11.2	10.6
50.0	15.5	14.9	14.3	13.7	13.2	12.6	12.0	11.4	10.8	10.2
51.0	14.9	14.4	13.8	13.2	12.7	12.1	11.5	10.9	10.4	9.8
52.0	14.4	13.8	13.3	12.7	12.2	11.6	11.1	10.5	9.9	9.3
53.0	13.8	13.3	12.7	12.2	11.7	11.1	10.6	10.0	9.5	8.9
54.0	13.2	12.7	12.2	11.7	11.2	10.6	10.1	9.6	9.0	8.5
55.0	12.7	12.2	11.7	11.2	10.7	10.1	9.6	9.1	8.6	8.1
56.0	12.1	11.6	11.1	10.6	10.1	9.6	9.1	8.6	8.1	7.6
57.0	11.5	11.1	10.6	10.1	9.6	9.1	8.7	8.2	7.7	7.2
58.0	10.9	10.5	10.0	9.6	9.1	8.6	8.2	7.7	7.2	6.7
59.0	10.4	9.9	9.5	9.0	8.6	8.1	7.7	7.2	6.7	6.3
60.0	9.8	9.3	8.9	8.5	8.1	7.6	7.2	6.7	6.3	5.8
61.0	9.2	8.8	8.4	7.9	7.5	7.1	6.7	6.2	5.8	5.4
62.0	8.6	8.2	7.8	7.4	7.0	6.6	6.2	5.7	5.3	4.9
63.0	8.0	7.6	7.2	6.8	6.4	6.1	5.7	5.3	4.9	4.4
64.0	7.4	7.0	6.7	6.3	5.9	5.5	5.1	4.8	4.4	4.0
65.0	6.8	6.4	6.1	5.7	5.4	5.0	4.6	4.3	3.9	3.5
66.0	6.2	5.8	5.5	5.2	4.8	4.5	4.1	3.8	3.4	3.0
67.0	5.6	5.3	4.9	4.6	4.3	3.9	3.6	3.3	2.9	2.6
68.0	5.0	4.7	4.4	4.0	3.7	3.4	3.1	2.8	2.4	2.1
69.0	4.4	4.1	3.8	3.5	3.2	2.9	2.6	2.3	1.9	1.6
70.0	3.7	3.5	3.2	2.9	2.6	2.3	2.0	1.7	1.4	1.1
71.0	3.1	2.9	2.6	2.3	2.1	1.8	1.5	1.2	1.0	0.7
72.0	2.5	2.3	2.0	1.8	1.5	1.3	1.0	0.7	0.5	0.2
73.0	1.9	1.7	1.4	1.2	1.0	0.7	0.5	0.2	–	–
74.0	1.3	1.1	0.8	0.6	0.4	0.2	–	–	–	–
75.0	0.7	0.5	0.3	0.0	–	–	–	–	–	–
76.0	0.1	–	–	–	–	–	–	–	–	–
77.0	–	–	–	–	–	–	–	–	–	–
78.0	–	–	–	–	–	–	–	–	–	–
79.0	–	–	–	–	–	–	–	–	–	–
80.0	–	–	–	–	–	–	–	–	–	–

Table A7.2 *Elevation angles*

| Latitude (°) | Longitude difference LS − LR (°) | | | | | | | | | |
	61.0	62.0	63.0	64.0	65.0	66.0	67.0	68.0	69.0	70.0
0.0	20.9	19.8	18.8	17.7	16.7	15.6	14.6	13.5	12.5	11.5
1.0	20.9	19.8	18.8	17.7	16.7	15.6	14.6	13.5	12.5	11.5
2.0	20.9	19.8	18.7	17.7	16.7	15.6	14.6	13.5	12.5	11.5
3.0	20.8	19.8	18.7	17.7	16.6	15.6	14.5	13.5	12.5	11.4
4.0	20.8	19.7	18.7	17.6	16.6	15.6	14.5	13.5	12.5	11.4
5.0	20.7	19.7	18.6	17.6	16.6	15.6	14.5	13.5	12.4	11.4
6.0	20.7	19.6	18.6	17.6	16.5	15.5	14.4	13.4	12.4	11.4
7.0	20.6	19.6	18.5	17.5	16.5	15.4	14.4	13.4	12.3	11.3
8.0	20.5	19.5	18.5	17.4	16.4	15.4	14.3	13.3	12.3	11.3
9.0	20.5	19.4	18.4	17.4	16.3	15.3	14.3	13.2	12.2	11.2
10.0	20.4	19.3	18.3	17.3	16.2	15.2	14.2	13.2	12.2	11.1
11.0	20.3	19.2	18.2	17.2	16.2	15.1	14.1	13.1	12.1	11.1
12.0	20.1	19.1	18.1	17.1	16.1	15.0	14.0	13.0	12.0	11.0
13.0	20.0	19.0	18.0	17.0	16.0	14.9	13.9	12.9	11.9	10.9
14.0	19.9	18.9	17.9	16.8	15.8	14.8	13.8	12.8	11.8	10.8
15.0	19.7	18.7	17.7	16.7	15.7	14.7	13.7	12.7	11.7	10.7
16.0	19.6	18.6	17.6	16.6	15.6	14.6	13.6	12.6	11.6	10.6
17.0	19.4	18.4	17.4	16.4	15.5	14.5	13.5	12.5	11.5	10.5
18.0	19.2	18.3	17.3	16.3	15.3	14.3	13.4	12.4	11.4	10.4
19.0	19.1	18.1	17.1	16.1	15.2	14.2	13.2	12.2	11.3	10.3
20.0	18.9	17.9	16.9	16.0	15.0	14.0	13.1	12.1	11.1	10.2
21.0	18.7	17.7	16.7	15.8	14.8	13.9	12.9	12.0	11.0	10.1
22.0	18.5	17.5	16.6	15.6	14.7	13.7	12.8	11.8	10.9	9.9
23.0	18.2	17.3	16.4	15.4	14.5	13.5	12.6	11.7	10.7	9.8
24.0	18.0	17.1	16.1	15.2	14.3	13.3	12.4	11.5	10.6	9.6
25.0	17.8	16.9	15.9	15.0	14.1	13.2	12.2	11.3	10.4	9.5
26.0	17.5	16.6	15.7	14.8	13.9	13.0	12.1	11.1	10.2	9.3
27.0	17.3	16.4	15.5	14.6	13.7	12.8	11.9	11.0	10.1	9.2
28.0	17.0	16.1	15.2	14.3	13.4	12.6	11.7	10.8	9.9	9.0
29.0	16.8	15.9	15.0	14.1	13.2	12.3	11.5	10.6	9.7	8.8
30.0	16.5	15.6	14.7	14.3	13.0	12.1	11.2	10.4	9.5	8.6
31.0	16.2	15.3	14.5	13.6	12.8	11.9	11.0	10.2	9.3	8.4
32.0	15.9	15.1	14.2	13.4	12.5	11.7	10.8	10.0	9.1	8.3
33.0	15.6	14.8	13.9	13.1	12.3	11.4	10.6	9.7	8.9	8.1
34.0	15.3	14.5	13.7	12.8	12.0	11.2	10.3	9.5	8.7	7.9
35.0	15.0	14.2	13.4	12.6	11.7	10.9	10.1	9.3	8.5	7.6
36.0	14.7	13.9	13.1	12.3	11.5	10.7	9.9	9.1	8.2	7.4
37.0	14.4	13.6	12.8	12.0	11.2	10.4	9.6	8.8	8.0	7.2
38.0	14.0	13.2	12.5	11.7	10.9	10.1	9.3	8.6	7.8	7.0
39.0	13.7	12.9	12.2	11.4	10.6	9.9	9.1	8.3	7.5	6.8
40.0	13.3	12.6	11.8	11.1	10.3	9.6	8.8	8.1	7.3	6.5
41.0	13.0	12.3	11.5	10.8	10.0	9.3	8.5	7.8	7.1	6.3
42.0	12.6	11.9	11.2	10.5	9.7	9.0	8.3	7.5	6.8	6.1
43.0	12.3	11.6	10.8	10.1	9.4	8.7	8.0	7.3	6.6	5.8
44.0	11.9	11.2	10.5	9.8	9.1	8.4	7.7	7.0	6.3	5.6
45.0	11.5	10.8	10.2	9.5	8.8	8.1	7.4	6.7	6.0	5.3
46.0	11.1	10.5	9.8	9.1	8.5	7.8	7.1	6.4	5.8	5.1
47.0	10.8	10.1	9.5	8.8	8.1	7.5	6.8	6.2	5.5	4.8
48.0	10.4	9.7	9.1	8.5	7.8	7.2	6.5	5.9	5.2	4.6
49.0	10.0	9.4	8.7	8.1	7.5	6.8	6.2	5.6	4.9	4.3
50.0	9.6	9.0	8.4	7.7	7.1	6.5	5.9	5.3	4.6	4.0
51.0	9.2	8.6	8.0	7.4	6.8	6.2	5.6	5.0	4.4	3.7
52.0	8.8	8.2	7.6	7.0	6.4	5.8	5.3	4.7	4.1	3.5
53.0	8.4	7.8	7.2	6.7	6.1	5.5	4.9	4.4	3.8	3.2
54.0	7.9	7.4	6.8	6.3	5.7	5.2	4.6	4.0	3.5	2.9
55.0	7.5	7.0	6.4	5.9	5.4	4.8	4.3	3.7	3.2	2.6

Table A7.2 *Elevation angles*

Latitude (°)	Longitude difference LS − LR (°)									
	61.0	62.0	63.0	64.0	65.0	66.0	67.0	68.0	69.0	70.0
56.0	7.1	6.6	6.1	5.5	5.0	4.5	3.9	3.4	2.9	2.3
57.0	6.7	6.2	5.7	5.1	4.6	4.1	3.6	3.1	2.6	2.0
58.0	6.2	5.7	5.3	4.8	4.3	3.8	3.3	2.8	2.3	1.7
59.0	5.8	5.3	4.9	4.4	3.9	3.4	2.9	2.4	1.9	1.4
60.0	5.4	4.9	4.4	4.0	3.5	3.0	2.6	2.1	1.6	1.1
61.0	4.9	4.5	4.0	3.6	3.1	2.7	2.2	1.8	1.3	0.8
62.0	4.5	4.1	3.6	3.2	2.8	2.3	1.9	1.4	1.0	0.5
63.0	4.0	3.6	3.2	2.8	2.4	1.9	1.5	1.1	0.7	0.2
64.0	3.6	3.2	2.8	2.4	2.0	1.6	1.2	0.8	0.3	–
65.0	3.1	2.8	2.4	2.0	1.6	1.2	0.8	0.4	0.0	–
66.0	2.7	2.3	1.9	1.6	1.2	0.8	0.4	0.1	–	–
67.0	2.2	1.9	1.5	1.2	0.8	0.4	0.1	–	–	–
68.0	1.8	1.4	1.1	0.8	0.4	0.1	–	–	–	–
69.0	1.3	1.0	0.7	0.3	0.0	–	–	–	–	–
70.0	0.8	0.5	0.2	–	–	–	–	–	–	–
71.0	0.4	0.1	–	–	–	–	–	–	–	–
72.0	–	–	–	–	–	–	–	–	–	–
73.0	–	–	–	–	–	–	–	–	–	–
74.0	–	–	–	–	–	–	–	–	–	–
75.0	–	–	–	–	–	–	–	–	–	–
76.0	–	–	–	–	–	–	–	–	–	–
77.0	–	–	–	–	–	–	–	–	–	–
78.0	–	–	–	–	–	–	–	–	–	–
79.0	–	–	–	–	–	–	–	–	–	–
80.0	–	–	–	–	–	–	–	–	–	–

Latitude (°)	Longitude difference LS − LR (°)									
	71.0	72.0	73.0	74.0	75.0	76.0	77.0	78.0	79.0	80.0
0.0	10.4	0.4	8.4	7.4	6.4	5.3	4.3	3.3	2.3	1.3
1.0	10.4	9.4	8.4	7.4	6.4	5.3	4.3	3.3	2.3	1.3
2.0	10.4	9.4	8.4	7.4	6.3	5.3	4.3	3.3	2.3	1.3
3.0	10.4	9.4	8.4	7.3	6.3	5.3	4.3	3.3	2.3	1.3
4.0	10.4	9.4	8.3	7.3	6.3	5.3	4.3	3.3	2.3	1.3
5.0	10.4	9.3	8.3	7.3	6.3	5.3	4.3	3.3	2.3	1.3
6.0	10.3	9.3	8.3	7.3	6.3	5.3	4.3	3.2	2.2	1.2
7.0	10.3	9.3	8.3	7.2	6.2	5.2	4.2	3.2	2.2	1.2
8.0	10.2	9.2	8.2	7.2	6.2	5.2	4.2	3.2	2.2	1.2
9.0	10.2	9.2	8.2	7.2	6.2	5.2	4.2	3.2	2.2	1.2
10.0	10.1	9.1	8.1	7.1	6.1	5.1	4.1	3.1	2.1	1.1
11.0	10.1	9.1	8.1	7.1	6.1	5.1	4.1	3.1	2.1	1.1
12.0	10.0	9.0	8.0	7.0	6.0	5.0	4.0	3.0	2.1	1.1
13.0	9.9	8.9	7.9	6.9	6.0	5.0	4.0	3.0	2.0	1.0
14.0	9.8	8.9	7.9	6.9	5.9	4.9	3.9	2.9	2.0	1.0
15.0	9.8	8.8	7.8	6.8	5.8	4.8	3.9	2.9	1.9	1.0
16.0	9.7	8.7	7.7	6.7	5.7	4.8	3.8	2.8	1.9	0.9
17.0	9.6	8.6	7.6	6.6	5.7	4.7	3.7	2.8	1.8	0.9
18.0	9.5	8.5	7.5	6.6	5.6	4.6	3.7	2.7	1.8	0.8
19.0	9.3	8.4	7.4	6.5	5.5	4.6	3.6	2.6	1.7	0.8
20.0	9.2	8.3	7.3	6.4	5.4	4.5	3.5	2.6	1.6	0.7
21.0	9.1	8.2	7.2	6.3	5.3	4.4	3.4	2.5	1.6	0.6
22.0	9.0	8.0	7.1	6.2	5.2	4.3	3.4	2.4	1.5	0.6
23.0	8.8	7.9	7.0	6.0	5.1	4.2	3.3	2.3	1.4	0.5
24.0	8.7	7.8	6.9	5.9	5.0	4.1	3.2	2.3	1.3	0.4
25.0	8.6	7.6	6.7	5.8	4.9	4.0	3.1	2.2	1.3	0.4
26.0	8.4	7.5	6.6	5.7	4.8	3.9	3.0	2.1	1.2	0.3

Table A7.2 *Elevation angles*

| Latitude (°) | Longitude difference LS − LR (°) | | | | | | | | | |
	71.0	72.0	73.0	74.0	75.0	76.0	77.0	78.0	79.0	80.0
27.0	8.3	7.4	6.5	5.6	4.7	3.8	2.9	2.0	1.1	0.2
28.0	8.1	7.2	6.3	5.4	4.5	3.7	2.8	1.9	1.0	0.1
29.0	7.9	7.0	6.2	5.3	4.4	3.5	2.7	1.8	0.9	0.0
30.0	7.8	6.9	6.0	5.1	4.3	3.4	2.5	1.7	0.8	–
31.0	7.6	6.7	5.9	5.0	4.1	3.3	2.4	1.6	0.7	–
32.0	7.4	6.5	5.7	4.8	4.0	3.2	2.3	1.5	0.6	–
33.0	7.2	6.4	5.5	4.7	3.9	3.0	2.2	1.3	0.5	–
34.0	7.0	6.2	5.4	4.5	3.7	2.9	2.1	1.2	0.4	–
35.0	6.8	6.0	5.2	4.4	3.6	2.7	1.9	1.1	0.3	–
36.0	6.6	5.8	5.0	4.2	3.4	2.6	1.8	1.0	0.2	–
37.0	6.4	5.6	4.8	4.0	3.2	2.4	1.7	0.9	0.1	–
38.0	6.2	5.4	4.6	3.9	3.1	2.3	1.5	0.7	–	–
39.0	6.0	5.2	4.5	3.7	2.9	2.1	1.4	0.6	–	–
40.0	5.8	5.0	4.3	3.5	2.7	2.0	1.2	0.5	–	–
41.0	5.6	4.8	4.1	3.3	2.6	1.8	1.1	0.3	–	–
42.0	5.3	4.6	3.9	3.1	2.4	1.7	0.9	0.2	–	–
43.0	5.1	4.4	3.7	2.9	2.2	1.5	0.8	0.0	–	–
44.0	4.9	4.2	3.5	2.7	2.0	1.3	0.6	–	–	–
45.0	4.6	3.9	3.2	2.5	1.8	1.2	0.5	–	–	–
46.0	4.4	3.7	3.0	2.3	1.7	1.0	0.3	–	–	–
47.0	4.2	3.5	2.8	2.1	1.5	0.8	0.1	–	–	–
48.0	3.9	3.2	2.6	1.9	1.3	0.6	–	–	–	–
49.0	3.7	3.0	2.4	1.7	1.1	0.4	–	–	–	–
50.0	3.4	2.8	2.1	1.5	0.9	0.2	–	–	–	–
51.0	3.1	2.5	1.9	1.3	0.7	0.1	–	–	–	–
52.0	2.9	2.3	1.7	1.1	0.5	–	–	–	–	–
53.0	2.6	2.0	1.4	0.8	0.3	–	–	–	–	–
54.0	2.3	1.8	1.2	0.6	0.0	–	–	–	–	–
55.0	2.1	1.5	1.0	0.4	–	–	–	–	–	–
56.0	1.8	1.3	0.7	0.2	–	–	–	–	–	–
57.0	1.5	1.0	0.5	–	–	–	–	–	–	–
58.0	1.2	0.7	0.2	–	–	–	–	–	–	–
59.0	1.0	0.5	–	–	–	–	–	–	–	–
60.0	0.7	0.2	–	–	–	–	–	–	–	–
61.0	0.4	–	–	–	–	–	–	–	–	–
62.0	0.1	–	–	–	–	–	–	–	–	–
63.0	–	–	–	–	–	–	–	–	–	–
64.0	–	–	–	–	–	–	–	–	–	–
65.0	–	–	–	–	–	–	–	–	–	–
66.0	–	–	–	–	–	–	–	–	–	–
67.0	–	–	–	–	–	–	–	–	–	–
68.0	–	–	–	–	–	–	–	–	–	–
69.0	–	–	–	–	–	–	–	–	–	–
70.0	–	–	–	–	–	–	–	–	–	–
71.0	–	–	–	–	–	–	–	–	–	–
72.0	–	–	–	–	–	–	–	–	–	–
73.0	–	–	–	–	–	–	–	–	–	–
74.0	–	–	–	–	–	–	–	–	–	–
75.0	–	–	–	–	–	–	–	–	–	–
76.0	–	–	–	–	–	–	–	–	–	–
77.0	–	–	–	–	–	–	–	–	–	–
78.0	–	–	–	–	–	–	–	–	–	–
79.0	–	–	–	–	–	–	–	–	–	–
80.0	–	–	–	–	–	–	–	–	–	–

Appendix 8

AZ/EL tables for major European towns and cities

Inclusion does not necessarily imply reception is possible in every case. Some locations may be outside usable footprint area for some satellites. Add local magnetic variation to true azimuth value for compass bearing. All values are decimalized degrees.

ALBANIA — Satellite longitude (°)

	26.5E	23.5E	19.2E	16.0E	13.0E	10.0E	1.0W	5.0W	19.0W	27.5W	31.0W
Tirana (41.33N,19.83E)											
AZ	170.0	174.5	181.0	185.8	190.3	194.7	209.9	215.0	230.6	238.7	241.7
EL	41.7	42.1	42.2	42.1	41.7	41.2	37.7	35.9	28.1	22.6	20.1

AUSTRIA
Satellite longitude (°)

	26.5E	23.5E	19.2E	16.0E	13.0E	10.0E	1.0W	5.0W	19.0W	27.5W	31.0W
Graz (47.08N,15.45E)											
AZ	165.1	169.1	174.9	179.2	183.3	187.4	202.0	207.0	223.1	231.8	235.2
EL	34.8	35.3	35.8	35.9	35.8	35.6	33.5	32.3	26.4	21.8	19.8
Innsbruck (47.27N,11.40E)											
AZ	159.8	163.7	169.4	173.7	177.8	181.9	196.7	201.8	218.6	227.7	231.2
EL	33.7	34.4	35.1	35.5	35.7	35.7	34.3	33.4	28.2	23.9	22.0
Linz (48.30N,14.30E)											
AZ	163.9	167.8	173.4	177.7	181.7	185.8	200.1	205.1	221.3	230.1	233.5
EL	33.3	33.8	34.3	34.5	34.5	34.4	32.6	31.5	26.0	21.6	19.7
Vienna (48.22N,16.33E)											
AZ	166.5	170.4	176.2	180.4	184.5	188.5	202.7	207.6	223.6	232.2	235.5
EL	33.8	34.2	34.6	34.6	34.5	34.3	32.1	30.9	25.1	20.6	18.6

BELGIUM
Satellite longitude (°)

	26.5E	23.5E	19.2E	16.0E	13.0E	10.0E	1.0W	5.0W	19.0W	27.5W	31.0W
Antwerp (51.22N,4.42E)											
AZ	152.5	156.1	161.3	165.3	169.0	172.9	186.9	192.0	209.1	218.6	222.4
EL	27.8	28.7	29.71	30.4	30.8	31.1	31.1	30.7	27.4	24.2	22.7
Brussels (50.83N,4.33E)											
AZ	152.3	155.9	161.1	165.1	168.9	172.7	186.9	192.0	209.1	218.7	222.4
EL	28.1	29.0	30.1	30.7	31.2	31.5	31.6	31.1	27.8	24.5	23.0

Liege (50.50N,5.50E)

AZ	153.6	157.2	162.5	166.5	170.3	174.2	188.4	193.5	210.6	220.1	223.8
EL	28.8	29.7	30.7	31.3	31.7	32.0	31.8	31.3	27.7	24.3	22.7

Ostende (51.22N,2.92E)

AZ	150.8	154.3	159.5	163.4	167.1	170.9	185.0	190.1	207.3	217.0	220.8
EL	27.3	28.2	29.4	30.1	30.6	31.0	31.2	30.9	27.8	24.8	23.3

BULGARIA
Satellite longitude (°)

	26.5E	23.5E	19.2E	16.0E	13.0E	10.0E	1.0W	5.0W	19.0W	27.5W	31.0W

Burgas (42.50N,27.47E)

	26.5E	23.5E	19.2E	16.0E	13.0E	10.0E	1.0W	5.0W	19.0W	27.5W	31.0W
AZ	181.4	185.9	192.1	196.7	200.9	205.0	218.7	223.3	237.3	244.7	247.5
EL	40.9	40.8	40.2	39.6	30.8	38.8	33.1	31.0	22.5	16.7	14.2

Ruse (43.83N,25.95E)

	26.5E	23.5E	19.2E	16.0E	13.0E	10.0E	1.0W	5.0W	19.0W	27.5W	31.0W
AZ	179.2	183.5	189.7	194.2	198.4	202.4	216.3	220.9	235.2	242.8	245.7
EL	39.5	39.4	39.0	38.5	37.8	37.0	32.7	30.7	22.7	17.1	14.8

Sofia (42.68N,23.32E)

	26.5E	23.5E	19.2E	16.0E	13.0E	10.0E	1.0W	5.0W	19.0W	27.5W	31.0W
AZ	175.3	179.7	186.1	190.7	195.0	199.2	213.7	218.5	233.3	241.1	244.0
EL	40.6	40.7	40.6	40.2	39.6	38.9	34.9	33.0	25.1	19.5	17.1

Varna (43.22N,27.92E)

	26.5E	23.5E	19.2E	16.0E	13.0E	10.0E	1.0W	5.0W	19.0W	27.5W	31.0W
AZ	182.1	186.4	192.6	197.1	201.3	205.3	218.9	223.4	237.4	244.7	247.6
EL	40.1	39.9	39.4	38.7	37.9	37.0	32.3	30.2	21.8	16.1	13.6

DENMARK
Satellite longitude (°)

	26.5E	23.5E	19.2E	16.0E	13.0E	10.0E	1.0W	5.0W	19.0W	27.5W	31.0W

Allorg (57.05N,9.93E)

	26.5E	23.5E	19.2E	16.0E	13.0E	10.0E	1.0W	5.0W	19.0W	27.5W	31.0W
AZ	160.5	164.0	169.0	172.8	176.3	179.9	193.0	197.6	213.4	222.4	225.9
EL	23.4	24.0	24.6	24.9	25.0	25.1	24.4	23.7	20.3	17.3	15.9

Arhus (56.15N,10.22E)

	26.5E	23.5E	19.2E	16.0E	13.0E	10.0E	1.0W	5.0W	19.0W	27.5W	31.0W
AZ	160.6	164.1	169.2	173.0	176.6	180.3	193.4	198.1	214.0	223.0	226.5
EL	24.4	24.9	25.5	25.8	26.0	26.0	25.3	24.6	21.0	17.9	16.4

Copenhagen (55.67N,12.58E)

	26.5E	23.5E	19.2E	16.0E	13.0E	10.0E	1.0W	5.0W	19.0W	27.5W	31.0W
AZ	163.3	166.9	172.0	175.9	179.5	183.1	196.3	201.0	216.7	225.5	229.1
EL	25.3	25.8	26.3	26.5	26.6	26.5	25.4	24.6	20.6	17.3	15.7

Esbjerg (55.47N,8.45E)

	26.5E	23.5E	19.2E	16.0E	13.0E	10.0E	1.0W	5.0W	19.0W	27.5W	31.0W
AZ	158.4	161.9	167.0	170.9	174.5	178.1	191.4	196.2	212.2	221.4	225.0
EL	24.7	25.3	26.0	26.4	26.6	26.8	26.2	25.6	22.1	19.1	17.7

Odense (55.40N,10.38E)

	26.5E	23.5E	19.2E	16.0E	13.0E	10.0E	1.0W	5.0W	19.0W	27.5W	31.0W
AZ	160.7	164.2	169.3	173.2	176.8	180.5	193.7	198.5	214.4	223.4	226.9
EL	25.2	25.7	26.3	26.6	26.8	26.8	26.0	25.3	21.6	18.4	16.9

FINLAND
Satellite longitude (°)

	26.5E	23.5E	19.2E	16.0E	13.0E	10.0E	1.0W	5.0W	19.0W	27.5W	31.0W

Helsinki (60.17N,24.97E)

	26.5E	23.5E	19.2E	16.0E	13.0E	10.0E	1.0W	5.0W	19.0W	27.5W	31.0W
AZ	178.2	181.7	186.6	190.3	193.7	197.1	209.3	213.6	228.0	236.3	239.6
EL	21.7	21.7	21.6	21.3	21.0	20.6	18.3	17.2	12.5	9.1	7.5

Tampere (61.50N,23.75E)

	26.5E	23.5E	19.2E	16.0E	13.0E	10.0E	1.0W	5.0W	19.0W	27.5W	31.0W
AZ	176.9	180.3	185.2	188.8	192.2	195.6	207.7	212.0	226.4	234.8	238.2
EL	20.3	20.3	20.2	20.0	19.8	19.4	17.4	16.4	12.0	8.8	7.4

Turku (60.45N,22.28E)

AZ	175.2	178.6	183.5	187.2	190.6	194.1	206.3	210.7	225.3	233.7	237.0
EL	21.4	21.4	21.4	21.3	21.0	20.7	18.7	17.7	13.3	10.0	8.5

Oulu (65.02N,25.47E)

AZ	178.9	182.2	186.9	190.4	193.7	197.0	208.8	213.0	227.3	235.6	239.0
EL	16.6	16.6	16.5	16.3	16.0	15.6	13.8	12.9	8.9	6.1	4.8

FRANCE
Satellite longitude (°)

	26.5E	23.5E	19.2E	16.0E	13.0E	10.0E	1.0W	5.0W	19.0W	27.5W	31.0W
Bordeaux (44.83N,0.57W)											
AZ	144.1	147.6	153.0	157.1	161.1	165.2	180.6	186.3	205.3	215.8	219.8
EL	31.8	33.1	34.7	35.8	36.6	37.3	38.4	38.2	35.2	31.8	30.2
Brest (48.40N,0.57W)											
AZ	141.2	144.6	149.6	153.5	157.2	160.9	175.3	180.7	199.1	209.6	213.7
EL	26.9	28.2	29.9	31.0	31.9	32.7	34.3	34.4	32.7	30.2	28.8
Clermont-Ferrand (45.78N,3.08E)											
AZ	148.9	152.6	158.0	162.3	166.3	170.4	185.7	191.2	209.5	219.5	223.4
EL	32.5	33.6	34.9	35.8	36.4	36.9	37.2	36.7	33.0	29.3	27.6
Dijon (47.32N,5.02E)											
AZ	151.8	155.5	161.0	165.2	169.2	173.2	188.2	193.5	211.2	220.9	224.7
EL	31.7	32.7	33.9	34.6	35.1	35.4	35.3	34.7	30.8	27.1	25.4
Le Havre (49.50N,0.13E)											
AZ	146.9	150.4	155.6	159.5	163.3	167.1	181.5	186.7	204.5	214.5	218.5
EL	27.9	29.0	30.4	31.2	31.9	32.4	33.2	33.0	30.3	27.4	26.0
Le Mans (49.00N,0.20E)											
AZ	146.8	150.3	155.5	159.4	163.2	167.1	181.6	186.9	204.8	214.8	218.7
EL	28.4	29.5	30.9	31.8	32.4	33.0	33.8	33.6	30.8	27.8	26.3
Lille (50.63N,3.07E)											
AZ	150.7	154.3	159.5	163.5	167.2	171.1	185.3	190.4	207.7	217.4	221.2
EL	27.9	28.9	30.0	30.7	31.2	31.6	31.9	31.5	28.4	25.2	23.7
Limoges (49.83N,1.27E)											
AZ	148.3	151.9	157.0	161.0	164.8	168.6	183.0	188.2	205.8	215.7	219.6
EL	28.0	29.1	30.4	31.2	31.8	32.3	32.8	32.6	29.7	26.7	25.2
Lyon (45.75N,4.85E)											
AZ	151.0	154.8	160.3	164.6	168.7	172.8	188.1	193.6	211.7	221.5	225.2
EL	33.2	34.2	35.5	36.2	36.7	37.1	37.0	36.4	32.3	28.5	26.7
Marseille (43.30N,5.40E)											
AZ	150.6	154.5	160.3	164.7	169.0	173.3	189.3	195.0	213.5	223.3	227.1
EL	35.7	36.8	38.1	38.9	39.5	39.8	39.6	39.0	34.3	30.1	28.2
Nancy (48.68N,6.20E)											
AZ	153.8	157.5	162.9	167.0	171.0	174.9	189.5	194.8	212.1	221.6	225.3
EL	30.8	31.7	32.7	33.3	33.7	34.0	33.7	33.1	29.1	25.5	23.8
Nantes (47.22N,1.55E)											
AZ	147.6	1151.2	156.6	160.7	164.6	168.6	183.5	188.9	207.1	217.1	221.0
EL	30.5	31.7	33.1	33.9	34.6	35.1	35.7	35.4	32.1	28.8	27.2
Nice (43.70N,7.25E)											
AZ	153.2	157.1	163.0	167.4	171.7	176.0	191.9	197.4	215.5	225.1	228.8
EL	36.0	37.0	38.2	38.8	39.3	39.5	38.9	38.1	33.1	28.8	26.8
Paris (48.87N,2.33E)											
AZ	149.2	152.8	158.1	162.1	166.0	169.9	184.4	189.7	207.4	217.3	221.1
EL	29.3	30.3	31.6	32.4	33.0	33.4	33.8	33.5	30.3	27.1	25.5

Reims (49.25N,4.03E)

AZ	151.4	155.0	160.3	164.4	168.2	172.1	186.6	191.9	209.3	219.0	222.8
EL	29.5	30.5	31.7	32.3	32.8	33.2	33.3	32.8	29.3	26.0	24.4

Rennes (48.08N,1.68W)

AZ	144.2	147.7	152.9	156.8	160.6	164.5	179.1	184.5	202.7	213.0	217.0
EL	28.4	29.6	31.2	32.2	33.0	33.6	34.8	34.7	32.3	29.4	27.9

Rouen (49.43N,1.08E)

AZ	148.0	151.5	156.7	160.7	164.5	168.3	182.7	188.0	205.7	215.6	219.5
EL	28.3	29.4	30.7	31.5	32.2	32.7	33.3	33.0	30.1	27.1	25.6

Strasbourg (48.58N,7.75E)

AZ	155.6	159.4	164.9	169.1	173.0	177.0	191.6	196.8	213.9	223.3	226.9
EL	31.4	32.2	33.1	33.7	34.0	34.2	33.6	32.9	28.6	24.8	23.1

Toulouse (43.60N,1.43E)

AZ	145.9	149.6	155.1	159.4	163.5	167.7	183.5	189.3	208.4	218.7	222.7
EL	33.8	35.0	36.6	37.6	38.4	39.0	39.7	39.3	35.7	32.0	30.2

GERMANY
Satellite longitude(°)

	26.5E	23.5E	19.2E	16.0E	13.0E	10.0E	1.0W	5.0W	19.0W	27.5W	31.0W
Berlin (52.50N,13.42E)											
AZ	163.7	167.4	172.7	176.7	180.5	184.3	198.0	202.8	218.7	227.5	231.0
EL	28.7	29.2	29.7	29.9	30.0	29.9	28.5	27.6	22.9	19.2	17.5
Bonn (50.73N,7.08E)											
AZ	155.5	159.2	164.5	168.5	172.4	176.2	190.4	195.5	212.3	221.7	225.3
EL	29.1	29.8	30.8	31.3	31.6	31.8	31.4	30.8	26.9	23.4	21.8
Bremen (53.08N,8.83E)											
AZ	158.3	161.9	167.1	171.1	174.8	178.5	192.2	197.1	213.4	222.6	226.2
EL	27.2	27.8	28.6	29.0	29.2	29.3	28.7	28.0	24.2	20.8	19.3
Dresden (51.05N,13.73E)											
AZ	163.8	167.5	173.0	177.1	180.9	184.8	198.7	203.6	219.6	228.4	231.9
EL	30.3	30.8	31.3	31.5	31.5	31.4	29.9	28.9	24.0	20.0	18.3
Essen (51.47N,7.02E)											
AZ	155.7	159.3	164.6	168.6	172.4	176.2	190.2	195.2	212.0	221.3	225.0
EL	28.3	29.1	30.0	30.5	30.8	31.0	30.6	30.0	26.2	22.9	21.3
Frankfurt (50.12N,8.67E)											
AZ	157.3	161.0	166.4	170.5	174.4	178.3	192.5	197.6	214.3	223.6	227.2
EL	30.1	30.8	31.7	32.1	32.4	32.5	31.8	31.1	26.8	23.2	21.5
Hamburg (53.55N,9.98E)											
AZ	159.8	163.4	168.6	172.5	176.3	180.0	193.6	198.4	214.6	223.6	227.2
EL	27.0	27.6	28.2	28.6	28.8	28.8	28.0	27.3	23.3	20.0	18.4
Hannover (52.40N,9.73E)											
AZ	159.2	162.8	168.1	172.1	175.9	179.7	193.5	198.4	214.7	223.8	227.4
EL	28.1	28.7	29.4	29.8	30.0	30.1	29.2	28.5	24.4	20.9	19.3
Kassel (51.32N,9.48E)											
AZ	158.6	162.3	167.6	171.7	175.5	179.3	193.3	198.3	214.8	224.0	227.6
EL	29.1	29.8	30.5	30.9	31.2	31.3	30.4	29.7	25.5	21.9	202
Kief (54.33N,10.13E)											
AZ	160.1	163.7	168.9	172.8	176.5	180.2	193.6	198.4	214.5	223.5	227.1
EL	26.2	26.8	27.4	27.8	27.9	28.0	27.2	26.5	22.6	19.3	17.8
Koln (50.93N,6.98E)											
AZ	155.5	159.1	164.4	168.4	172.3	176.1	190.2	195.3	212.1	221.5	225.2
EL	28.8	29.6	30.5	31.0	31.4	31.6	31.2	30.6	26.7	233	21.7

Leipzig (51.32N,12.33E)

AZ	162.1	165.8	171.2	175.3	179.1	183.0	196.9	201.8	218.0	226.9	230.4
EL	29.8	30.3	30.9	31.2	31.2	31.2	29.9	29.0	24.3	20.5	18.8

Magdeburg (52.12N,11.63E)

AZ	161.4	165.1	170.4	174.5	178.3	182.1	195.9	200.7	216.9	225.9	229.4
EL	28.8	29.4	30.0	30.2	30.4	30.4	29.2	28.4	23.9	20.3	18.6

Manheim (49.48N,8.48E)

AZ	156.8	160.6	166.0	170.2	174.1	178.0	192.4	197.5	214.4	223.7	227.3
EL	30.7	31.5	32.3	32.8	33.1	33.2	32.5	31.8	27.5	23.8	22.0

Munich (48.13N,11.57E)

AZ	160.3	164.2	169.8	174.1	178.1	182.1	196.7	201.8	218.4	227.5	231.0
EL	32.9	33.5	34.2	34.6	34.7	34.7	33.4	32.4	27.4	23.2	21.3

Nuremburg (49.45N,11.07E)

AZ	160.0	163.8	169.3	173.5	177.5	181.4	195.7	200.8	217.3	226.4	229.9
EL	31.4	32.0	32.7	33.1	33.3	33.3	32.1	31.2	26.5	22.5	20

Rostock (54.08N,12.12E)

AZ	162.4	166.0	171.3	175.2	178.9	182.6	196.1	200.8	216.7	225.6	229.1
EL	26.9	27.4	27.9	28.2	28.3	28.2	27.1	26.3	22.1	18.6	17.0

Stuttgart (48.77N,9.18E)

AZ	157.5	161.3	166.8	171.0	174.9	178.9	193.4	198.6	215.5	224.7	228.3
EL	31.6	32.3	33.2	33.6	33.9	34.0	33.2	32.4	27.8	24.0	22.2

GREECE
Satellite longitude (°)

	26.5E	23.5E	19.2E	16.0E	13.0E	10.0E	1.0W	5.0W	19.0W	27.5W	31.0W

Athens (37.97N,23.72E)

AZ	175.5	180.4	187.3	192.4	197.1	201.6	216.8	221.7	236.3	243.7	246.5
EL	45.9	46.0	45.7	45.3	44.6	43.7	39.0	36.8	27.7	21.5	18.9

Irraklion (35.33N,25.15E)

AZ	177.7	182.9	190.2	195.6	200.4	205.1	220.3	225.1	239.2	246.2	248.8
EL	48.9	48.9	48.5	47.8	47.0	45.9	40.5	38.0	28.2	21.6	18.8

Patrai (38.25N,21.73E)

AZ	172.3	177.1	184.1	189.2	193.9	198.5	214.1	219.1	234.3	241.9	244.8
EL	45.4	45.6	45.6	45.3	44.7	44.0	39.7	37.7	28.9	22.8	20.2

Thessaloniki (40.63N,22.93E)

AZ	174.5	179.1	185.7	190.6	195.1	199.4	214.3	219.2	234.1	241.7	244.6
EL	42.9	43.0	42.9	42.5	41.9	41.2	37.0	35.0	26.6	20.8	18.3

HOLLAND
Satellite longitude (°)

	26.5E	23.5E	19.2E	16.0E	13.0E	10.0E	1.0W	5.0W	19.0W	27.5W	31.0W

Amsterdam (52.37N,4.90E)

AZ	153.4	157.0	162.2	166.1	169.8	173.6	187.4	192.4	209.2	218.7	222.4
EL	26.8	27.7	28.6	29.2	29.6	29.9	29.9	29.4	26.1	23.0	21.6

Arnhem (51.98N,5.92E)

AZ	154.5	158.1	163.3	167.3	171.0	174.8	188.8	193.8	210.5	219.9	223.6
EL	27.5	28.3	29.2	29.8	30.2	30.4	30.2	29.7	26.2	22.9	21.4

Groningen (53.22N,6.55E)

AZ	155.6	159.2	164.3	168.3	172.0	175.7	189.4	194.3	210.8	220.2	223.8
EL	26.5	27.2	28.1	28.6	28.9	29.1	28.8	28.3	24.8	21.7	20.2

Rotterdam (51.92N,4.47E)

AZ	152.8	156.3	161.5	165.5	169.2	173.0	186.9	192.0	208.9	218.4	222.1
EL	27.1	28.0	29.0	29.6	30.1	30.4	30.4	29.9	26.7	23.6	22.1

Utrecht (52.08N,5.13E)

AZ	153.6	157.2	162.4	166.3	170.1	173.8	187.8	192.8	209.6	219.1	222.8
EL	27.2	28.0	29.0	29.6	30.0	30.2	30.1	29.7	26.3	23.2	21.7

HUNGARY
Satellite longitude (°)

	26.5E	23.5E	19.2E	16.0E	13.0E	10.0E	1.0W	5.0W	19.0W	27.5W	31.0W

Budapest (47.50N,19.08E)

	26.5E	23.5E	19.2E	16.0E	13.0E	10.0E	1.0W	5.0W	19.0W	27.5W	31.0W
AZ	170.0	174.0	179.8	184.2	181.2	192.2	206.4	211.2	226.7	235.1	238.3
EL	34.9	35.2	35.4	35.3	35.1	34.7	32.0	30.6	24.2	19.5	17.4

Debrecen (47.53N,21.63E)

	26.5E	23.5E	19.2E	16.0E	13.0E	10.0E	1.0W	5.0W	19.0W	27.5W	31.0W
AZ	173.4	177.5	183.3	187.6	191.6	195.6	209.5	214.2	229.3	237.5	240.6
EL	35.2	35.4	35.3	35.1	34.7	34.2	31.1	29.6	22.8	17.9	15.8

Pecs (46.08N,18.22E)

	26.5E	23.5E	19.2E	16.0E	13.0E	10.0E	1.0W	5.0W	19.0W	27.5W	31.0W
AZ	168.6	172.7	178.6	183.1	187.2	191.3	205.8	210.8	226.5	234.9	238.1
EL	36.3	36.7	37.0	36.9	36.7	36.4	33.7	32.3	25.7	20.8	18.7

Szeged (46.25N,20.15E)

	26.5E	23.5E	19.2E	16.0E	13.0E	10.0E	1.0W	5.0W	19.0W	27.5W	31.0W
AZ	171.2	175.4	181.3	185.7	189.9	193.9	208.2	213.0	228.4	236.6	239.8
EL	36.4	36.7	36.8	36.6	36.3	35.9	32.9	31.3	24.5	19.6	17.4

ICELAND
Satellite longitude (°)

	26.5E	23.5E	19.2E	16.0E	13.0E	10.0E	1.0W	5.0W	19.0W	27.5W	31.0W

Reykiavick (64.15N,21.85W)

	26.5E	23.5E	19.2E	16.0E	13.0E	10.0E	1.0W	5.0W	19.0W	27.5W	31.0W
AZ	128.7	131.6	135.9	139.2	142.3	145.4	157.1	161.4	176.8	186.3	190.1
EL	8.2	9.3	10.6	11.6	12.5	13.3	15.7	16.3	17.5	17.4	17.2

ITALY
Satellite longitude (°)

	26.5E	23.5E	19.2E	16.0E	13.0E	10.0E	1.0W	5.0W	19.0W	27.5W	31.0W

Bari (47.12N,16.87E)

	26.5E	23.5E	19.2E	16.0E	13.0E	10.0E	1.0W	5.0W	19.0W	27.5W	31.0W
AZ	167.0	171.0	176.8	181.2	185.3	189.3	203.7	208.7	224.6	233.2	236.5
EL	35.0	35.4	35.8	35.8	35.7	35.4	33.1	31.8	25.6	21.0	18.9

Bologna (44.48N,11.33E)

	26.5E	23.5E	19.2E	16.0E	13.0E	10.0E	1.0W	5.0W	19.0W	27.5W	31.0W
AZ	158.9	162.9	168.8	173.4	177.6	181.9	197.3	202.7	219.9	229.0	232.4
EL	36.5	37.3	38.1	38.5	38.7	38.7	37.3	36.2	30.5	25.9	23.9

Cagliari (39.22N,9.12E)

	26.5E	23.5E	19.2E	16.0E	13.0E	10.0E	1.0W	5.0W	19.0W	27.5W	31.0W
AZ	153.7	157.9	164.3	169.2	173.9	178.6	195.8	201.7	220.2	229.6	233.1
EL	41.1	42.2	43.4	44.0	44.4	44.6	43.4	42.3	36.1	31.0	28.7

Catania (37.50N,15.10E)

	26.5E	23.5E	19.2E	16.0E	13.0E	10.0E	1.0W	5.0W	19.0W	27.5W	31.0W
AZ	161.7	166.4	173.3	178.5	183.4	188.3	205.4	211.0	228.0	236.5	239.6
EL	44.9	45.6	46.3	46.5	46.5	46.2	43.3	41.7	33.9	28.1	25.5

Florence (43.77N,11.25E)

	26.5E	23.5E	19.2E	16.0E	13.0E	10.0E	1.0W	5.0W	19.0W	27.5W	31.0W
AZ	158.5	162.6	168.6	173.2	177.5	181.8	197.4	202.8	220.1	229.2	232.7
EL	37.2	38.0	38.9	39.3	39.5	39.5	38.0	36.9	31.2	26.5	24.4

Foggia (43.45N,15.57E)

	26.5E	23.5E	19.2E	16.0E	13.0E	10.0E	1.0W	5.0W	19.0W	27.5W	31.0W
AZ	165.3	169.3	175.1	179.4	183.5	187.5	202.0	207.0	223.1	231.8	235.1
EL	34.4	34.9	35.4	35.5	35.4	35.2	33.1	31.9	26.0	21.5	19.5

Genova (44.42N,8.95E)

AZ	155.7	159.7	165.5	170.0	174.2	178.5	194.1	199.5	217.2	226.5	230.1
EL	35.9	36.8	37.8	38.3	38.7	38.8	37.9	36.9	31.7	27.3	25.3

Milan (45.47N,9.20E)

AZ	156.4	160.3	166.1	170.5	174.7	178.9	194.2	199.5	217.0	226.3	229.9
EL	34.9	35.8	36.7	37.2	37.5	37.6	36.7	35.8	30.7	26.4	24.5

Naples (40.85N,14.28E)

AZ	161.7	166.1	172.5	177.4	182.0	186.5	202.7	208.1	225.1	233.8	237.1
EL	41.1	41.8	42.5	42.7	42.8	42.6	40.2	38.8	31.8	26.6	24.2

Palermo (38.12N,13.35E)

AZ	159.3	163.8	170.6	175.7	180.6	185.4	202.5	208.3	225.7	234.5	237.7
EL	43.7	44.6	45.4	45.7	45.8	45.7	43.3	41.8	34.5	28.9	26.5

Reggio di Calabria (38.12N,15.65E)

AZ	162.8	167.4	174.3	179.4	184.3	189.1	205.9	211.4	228.2	236.6	239.8
EL	44.4	45.1	45.7	45.8	45.7	45.4	42.5	40.8	33.0	27.3	24.8

Rome (41.90N,12.48E)

AZ	159.5	163.7	170.0	174.7	179.2	183.7	199.7	205.3	222.5	231.5	234.8
EL	39.5	40.3	41.1	41.5	41.6	41.5	39.7	38.4	32.0	27.0	24.8

Turin (45.05N,7.67E)

AZ	154.3	158.2	163.9	168.3	172.5	176.7	192.2	197.6	215.4	224.9	228.5
EL	34.8	35.8	36.9	37.5	37.8	38.1	37.4	36.6	31.8	27.6	25.6

Venice (45.45N,12.35E)

AZ	160.5	164.5	170.4	174.9	179.1	183.3	198.4	203.7	220.5	229.5	233.0
EL	35.8	36.5	37.2	37.5	37.7	37.6	36.0	34.9	29.2	24.7	22.6

LUXEMBOURG
Satellite longitude (°)

	26.5E	23.5E	19.2E	16.0E	13.0E	10.0E	1.0W	5.0W	19.0W	27.5W	31.0W
Luxembourg (49.75N,6.08E)											
AZ	154.0	157.7	163.0	167.1	171.0	174.9	189.2	194.4	211.5	221.0	224.7
EL	29.7	30.6	31.6	32.2	32.6	32.8	32.6	32.0	28.2	24.7	23.0

NORWAY
Satellite longitude (°)

	26.5E	23.5E	19.2E	16.0E	13.0E	10.0E	1.0W	5.0W	19.0W	27.5W	31.0W
Bergen (60.38N,5.33E)											
AZ	156.0	159.3	164.1	167.8	171.2	174.6	187.3	191.8	207.5	216.6	220.2
EL	19.2	19.8	20.5	20.9	21.2	21.4	21.3	21.0	18.5	16.2	15.1
Oslo (59.92N,10.75E)											
AZ	161.9	165.3	170.3	173.9	177.4	180.9	193.5	198.1	213.4	222.3	225.9
EL	20.7	21.2	21.6	21.9	22.0	22.0	21.3	20.7	17.5	14.8	13.5
Stravenger (58.97N,5.75E)											
AZ	156.1	159.5	164.4	168.1	171.6	175.0	187.9	192.5	208.3	217.4	221.1
EL	20.7	21.3	22.0	22.4	22.7	22.9	22.8	22.4	19.7	17.2	16.0
Trondheim (65.42N,10.42E)											
AZ	162.4	165.7	170.4	173.9	177.2	180.5	192.5	196.9	211.8	220.6	224.1
EL	15.2	15.5	15.9	16.1	16.2	16.2	15.7	15.3	12.8	10.6	9.6

POLAND
Satellite longitude (°)

	26.5E	23.5E	19.2E	16.0E	13.0E	l0.0E	1.0W	5.0W	19.0W	27.5W	31.0W
Gdansk (54.38N,18.67E)											
AZ	170.4	174.1	179.3	183.3	187.0	190.6	203.7	208.3	223.5	232.0	235.4
EL	27.5	27.8	27.9	27.9	27.7	27.4	25.4	24.3	19.2	15.4	13.7
Krakow (50.05N,19.97E)											
AZ	171.5	177.0	181.0	185.2	189.1	192.9	206.6	211.3	226.5	234.9	238.1
EL	32.3	32.5	32.6	32.5	32.2	31.8	29.3	27.9	21.9	17.4	15.5
Lodz (51.77N,19.50E)											
AZ	171.1	174.9	180.4	184.5	188.3	192.0	205.5	210.1	225.4	233.8	237.1
EL	30.4	30.6	30.8	30.7	30.5	30.1	27.7	26.5	20.8	16.6	14.8
Lubin (51.25N,22.58E)											
AZ	175.0	178.8	184.3	188.4	192.2	196.0	209.2	213.8	228.7	236.9	240.1
EL	31.2	31.3	31.2	31.0	30.6	30.1	27.3	25.9	19.7	15.3	13.3
Poznan (52.42N,16.92E)											
AZ	168.0	171.7	177.1	181.2	184.9	188.7	202.2	206.9	222.4	231.0	234.4
EL	29.4	29.7	30.0	30.1	29.9	29.7	27.8	26.7	21.5	17.5	15.8
Szczecin (53.40N,14.53E)											
AZ	165.2	168.9	174.2	178.2	181.9	185.6	199.1	203.8	219.5	228.3	231.8
EL	28.0	28.4	28.8	29.0	29.0	28.9	27.3	26.4	21.7	18.0	16.3
Warsaw (52.25N,21.00E)											
AZ	173.1	176.8	182.3	186.3	190.1	193.8	207.1	211.7	226.7	235.0	238.3
EL	30.0	30.2	30.2	30.1	29.8	29.4	26.8	25.5	19.8	15.6	13.7

PORTUGAL
Satellite longitude (°)

	26.5E	23.5E	19.2E	16.0E	13.0E	10.0E	1.0W	5.0W	19.0W	27.5W	31.0W
Lisbon (38.72N,9.13W)											
AZ	131.1	134.3	139.2	143.1	147.0	151.0	167.1	173.4	195.5	208.0	212.7
EL	32.0	33.9	36.4	38.1	39.6	40.9	44.4	45.0	44.0	41.2	39.7
Portimao (37.13N,8.53W)											
AZ	130.7	134.0	138.9	142.9	146.8	151.0	167.6	174.2	197.0	209.7	214.4
EL	33.5	35.4	38.0	39.8	41.3	42.7	46.2	46.8	45.5	42.5	40.9
Porto (41.18N,8.60W)											
AZ	133.1	136.4	141.3	145.2	149.0	152.9	168.5	174.5	195.6	207.5	212.0
EL	30.5	32.3	34.6	36.2	37.5	38.7	41.8	42.3	41.2	38.6	37.2

ROMANIA
Satellite longitude (°)

	26.5E	23.5E	19.2E	16.0E	13.0E	10.0E	1.0W	5.0W	19.0W	27.5W	31.0W
Bucharest (44.43N,26.10E)											
AZ	179.4	183.7	189.8	194.3	198.4	202.4	216.2	220.8	235.1	242.7	245.6
EL	38.8	38.7	38.3	37.8	37.1	36.3	32.1	30.2	22.2	16.7	14.4
Cluj-Napoca (46.78N,23.60E)											
AZ	176.0	180.1	186.0	190.4	194.4	198.4	212.1	216.8	231.6	239.5	242.6
EL	36.1	36.2	36.0	35.7	35.2	34.6	31.1	29.4	22.2	17.2	15.0
Constanta (44.18N,28.65E)											
AZ	183.1	187.4	193.4	197.9	201.9	205.8	219.2	223.7	237.6	244.9	247.8
EL	39.0	38.8	38.2	37.5	36.7	35.7	31.1	29.1	20.8	15.1	12.8

Lasi (47.17N,27.58E)

AZ	181.5	185.6	191.4	195.6	199.5	203.4	216.6	221.1	235.2	242.9	245.9
EL	35.8	35.6	35.2	34.6	33.9	33.1	29.1	27.2	19.7	14.5	12.3

Timisoara (45.75N,25.22E)

AZ	178.2	182.4	188.4	192.8	196.8	200.8	214.5	219.1	233.6	241.4	244.4
EL	37.3	37.3	37.0	36.6	36.0	35.2	31.3	29.5	21.9	16.7	14.4

SPAIN
Satellite longitude (°)

	26.5E	23.5E	19.2E	16.0E	13.0E	10.0E	1.0W	5.0W	19.0W	27.5W	31.0W
Barcelona (41.38N,2.18E)											
AZ	145.6	149.4	155.2	159.6	163.9	168.3	184.8	190.8	210.4	220.8	224.7
EL	36.1	37.4	39.1	40.1	40.9	41.5	42.1	41.6	37.5	33.4	31.5
Bilbao (43.25N,2.97W)											
AZ	140.5	144.0	149.3	153.4	157.3	161.4	177.1	183.0	202.8	213.7	217.9
EL	32.0	33.4	35.3	36.6	37.6	38.4	40.1	40.1	37.5	34.3	32.7
Cadiz (36.53N,6.30W)											
AZ	132.7	136.1	141.3	145.4	149.5	153.8	171.1	177.8	200.7	213.1	217.7
EL	35.4	37.3	39.8	41.5	43.0	44.2	47.2	47.6	45.5	42.1	40.3
Cordoba (37.88N,4.77W)											
AZ	135.3	138.8	144.1	148.3	152.4	156.8	173.9	180.4	202.4	214.3	218.7
EL	35.3	37.1	39.4	41.0	42.3	43.4	45.9	46.1	43.6	40.1	38.2
Gibraltar (UK) (36.13N,5.35W)											
AZ	133.5	136.9	142.2	146.5	150.6	155.0	172.6	179.4	202.4	214.6	219.2
EL	36.3	38.2	40.7	42.4	43.8	45.0	47.8	48.1	45.6	42.0	40.1
Madrid (40.40N,3.68W)											
AZ	138.1	141.6	146.9	151.1	155.2	159.4	175.9	182.0	202.9	214.3	218.6
EL	34.0	35.6	37.7	39.1	40.2	41.2	43.2	43.3	40.7	37.2	35.5
Malaga (36.72N,4.42W)											
AZ	135.0	138.5	143.8	148.1	152.3	156.7	174.3	181.0	203.5	215.5	219.9
EL	36.53	38.3	40.7	42.3	43.6	44.8	47.3	47.4	44.7	41.0	39.0
Oviedo (43.37N,5.83W)											
AZ	137.3	140.7	145.8	149.7	153.6	157.6	173.0	178.8	198.8	210.1	214.4
EL	30.4	32.0	34.0	35.4	36.5	37.5	39.7	40.0	38.2	35.4	33.9
Palma (39.57N,2.65E)											
AZ	145.2	149.1	155.0	159.6	164.0	168.6	185.7	191.9	211.9	222.4	226.3
EL	38.0	39.4	41.1	42.2	43.0	43.6	44.1	43.5	39.0	34.7	32.6
Sevilla (37.38N,5.98W)											
AZ	133.6	137.0	142.2	146.4	150.5	154.7	171.8	178.4	200.8	213.0	217.5
EL	35.0	36.8	39.3	40.9	42.3	43.5	46.3	46.6	44.5	41.1	39.3
Valencia (39.47N,0.37W)											
AZ	141.4	145.2	150.8	155.2	159.5	163.9	181.0	187.3	207.9	218.9	223.0
EL	36.5	38.1	40.0	41.3	42.3	43.1	44.3	44.1	40.4	36.4	34.5
Valladolid (41.65N,4.72W)											
AZ	137.6	141.1	146.3	150.4	154.3	158.4	174.4	180.4	201.0	212.3	216.6
EL	32.4	34.0	36.1	37.4	38.6	39.6	41.7	41.9	39.7	36.6	34.9
Vigo (42.23N,8.72W)											
AZ	133.6	136.8	141.8	145.6	149.4	153.2	168.6	174.5	195.1	206.8	211.4
EL	29.7	31.4	33.6	35.2	36.5	37.6	40.6	41.1	40.1	37.6	36.2

SWEDEN
Satellite longitude (°)

	26.5E	25.5E	19.2E	16.0E	13.0E	10.0E	1.0W	5.0W	19.0W	27.5W	31.0W
Gothenburg (57.72N,11.97E)											
AZ	163.0	166.4	171.5	175.2	178.8	182.3	195.2	199.8	215.4	224.2	227.8
EL	23.1	23.6	24.1	24.3	24.4	24.3	23.4	22.7	19.0	16.0	14.6
Malmo (55.60N,13.00E)											
AZ	163.8	167.3	172.5	176.4	180.0	183.6	196.8	201.5	217.1	226.0	229.5
EL	25.5	25.9	26.4	26.6	26.6	26.6	25.4	24.6	20.5	17.1	15.6
Stockholm (59.33N,18.05E)											
AZ	170.2	173.7	178.7	182.4	185.9	189.3	201.9	206.3	221.3	229.8	233.3
EL	22.3	22.5	22.6	22.6	22.5	22.3	20.7	19.8	15.6	12.4	11.0
Sundsvall (62.38N,17.30E)											
AZ	169.6	173.0	177.9	181.5	184.9	188.2	200.5	204.8	219.7	228.3	231.7
EL	19.0	19.2	19.4	19.4	19.3	19.2	17.8	17.1	13.5	10.7	9.4

SWITZERLAND
Satellite longitude (°)

	26.5E	23.5E	19.2E	16.0E	13.0E	10.0E	1.0W	5.0W	19.0W	27.5W	31.0W
Basel (47.55N,7.58E)											
AZ	155.1	158.9	164.4	168.7	172.7	176.7	191.6	196.8	214.1	223.6	227.2
EL	32.3	33.2	34.2	34.7	35.1	35.3	34.7	34.0	29.6	25.7	23.9
Bern (46.95N,7.43E)											
AZ	154.7	158.5	164.1	168.4	172.4	176.5	191.5	196.8	214.2	223.7	227.4
EL	32.9	33.8	34.8	35.4	35.7	36.0	35.4	34.7	30.2	26.2	24.4
Chur (46.85N,9.53E)											
AZ	157.3	161.2	166.9	171.2	175.3	179.4	194.3	199.6	216.7	226.0	229.5
EL	33.6	34.4	35.3	35.8	36.0	36.1	35.1	34.3	29.4	25.2	23.3
Geneva (46.20N,6.15E)											
AZ	152.8	156.6	162.2	166.5	170.6	174.7	189.9	195.3	213.0	222.7	226.4
EL	33.2	34.2	35.3	36.0	36.4	36.7	36.4	35.7	31.4	27.5	25.6
Zurich (47.38N,8.53E)											
AZ	156.2	160.0	165.6	169.9	173.9	178.0	192.9	198.1	215.3	224.7	228.3
EL	32.8	33.6	34.6	35.1	35.4	35.5	34.8	34.0	29.3	25.3	23.5

CZECHOSLOVAKIA
Satellite longitude (°)

	26.5E	23.5E	19.2E	16.0E	13.0E	10.0E	1.0W	5.0W	19.0W	27.5W	31.0W
Bratislava (48.15N,17.12E)											
AZ	167.5	171.5	177.2	181.5	185.5	189.5	203.7	208.6	224.4	232.9	236.3
EL	34.0	34.4	34.7	34.7	34.6	34.3	32.0	30.7	24.7	20.2	18.2
Brno (49.20N,16.62E)											
AZ	167.0	170.9	176.6	180.8	184.8	188.7	202.8	207.6	223.4	232.0	235.4
EL	32.8	33.2	33.5	33.6	33.5	33.2	31.1	29.9	24.2	19.8	17.9
Kosice (48.72N,21.25E)											
AZ	173.0	177.0	182.7	187.0	190.9	194.8	208.6	213.3	228.4	236.6	239.8
EL	33.9	34.0	34.1	33.9	33.5	33.0	30.1	28.7	22.2	17.5	15.4
Prague (50.08N,14.43E)											
AZ	164.4	168.2	173.8	178.0	181.9	185.8	199.8	204.7	220.7	229.5	232.9
EL	31.5	31.9	32.4	32.6	32.6	32.4	30.7	29.7	24.5	20.4	18.5

YUGOSLAVIA
Satellite longitude (°)

	26.5E	23.5E	19.2E	16.0E	13.0E	10.0E	1.0W	5.0W	19.0W	27.5W	31.0W
Belgrade (44.83N,20.50E)											
AZ	171.5	175.7	181.8	186.4	190.6	194.7	209.2	214.1	229.5	237.6	240.7
EL	38.0	38.3	38.3	38.2	37.8	37.3	34.1	32.5	25.3	20.2	17.9
Skopje (41.98N,21.43E)											
AZ	172.4	176.9	183.3	188.1	192.5	196.8	211.7	216.6	231.9	239.8	242.8
EL	41.2	41.5	41.5	41.2	40.8	40.1	36.4	34.6	26.7	21.1	18.7
Split (43.52N,16.45E)											
AZ	165.6	169.8	176.0	180.7	185.0	189.3	204.5	209.7	226.0	234.5	237.7
EL	38.8	39.3	39.7	39.8	39.7	39.4	36.8	35.4	28.6	23.5	21.3
Zagreb (45.80N,15.97E)											
AZ	165.5	169.5	175.5	180.0	184.1	188.3	203.1	208.1	224.3	232.9	236.2
EL	36.3	36.8	37.2	37.3	37.2	37.0	34.7	33.4	27.1	22.4	20.3

Appendix 9

Useful constants and equivalents

Boltzmann's constant	1.3806×10^{-23} J/K
10 log (Boltzmann's constant)	-228.6 dB J/K
Altitude of geostationary orbit above sea level	35 784 km (22 235 miles)
Earth equatorial radius	6378.16 km
Effective Earth radius, R_e (due to atmospheric refractive effects)	8500 km
Ratio of geostationary orbital radius to equatorial Earth radius	6.61
Power ratio, P	10 log (P) dB
Voltage ratio, V	20 log (V) dB
1 MHz	1 000 000 Hz
1 GHz	1000 MHz
exp (x)	e^x where $e = 2.718$
antilog (base 10)	$\log^{-1}$ or 10^x
arcsin, arccos, arctan	$\sin^{-1}$, $\cos^{-1}$, $\tan^{-1}$

Index

Printed and bound by CPI Group (UK) Ltd, Croydon, CR0 4YY

03/10/2024

01040434-0017